Phase Change Materials for Thermal Energy Management and Storage

Phase Change Materials for Thermal Energy Management and Storage: Fundamentals and Applications provides the latest advances in thermal energy applications of phase change materials (PCMs). It introduces definitions and offers a brief history, and then delves into preparation techniques, thermophysical properties, and heat transfer characteristics with mathematical models, performance-affecting factors, and applications and challenges of PCMs.

FEATURES

- Provides key heat transfer enhancement and thermophysical properties features for a wide range of PCMs.
- Presents detailed parameter selection procedures impacting heat transfer.
- Reviews available prediction methods for heat transfer and thermophysical properties of PCMs.
- Discusses practical applications for enhanced thermal control.
- Explores challenges and potential opportunities for heat transfer enhancement.

This reference offers a comprehensive overview of the fundamentals, technologies, and current and near-future applications of PCMs for thermal energy management and storage for researchers and advanced students in materials, mechanical, and related fields of engineering.

Hafiz Muhammad Ali is Associate Professor of Mechanical Engineering at King Fahd University of Petroleum and Minerals, Saudi Arabia. He received his doctoral degree in mechanical engineering from the School of Engineering and Materials Science, Queen Mary University of London, United Kingdom, in 2011. He was a postdoc at the Water and Energy Laboratory of the University of California at Merced, United States, in 2016.

Emerging Materials and Technologies

Series Editor: Boris I. Kharissov

The *Emerging Materials and Technologies* series is devoted to highlighting publications centered on emerging advanced materials and novel technologies. Attention is paid to those newly discovered or applied materials with potential to solve pressing societal problems and improve quality of life, corresponding to environmental protection, medicine, communications, energy, transportation, advanced manufacturing, and related areas.

The series takes into account that, under present strong demands for energy, material, and cost savings, as well as heavy contamination problems and worldwide pandemic conditions, the area of emerging materials and related scalable technologies is a highly interdisciplinary field, with the need for researchers, professionals, and academics across the spectrum of engineering and technological disciplines. The main objective of this book series is to attract more attention to these materials and technologies and invite conversation among the international R&D community.

Advanced Synthesis and Medical Applications of Calcium Phosphates
Edited by S.S. Nanda, Jitendra Pal Singh, Sanjeev Gautam, and Dong Kee Yi

Non-Metallic Technical Textiles
Materials and Technologies
Mukesh Kumar Sinha and Ritu Pandey

Smart Micro- and Nanomaterials for Drug Delivery: Two-Volume Set
Edited by Ajit Behera, Arpan Kumar Nayak, Ranjan K. Mohapatra, and Ali Ahmed Rabaan

Smart Micro- and Nanomaterials for Drug Delivery: Volume One
Edited by Ajit Behera, Arpan Kumar Nayak, Ranjan K. Mohapatra, and Ali Ahmed Rabaan

Smart Micro- and Nanomaterials for Pharmaceutical Applications
Edited by Ajit Behera, Arpan Kumar Nayak, Ranjan K. Mohapatra, and Ali Ahmed Rabaan

Phase Change Materials for Thermal Energy Management and Storage
Fundamentals and Applications
Edited by Hafiz Muhammad Ali

For more information about this series, please visit: www.routledge.com/Emerging-Materials-and-Technologies/book-series/CRCEMT

Phase Change Materials for Thermal Energy Management and Storage

Fundamentals and Applications

Edited by Hafiz Muhammad Ali

CRC Press
Taylor & Francis Group
Boca Raton London New York

CRC Press is an imprint of the
Taylor & Francis Group, an **informa** business

Designed cover image: © Shutterstock

First edition published 2025
by CRC Press
2385 NW Executive Center Drive, Suite 320, Boca Raton FL 33431

and by CRC Press
4 Park Square, Milton Park, Abingdon, Oxon, OX14 4RN

CRC Press is an imprint of Taylor & Francis Group, LLC

ISBN: 978-1-032-35993-9 (hbk)
ISBN: 978-1-032-36438-4 (pbk)
ISBN: 978-1-003-33195-7 (ebk)

DOI: 10.1201/9781003331957

Typeset in Times
by Apex CoVantage, LLC

Dr. Hafiz Muhammad Ali dedicates this work to his beloved sisters, Mrs. Naheed Waseem, Mrs. Nadia Khurram, and Mrs. Attia Rashid, for their continuous support and sincere wishes throughout his life.

.

Contents

Chapter 11   Passive Thermal Regulation of Batteries 261

Bilal Lamrani, Badr Eddine Lebrouhi, and
Tarik Kousksou

Chapter 12   Application of a Solid–Solid Nanocomposite PCM for
Thermal Management of a Solar PV Panel...................................... 275

Praveen Bhaskaran Pillai, K. P. Venkitaraj, S. Suresh,
Hafiz Muhammad Ali, Aswin G., Abhishek R., Aravind S.,
Arjun P. Suresh, and Arun Kumar C. S.

*Ankit Bisariya, Rajan Kumar, Dwesh Kumar Singh,
Shailendra Kumar Shukla, Pushpendra Kumar Singh Rathore,
and Hafiz Muhammad Ali*

Contributors

Sanaz Akbarzadeh
Semnan University
Semnan, Iran

Hafiz Muhammad Ali
Mechanical Engineering Department
and
Interdisciplinary Research Center
   for Sustainable Energy Systems
   (IRC-SES)
King Fahd University of Petroleum and
   Minerals
Dhahran, Saudi Arabia

Tehmina Ambreen
School of Engineering
The University of Sheffield
Sheffield, UK

Pratyush Anand
Energy Institute Bengaluru
Centre of Rajiv Gandhi Institute of
   Petroleum Technology
Bengaluru, Karnataka, India

Mohamed M. Awad
Mechanical Power Engineering
   Department, Faculty of Engineering
Mansoura University
Mansoura, Egypt

Ankit Bisariya
Department of Mechanical
   Engineering
Dr. B. R. Ambedkar National Institute
   of Technology
Jalandhar, Punjab, India
and
Department of Mechanical
   Engineering
Raja Balwant Singh Engineering
   Technical Campus
Agra, Uttar Pradesh, India

Arun Kumar C. S.
Department of Mechanical Engineering
College of Engineering Adoor
Adoor, Kerala, India

Desai Maharshi D.
Sardar Vallabhbhai Nation Institute of
   Technology
Surat, Gujarat, India

Maziar Dehghan
Materials and Energy Research Center
Karaj, Iran

Benjamin Duraković
International University of Sarajevo
Sarajevo, Bosnia

Saeed Esfandeh
Department of Mechanical Engineering
Jundi-Shapur University of Technology
Dezful, Khuzestan, Iran

Aswin G.
Department of Mechanical Engineering
College of Engineering Adoor
Adoor, Kerala, India

Umit Gunes
Department of Naval Architecture and
   Marine Engineering
Yildiz Technical University
Istanbul, Turkey

S. Harikrishnan
Kings Engineering College Chennai
Tamilnadu, India

Mohd Naqueeb Shaad Jagirdar
Energy Institute Bengaluru
Centre of Rajiv Gandhi Institute of
   Petroleum Technology
Bengaluru, Karnataka, India

Mohammad Hassan Kamyab
Department of Mechanical Engineering
University of Kashan
Isfahan, Kashan, Iran

Tarik Kousksou
Université de Pau et des Pays de l'Adour
Pau, France

Anirudh Kulkarni
Department of Mechanical Engineering
School of Technology, Pandit
    Deendayal Energy University
Gandhinagar, Gujarat, India

Rajan Kumar
Department of Mechanical Engineering
Dr. B. R. Ambedkar National Institute
    of Technology
Jalandhar, Punjab, India

Bilal Lamrani
Université Mohammed V. de Rabat
Rabat, Morocco

Badr Eddine Lebrouhi
Université de Pau et des Pays de l'Adour
Pau, France

S. Shankara Narayanan
Nanomaterials Laboratory, Department
    of Physics
Sharda School of Basic Sciences and
    Research, Sharda University
Greater Noida, Uttar Pradesh, India

Hakeem Niyas
Energy Institute Bengaluru
Centre of Rajiv Gandhi Institute of
    Petroleum Technology
Bengaluru, Karnataka, India

Cheol Woo Park
School of Mechanical Engineering
Kyungpook National University
Bukgu Daegu, South Korea

Praveen Bhaskaran Pillai
School of Engineering & Physical
    Sciences
Heriot-Watt University
Edinburgh, Scotland, UK

Venkata Reddy Poluru
School of Engineering, Architecture
    and Interior Design
Amity University Dubai
Dubai, United Arab Emirates

Abhishek R.
Department of Mechanical
    Engineering
College of Engineering Adoor
Adoor, Kerala, India

Mohammed Fareed Rahi
Energy Institute Bengaluru
Centre of Rajiv Gandhi Institute of
    Petroleum Technology
Bengaluru, Karnataka, India

Pushpendra Kumar Singh Rathore
Department of Mechanical
    Engineering
Amity University
Uttar Pradesh, Noida, India

Rathod Manish K.
Sardar Vallabhbhai Nation Institute of
    Technology
Surat, Gujarat, India

Paula Ruiz-Hincapie
School of Physics, Engineering, and
    Computer Science
University of Hertfordshire
Hatfield, UK

Aravind S.
Department of Mechanical
    Engineering
College of Engineering Adoor
Adoor, Kerala, India

Arslan Saleem
Cardiff School of Engineering
Cardiff University
Cardiff, Wales, UK

S. Senthilraja
Department of Mechatronics
  Engineering
SRM Institute of Science and
  Technology
Kattankulathur, India

Shailendra Kumar Shukla
Department of Mechanical Engineering
Indian Institute of Technology (BHU)
  Varanasi
Uttar Pradesh, India

Dwesh Kumar Singh
Department of Mechanical
  Engineering
Dr. B. R. Ambedkar National Institute
  of Technology
Jalandhar, Punjab, India

P. Sivasamy
PSR Engineering College
Sivakasi, Tamilnadu, India

Arjun P. Suresh
Department of Mechanical
  Engineering
College of Engineering Adoor
Adoor, Kerala, India

S. Suresh
National Institute of Technology
  Tiruchirappalli
Tiruchirappalli, Tamil Nadu,
  India

H. M. Teamah
Algonquin College Center for
  Construction Excellence
Ottawa, ON, Canada

K. P. Venkitaraj
Department of Mechanical
  Engineering
College of Engineering Adoor
Adoor, Kerala, India

Apurv Yadav
School of Engineering
Architecture and Interior Design,
  Amity University Dubai
Dubai, United Arab Emirates

Acknowledgment

Dr. Hafiz Muhammad Ali would like to thankfully acknowledge the support provided by King Fahd University of Petroleum and Minerals, Dhahran, 31261, Saudi Arabia.

Preface

Phase Change Materials for Thermal Energy Management and Storage: Fundamentals and Applications is the new book that aims to comprehensively present the fundamental advances including experimental, numerical, theoretical, and empirical as well as the applied advances of phase change materials in various fields such as thermal management of electronics, batteries, etc., energy storage in buildings and constructional materials, desalination, and many allied fields.

This book consists of 13 distinctive and unique chapters; at first place each chapter provides a complete detailed overview and in-depth understanding on a unique aspect of phase change materials and at second place is well integrated with the forthcoming chapters of the book to develop step by step knowledge. Chapter 1 presents the detailed bibliometric research advances and historical advances of phase change materials. Chapter 2 is dedicated to the systematical classifications of phase change materials and unique preparation methods for the development of phase change materials. Chapter 3 further extends the discussion on the preparation and development methods of micro/nano encapsulated phase change materials including physical, chemical, and physicochemical processes. The characterization techniques of micro-encapsulated phase change materials are discussed as well in this chapter. Thermophysical properties and measurement techniques are precisely presented in Chapter 4 with special emphasis on melting point, latent heat and specific heat, and thermal conductivity measurements. Chapter 5 further develops the understanding into thermal characterization techniques. The importance of thermal energy storage and role of phase change materials to capture thermal energy are explored in Chapter 6. Phase change materials have high thermal storage capacity; however, they are not good conductors of thermal energy due to lower thermal conductivity. Therefore, detailed heat transfer augmentation methods are introduced in chapter sever that includes the use of extended surfaces, multiple (or cascaded) phase change materials, heat pipes, composite materials, encapsulation, and external fields. Chapter 8 further furnishes the use of nanomaterials into phase change materials to enhance thermal characteristics. Chapters 9–13 are fully dedicated to various applications of phase change materials including the thermal management of electronics and electrical systems, construction and building materials, passive thermal regulation of batteries, thermal management of a solar photovoltaic panel, and thermal energy storage systems.

In conclusion, this book provides a ready reference and complete guidelines to the experts working in the field of phase change materials and are willing to update their skill sets with advances in fundamentals and applications of phase change materials. At the same time, this book is a step-by-step guideline for beginners who are interested to understand and develop their knowledge into phase change materials. Therefore, it is hoped that book will prove itself an

asset for engineers, researchers, scientists, academicians, and scholars throughout the globe.

2 March 2024
Hafiz Muhammad Ali
Mechanical Engineering Department
King Fahd University of Petroleum and Minerals
Dhahran, Saudi Arabia
Interdisciplinary Research Center for Sustainable Energy Systems (IRC-SES)
King Fahd University of Petroleum and Minerals
Dhahran, Saudi Arabia

1 Research Trends and Perspectives on Phase Change Material
Bibliometric Analysis

Rathod Manish K. and Desai Maharshi D.

HIGHLIGHTS

- Thermal energy storage using latent heat is promising for improving energy system performance and addressing renewable energy's intermittent nature.
- This paper outlines the most critical trends and unmet research needs in thermal energy storage materials, i.e., phase change materials (PCMs), to guide future researchers and entrepreneurs.
- The results of the study indicate that China, the United States, and India are the primary contributors to research in the field of PCM on a worldwide scale.
- PCMs have been utilised in many applications such as photovoltaics, battery thermal management systems (BTMSs), and building temperature control.

ABBREVIATIONS

BTMS   Battery thermal management system
CFD   Computational fluid dynamics
LHSU   Latent heat storage unit
PCM   Phase change material
PV/T   Photovoltaic/thermal
TES   Thermal energy storage

1.1 INTRODUCTION

Energy is crucial to a country's economic growth and technical prowess. The growing demand for readily available energy sources, such as fossil fuels, can be attributed to various factors, including population growth, rapid urbanisation, and economic development. These factors have collectively contributed to a significant rise in the demand for energy, which is growing at an almost exponential rate. Even with increased resources and a larger energy supply, the rising demand for power is still

DOI: 10.1201/9781003331957-1

not being met. The fundamental challenge of the present is climate change, which is accelerated by the widespread use of fossil fuels for energy. Therefore, renewable energy is becoming increasingly significant as a means to supply environmentally friendly energy rather than having to constantly seek for new energy supplies. This has the potential to minimise the use of fossil fuels as a major energy source, thereby lowering greenhouse gas emissions. It is anticipated that renewables will continue to be the power source with the highest rate of increase (about 6%) through the year 2035 [1] and that efficiency improvements would lower demand growth to 60% in Organisation for Economic Co-operation and Development nations [2]. The world urgently requires a revolution that makes reliable, cheap power widely available. The importance of energy efficiency and conservation in this movement toward change cannot be overstated. Therefore, researchers and the energy sector have made it a top priority to find techniques for storing and efficiently using energy in order to address the gap between supply and demand. Different energy storage systems and their current stages of development are summarised in Figure 1.1. According to their level of development, these technologies fall into one of three categories: (i) mature, (ii) developed, or (iii) developing. Although thermal energy storage (TES) has been developed technically and is commercially available, it is still not widely

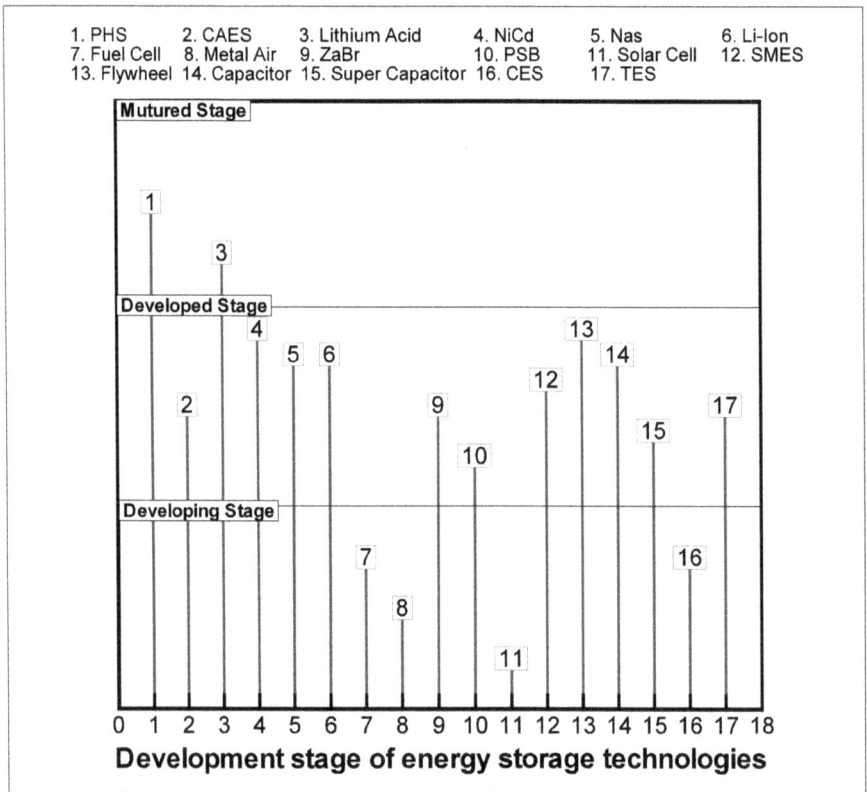

FIGURE 1.1   Technical and developmental stages of different energy storage technologies [3].

used, particularly for large-scale utilities. More trials by the industry and market are required to determine its competitiveness and reliability.

By utilising TES materials, one may decrease their overall energy consumption, include renewable energy sources in the production of electricity, and store large amounts of thermal energy at either high or low temperatures. It is possible to categorise TES systems according to the form of thermal energy that is being stored, which can be sensible, latent, or thermochemical. The process of storing sensible heat causes a temperature change. It is the most common and basic method of storing heat. Latent heat storage is a low-cost option for TES. The technique is based on the material's phase transition properties, namely its capacity to absorb or release heat when it changes from a solid, liquid, or gaseous state. One novel and fascinating alternative to traditional methods of sensible heat storage is thermochemical heat storage.

The use of a latent heat storage unit (LHSU) based on a phase change material (PCM) is preferable to the use of a sensible heat storage unit in TES because of the LHSU's greater energy density and its ability to store or extract energy in an isothermal manner. As a result, less material, in terms of mass and volume, is needed to achieve excellent energy efficiency. The phase change occurs at a constant temperature, eliminating hassles associated with corrosion caused by temperature and other handling difficulties. PCMs are materials that can change between solid and liquid phases while absorbing or releasing large amounts of energy [4]. This property makes them useful in a variety of applications, including thermal energy storage [5], temperature regulation [6], and phase change memory [7]. PCMs that exhibit a broad temperature range between melting and freezing have been found and widely explored. Materials can be categorised into different classes based on their composition and properties. One common classification scheme involves categorising materials as organic, inorganic, or eutectic. Organic materials include substances like paraffins and fatty acids, which are primarily composed of carbon and hydrogen atoms. Inorganic materials, on the other hand, consist of compounds such as salt hydrates and metals, which do not contain carbon–hydrogen bonds. Eutectic materials represent a combination of both organic and inorganic components, exhibiting unique properties resulting from their mixture. Figure 1.2 provides a categorisation of PCMs for the solid–liquid phase

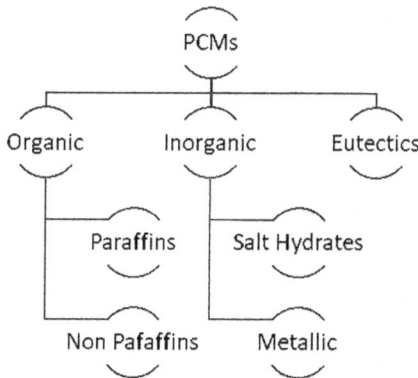

FIGURE 1.2 Categorisation of phase change materials (PCMs).

transition. Each class of PCMs has been extensively documented in the literature, along with its characteristics, benefits, and drawbacks.

In recent years, there has been a lot of research on PCMs because they have the potential to improve the efficiency of many systems and processes. For example, PCMs can be used to store excess thermal energy that would otherwise be wasted and then release that energy when it is needed. This can help reduce energy consumption and costs in buildings, vehicles, and industrial processes.

From review articles [4] and [5], several trends are observed in research on PCMs. They are:

1. *Development of new PCMs:*
   New PCMs with enhanced qualities are currently being developed by researchers. These new PCMs will have higher energy densities, faster phase change rates, and greater stability. These materials have the potential to be utilised in a variety of applications, including temperature management, phase change memory, and thermal equilibrium storage.

2. *Optimisation of existing PCMs:*
   In addition to ongoing investigations, researchers are actively engaged in enhancing the performance of currently available PCMs in various ways, such as by boosting their thermal conductivity or lowering their melting point. These efforts could result in PCM-based systems that are more effective both in terms of efficiency and cost.

3. *Integration of PCMs into various applications:*
   Investigations into the application of PCMs in a wide range of contexts, such as the insulation of buildings, cooling systems, and electric vehicles, are still in progress. The purpose of these initiatives is to discover methods for using PCMs in these systems in order to improve the systems' functionality and make them more efficient.

4. *Modelling and simulation of PCM behaviour:*
   Researchers use modelling and simulation techniques in order to obtain a deeper comprehension of the behaviour of PCMs and to more accurately predict how they will function in a variety of contexts. This effort contributes to the development of novel PCMs as well as the optimisation of their use in a variety of systems.

To gain a better insight from papers, articles, and book chapters, thorough quantitative and qualitative data on the subject are required. In this context, the term "quantitative information" pertains to the quantity of bibliometric data that has been enriched with statistical analysis. On the other hand, "qualitative information" pertains to the specific subtopics and keywords that have been investigated within a given subject [8]. With the use of a statistical technique called bibliometric analysis [9], a close look at the current state of the art, the most important contributions, the gaps in the research, and the potential future directions of PCMs and their applications can be investigated. It also helps to identify the most influential researchers and research groups in the field, as well as the most highly cited papers and articles. This is useful for researchers seeking to collaborate with leading experts in the field or for policymakers and funding agencies seeking to prioritise resources for further research. It also highlights the most promising areas for future research and development. Overall, bibliometric analysis provides valuable insights into the state of the field and the direction of future research on PCMs, which helps researchers make more informed decisions and improve the efficiency and impact of their work.

In short, with a bibliometric methodology, this chapter goes beyond the standard literature review to trace the development of the knowledge structure around the topic of PCMs.

1.2 METHODOLOGY

The main objective of the present work was to find research trends in the field of PCMs. To fulfil this objective, the Web of Science database was used. Research articles were selected and screened by condition (TI = "Phase Change Material" OR AK = "Phase Change Material" OR TI = "Latent Heat Storage" OR AK = "Latent Heat Storage" OR TI = "PCM" OR TI = "Phase Change Materials" OR AK = "Phase Change Materials") where TI and AK represent the title and author keywords, respectively. This indicates that papers with the desired search strings were retrieved from either the title or the keywords section. However, the queries did not search for the keywords from the introduction part of the article, since this area typically includes general information and might produce an inaccurate data sample for the study. A total of 16,165 research articles, book chapters, conference proceedings, review articles, etc., were found and analysed using the bibliometric network software VOSviewer [10] due to its ability to display large networks in an easy-to-understand way. As suggested by the methodologies in the articles [11], [12], and [13], analysis was carried out by analysing the type of document; year-by-year trends; and distribution by country, journal, and institute; as well as keywords and authors. It is to be noted that the database is reflected as of December 31st, 2022.

1.3 RESULTS AND DISCUSSION

This section presents the outcomes of the bibliometric and keyword analyses conducted on the data extracted from the database in response to the research query.

1.3.1 Bibliometric Analysis of PCMs

1.3.1.1 Number of Publications

Figure 1.3 shows the quantitative trend for total annual publications. The analysis was conducted using the available data from the Web of Science database, which yielded a total of 16,165 documents. The publication of the first article in 1948 was noted in the observation. Further, it was observed that number of documents published per year remained below 100 for the years 1980 to 2003, after which, an approximate linear trend was observed until the year 2009. A notable trend observed between 2010 and 2021 was the exponential growth in the number of publications. This indicates that the field is going through a period of rapid expansion, which is

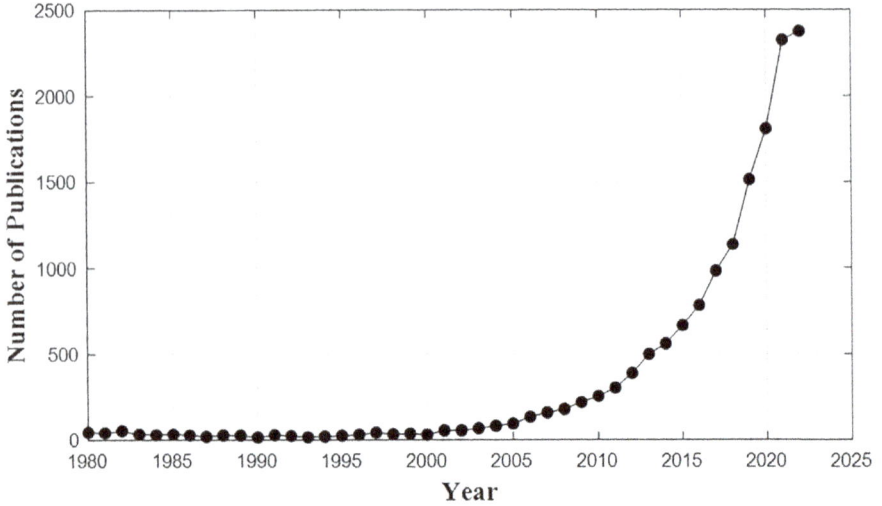

FIGURE 1.3 Total annual publications.

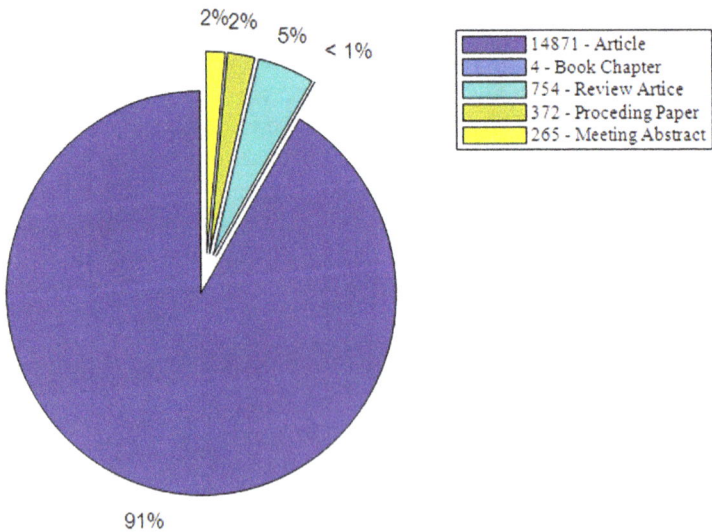

FIGURE 1.4 Types of documents.

also being driven by the need for sophisticated solutions that are already market-ready in the coming years.

Further, Figure 1.4 shows the distribution of types of documents. It can be observed that 91% of documents are research articles, while 5% of documents are review articles. About 2% of documents are proceedings papers or meeting abstracts. The distribution of documents related to PCM across different journals is shown in Figure 1.5. It can be observed that the majority of documents are published in

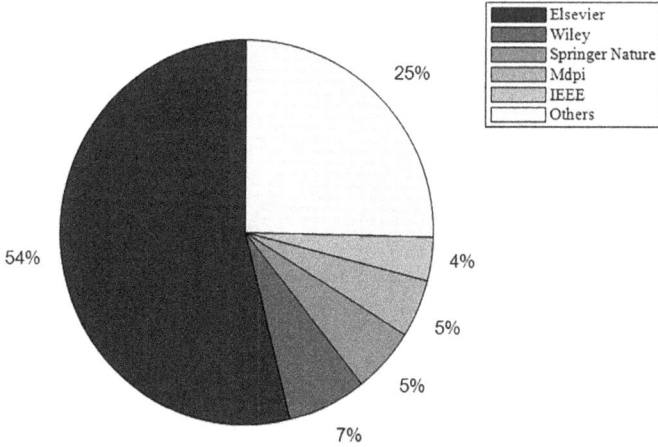

FIGURE 1.5 Publication distribution across journals.

Elsevier journals (54%), followed by Wiley (7%) and Springer Nature (5%), with about a quarter of the documents published in other journals.

1.3.1.2 Geographic Distribution of Publications

Figures 1.6 and 1.7 illustrate the global breakdown of the top 15 nations in terms of total number of publications, normalised publications per million residents, and average citations per document. With the aid of this study, researchers are able to determine which countries are seeing a surge in research activity and which counties have experienced a stagnation in research activity. It can be observed that China, with 5669 documents published, has the highest number of publications amongst all the countries and accounts for approximately 35% of all documents on PCMs. It is then followed by the USA (1759), followed by India (1286). Further, France (649), Germany (645), and Spain (624) have similar numbers of publications. However, upon normalising the documents published with respect to the population of each country, it can be observed that Australia (18.14) has one of the best publications-to-population ratios amongst the top 15 countries, followed by England and Saudi Arabia. Further, it is observed that the USA (5.29), China (4.01), and India (0.92) have low publications when normalised. This provides a general indication of a nation's research infrastructure, population, literacy rate, and the number of persons actively engaged in research within that country.

Further, it can be observed that, when normalised with respect to citations, Germany (48.41) has one of the best ratios of citations to publications, followed by Italy (45.42) and Spain (42.54). Further, Japan (24.90), Saudi Arabia (24.60), and South Korea (20.09) have low ratios of citations to publications. This serves as a rough indicator of the quality of the research articles that are published.

Further, top countries were selected based on total publications, and their publication trends were analysed and shown in Figure 1.8. It was observed that the

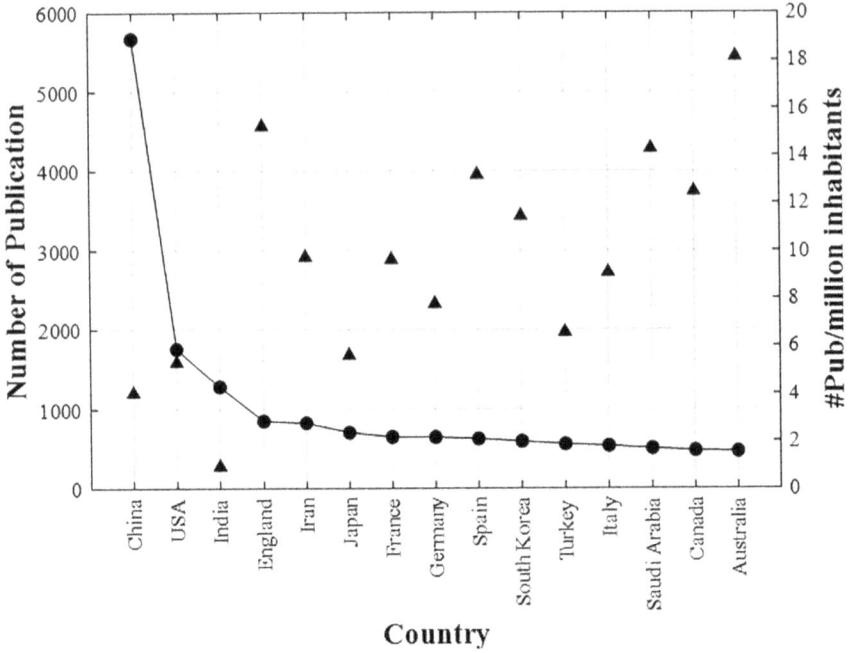

FIGURE 1.6   Country-wise number of publications and publications per million inhabitants.

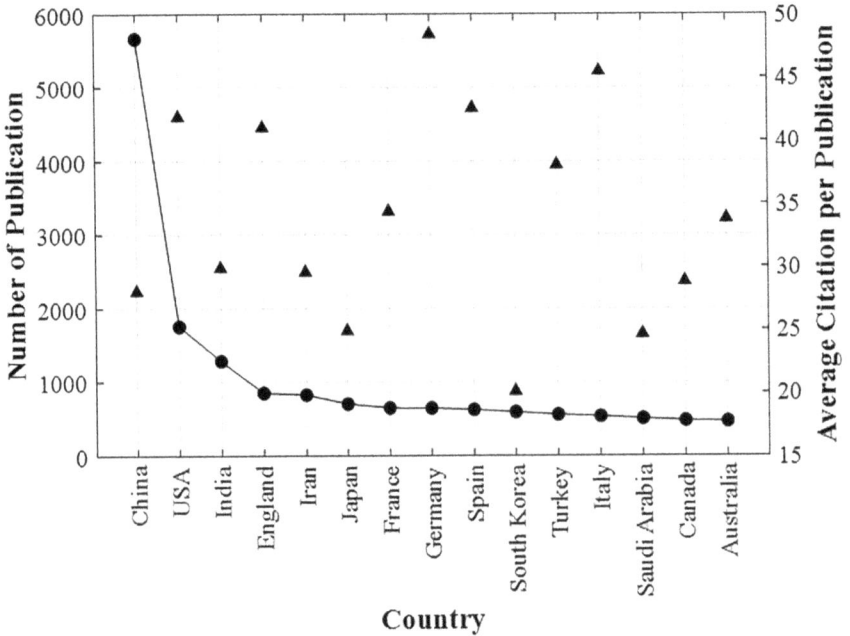

FIGURE 1.7   Country-wise number of publications and average citations per document.

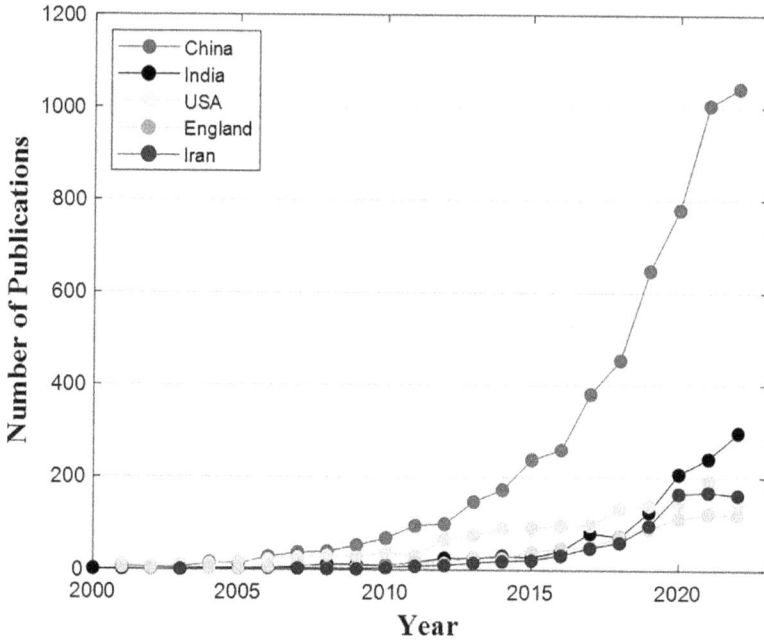

FIGURE 1.8 Publication trends in top five countries.

USA pioneered research on PCMs in 1948 (not shown in the graph) but has fallen behind China, India, and Iran in the number of articles published per year. China has shown tremendous growth in publications per year, crossing 1000 publications per year. It is observed that China started to grow exponentially from 2013 onward.

Figure 1.9 depicts the relationship between countries that publish on PCMs. Countries with a minimum of 100 published documents were selected for the analysis, amounting to 32 countries. Every connection between two nations is represented by a link, and the degree to which those countries are connected may be inferred from the thickness of the link. A threshold criterion was also set at a link strength of 800 to increase readability. Furthermore, the size of the node for each country corresponds to the number of publications. It can be seen that the USA and Japan have the lowest average year of publication, with India, China, and Australia having an average year of publication in the neighbourhood of 2018 and Saudi Arabia, Iraq, Pakistan, and Vietnam having the vast majority of their documents recently published.

The assessment of publications based on researchers' affiliations has been recognised as a valuable approach for identifying institutes or research organisations that focus on similar research areas. This practise can facilitate collaboration amongst researchers and expedite the generation of research outcomes. Figure 1.10 depicts the top ten institutes in terms of publication quantity. It can be seen that the Chinese Academy of Sciences (669) is a stand-alone institute in

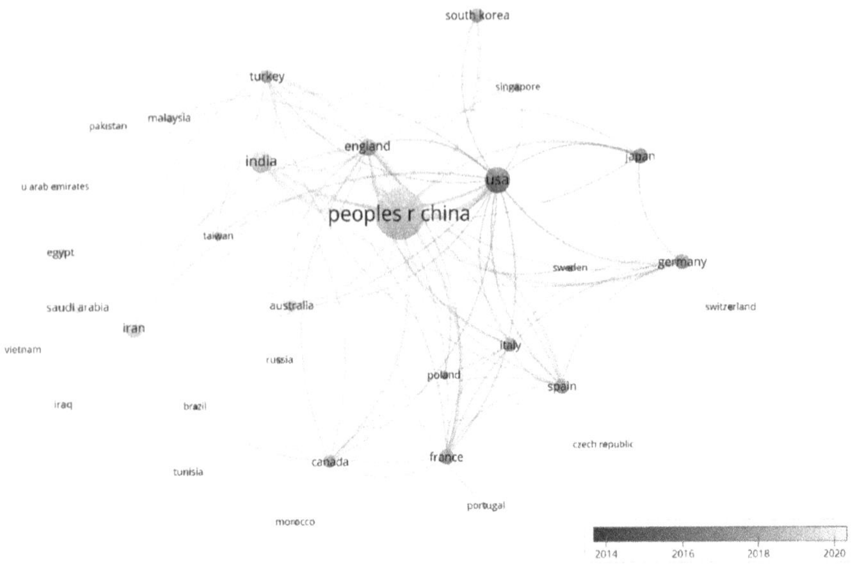

FIGURE 1.9   Relationship between countries.

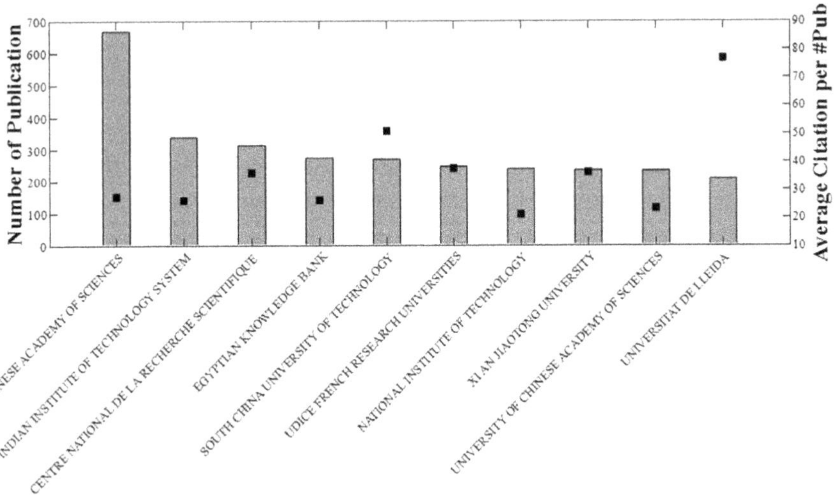

FIGURE 1.10   Top 10 institutes by publications.

terms of publications, with nearly double the number of publications as the next best institute. It is followed by the Indian Institute of Technology (339), closely followed by the Centre National De La Recherche Scientifique (314). Further, it can be observed that Universitat De Lleida (76.73) has the highest average citation per publication, followed by South China University of Technology (50.87) and Udice French Research Universities (37.63).

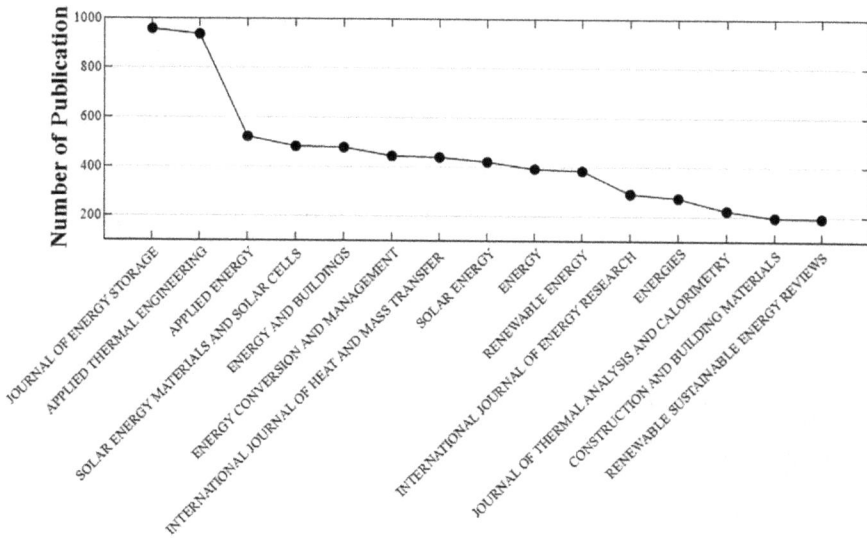

FIGURE 1.11 Top 15 journals by publication.

1.3.1.3 Prominent Publication Journal

Figure 1.11 presents a visual representation of the aggregate quantity of articles that have been published across the top 15 most influential journals within the specific field of investigation. It can be noted that the *Journal of Energy Storage* has the highest number of publications (958), having an impact factor of 8.78, closely followed by *Applied Thermal Engineering* (937) with an impact factor of 6.465, followed by *Applied Energy* (522) with an impact factor of 9.74. Further, it is to be noted that *Renewable Sustainable Energy Reviews* has the highest impact factor of all the top 15 journals at 16.8.

1.3.1.4 Top Researchers Based on Average Citations

Access to research facilities, financing possibilities, the exchange of ideas, and the creation of novel information are all crucial to the success of modern research, making it imperative that authors and academic institutions work together [14]. In this part, visual mapping is used to explore the publication landscape of publications with many authors. To identify the top researchers in the field of PCMs, a bibliometric analysis was performed. To identify top researchers, authors were screened based on the criteria of having a minimum of 10 research articles and a minimum of 2000 citations. Thus, out of 32,231 authors, 50 authors were screened. The density visualisation based on normalised citation is shown in Figure 1.12. It can be observed that Zhang, Zhengguo (195.51), had the highest normalised citation quantity, followed by Cabeza, L.F., (191.25), and Wuttig and Matthias (184.55). Further, it was observed that the top authors based on average citation quantity were Mehling, H., (423.6), followed by Cabeza, L.F., (349.0769), and then Yamada and Noboru (305.1).

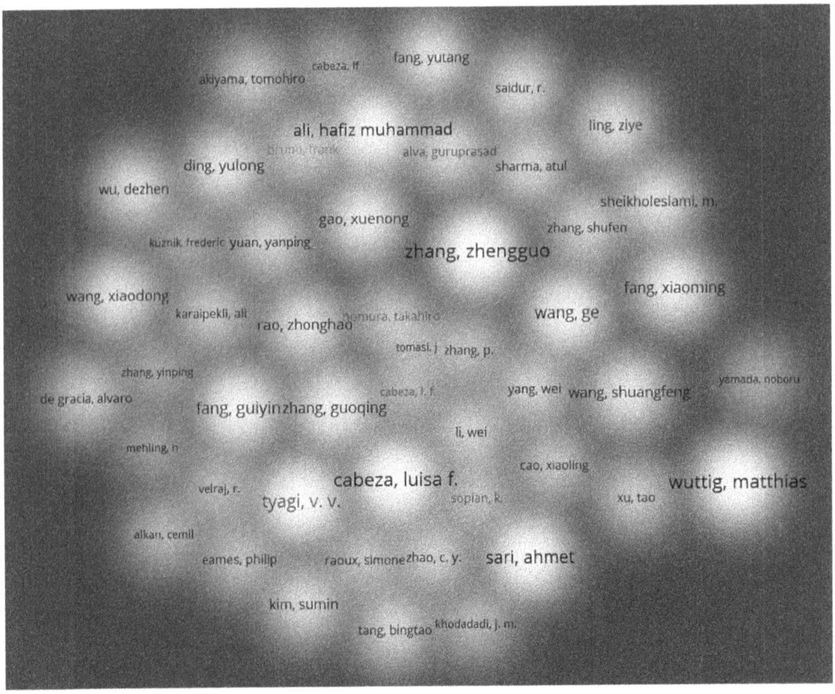

FIGURE 1.12 Density visualisation of authors based on normalised citations.

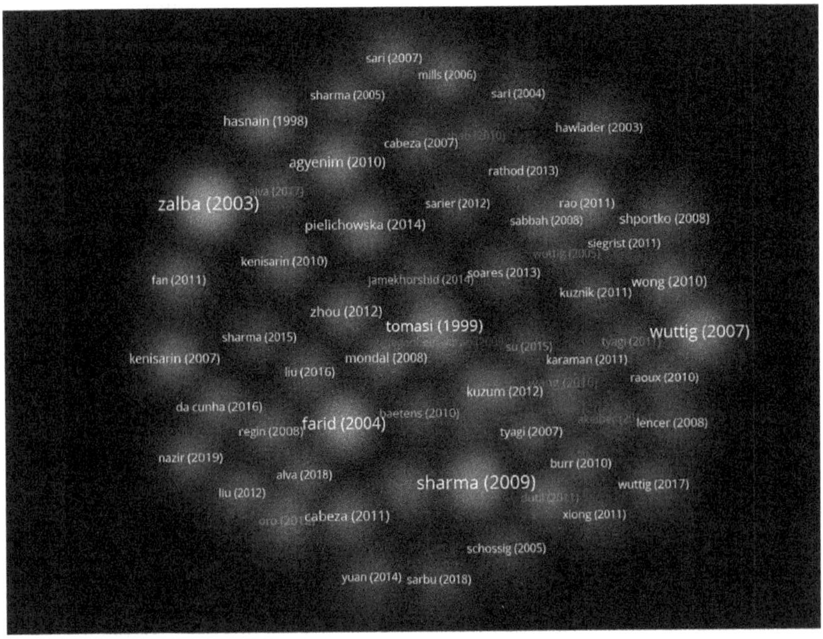

FIGURE 1.13 Density visualisation of top research articles.

1.3.1.5 Top Articles Cited by Researchers

A density visualisation, as shown in Figure 1.13, was created to identify the top research articles cited by researchers. The articles were chosen based on a citation count of at least 400. Thus, out of 16,165 articles, 65 were identified. Here, the size of each node reflects the number of times an article has been cited. The article "Review on Thermal Energy Storage with Phase Change: Materials, Heat Transfer Analysis, and Applications" by Sharma A. received the most citations, with 3383, followed by "Review on Thermal Energy Storage with Phase Change: Materials, Heat Transfer Analysis, and Applications" by Zalba B.

1.3.2 KEYWORD NETWORK ANALYSIS

1.3.2.1 Methods

A keyword network analysis was also performed to better understand the connection between the research's substantive scientific value and the bibliometric tendencies highlighted so far. From the data obtained from the Web of Science, appropriate corrections for repeated words such as "phase-change" vs "phase change" were made to increase the readability of the chart. A minimum repetition threshold of 20 was used to screen 21,190 keywords. A total of 327 keywords were obtained and screened down to 60 based on link strength. The primary keywords were analysed and then arranged into clusters with other terms that appeared at the same time. An analysis based on author keywords was carried out and is shown in Figure 1.14. The size of each node is indicative of the occurrence of each keyword.

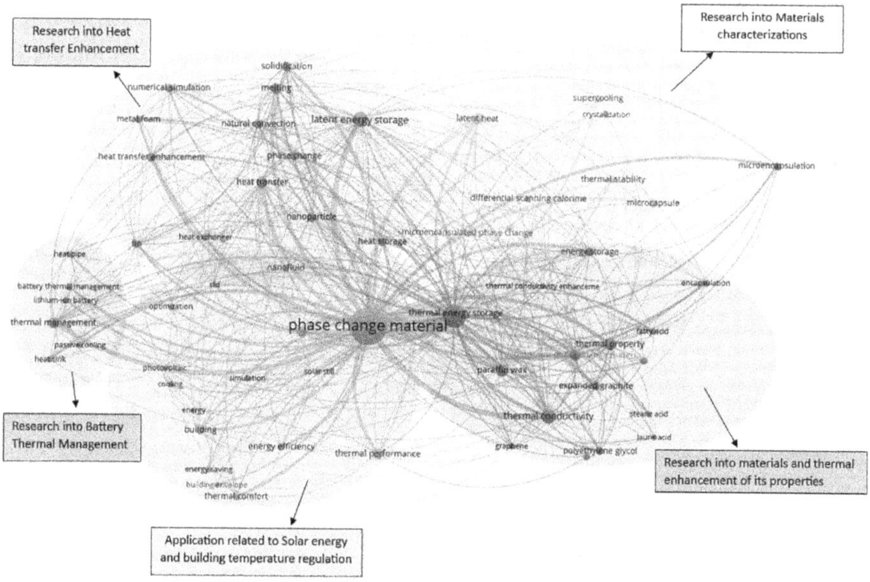

FIGURE 1.14 Bibliometric analysis based on author keywords.

1.3.2.2 Results and Discussion

In the analysis of author keywords, it is observed that the term "phase change material" has the highest frequency of occurrences, with a total of 9346 instances. Following closely behind is the term "thermal energy storage" with a count of 2738 occurrences. This data are depicted in Figure 1.14. The presented literature map illustrates the clustering of co-occurring keywords into five distinct groups, denoted by the colours red, blue, green, purple, and yellow. The analysis of the literature map reveals that a significant portion of the articles found in this search pertain to the investigation of heat transfer phenomena observed during the process of melting and solidification across diverse materials. Additionally, a notable focus is placed on enhancing the thermal performance of various materials and systems. These findings are visually represented by the presence of red and blue clusters on the literature map. In addition, it has been noted that the most researched uses of PCMs are related to solar energy, namely photovoltaic–thermal (PV/T) systems, solar stills, etc.; building cooling/passive systems; and battery thermal management systems. Microencapsulation of PCM appears to be a growing trend in the scientific community, which is interesting to observe. It is also noted that the thermal stability of a PCM, measured with a differential scanning calorimeter, is deemed to be the most crucial component of the PCM's characterisation.

Table 1.1 displays the frequency of the primary keywords in each cluster, which may be used for a more thorough analysis of the data. It is seen that research has been focused on mostly paraffin wax (431), polyethylene glycol (167), fatty acids (111), stearic acid (80), and lauric acid (60), and mainly proposes the use of expanded graphite (206) and graphene (89) for the enhancement of thermal conductivity. Further, it is observed that applications of PCMs include building cooling (192), battery thermal management (124), and photovoltaics (82) and that researchers have mainly resorted to using CFD (79) to carry out simulations. Research is also focused on heat transfer enhancement (214) by metal foam (133), heat pipe (92), fin (140), encapsulation (118), and microencapsulation (183).

1.3.2.3 Material-Oriented Keyword Network Analysis: A Critique

To identify different PCMs used by researchers along with different nanoparticles, a bibliographic analysis was carried out. To create a network, first, keywords were screened based on number of repetitions (minimum 20 repetitions), and 329 keywords were identified. Further, keywords were manually screened to match the perspective, and 38 keywords were identified and shown in Figure 1.15. It is to be noted that the larger the node, the more times the keyword appears, and the thicker the lines between them, the stronger the connection. It can be observed that paraffin wax (431) has been extensively used by the researchers. Further, it is also observed that extended graphene (206) has been extensively used as a nanoparticle for performance enhancement.

1.3.2.4 Application-Oriented Keyword Network Analysis: A Critique

A wide variety of fields have benefited from PCMs throughout the years. As a means of improved comprehension, a bibliometric study was performed through the use of manual identification. First, 239 keywords were found after being filtered based on

TABLE 1.1
Prominence of Keywords in Different Clusters

Cluster	Keyword	Occurrence	Avg. pub. year
1 (green)	Phase change material	9346	2018
	Building	192	2019
	Photovoltaic	82	2020
	Solar still	72	2020
	Computational fluid dynamics (CFD)	79	2019
	Solar energy	307	2017
2 (red)	Thermal energy storage	2738	2018
	Paraffin wax	431	2017
	Fatty acid	111	2016
	Graphene	89	2019
	Expanded graphite	206	2018
	Encapsulation	118	2017
	Microencapsulation	183	2017
	Stearic acid	80	2017
	Lauric acid	60	2017
	Composite PCM	329	2019
3 (yellow)	Differential scanning calorimetry	160	2015
	Latent heat	417	2016
	Supercooling	160	2018
	Crystallisation	101	2016
	Thermal stability	157	2017
4 (blue)	Latent energy storage	774	2017
	Numerical simulation	181	2018
	Heat exchanger	110	2018
	Metal foam	133	2020
	Fin	140	2020
5 (purple)	Thermal management	395	2019
	Battery thermal management	124	2020
	Lithium-ion battery	92	2020
	Heat sink	106	2018
	Heat pipe	92	2018

the number of times they appeared (a threshold of 25 repetitions was set). Further, keywords were manually screened to match the perspective, and 40 keywords were identified and shown in Figure 1.16. Researchers have looked at a wide variety of possible uses, but thermal management (395), solar energy (307), and heat storage (221) stand out as some of the most studied. Furthermore, the most studied areas of application include thermal management of batteries (118), photovoltaics (82), and cold storage (62).

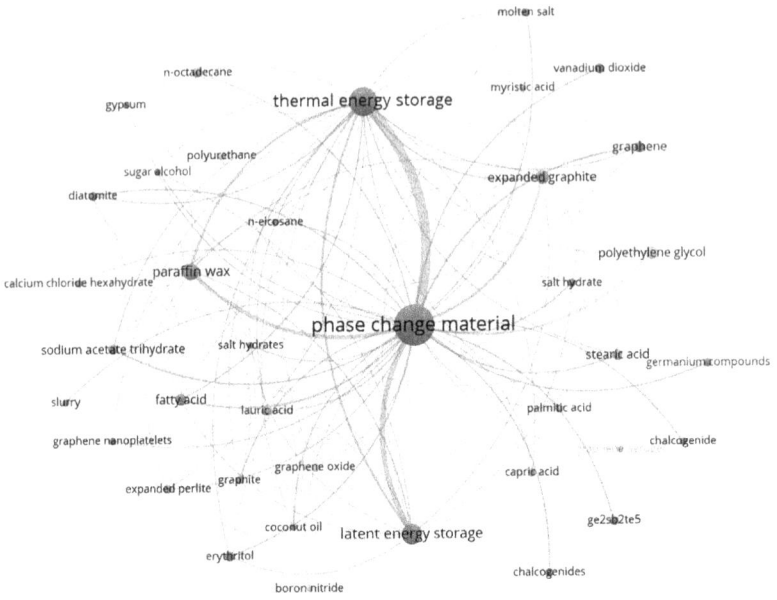

FIGURE 1.15   Bibliographic analysis of PCMs.

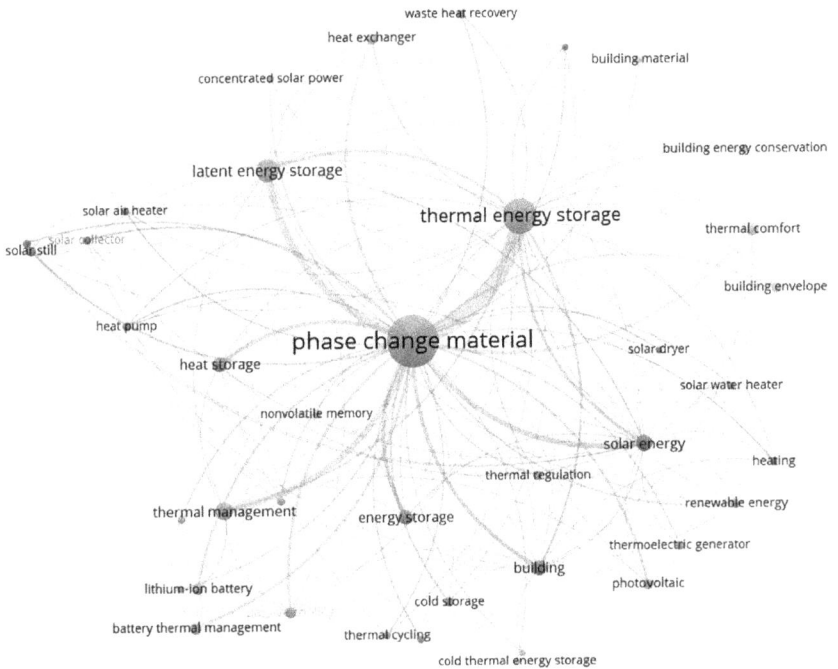

FIGURE 1.16   Bibliographic analysis of PCM applications.

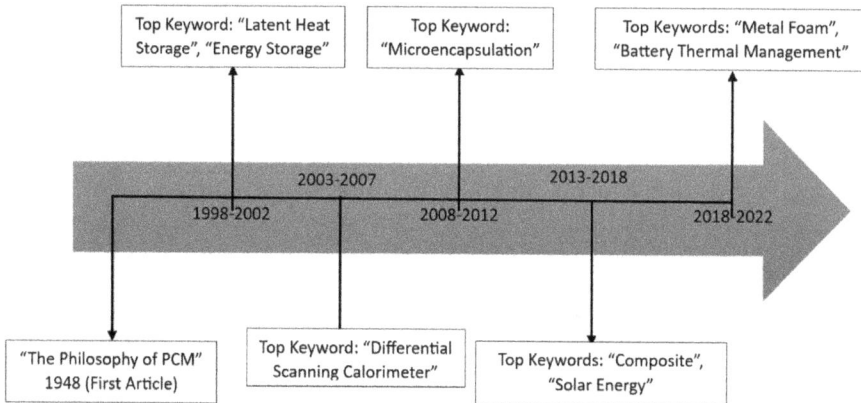

FIGURE 1.17 Year-wise analysis of most frequently occurring keywords.

1.3.3 Current Developments and Prospects for the Future

Following the pattern seen in Figure 1.3, the overlay visualisation by year of the keyword analysis (Figures 1.17–1.18) reveals that the majority of keywords are no older than eight years. The keywords "phase change," "latent heat," "heat transfer," and "differential scanning calorimeter" are the ones with the longest publication histories, with an average publication year of 2015–16. This leads one to believe that a significant portion of the study was motivated by the development of thermophysics linked with phase change processes during the melting and solidification processes. The most recent keywords identified in the literature review include "metal foam," "nanoparticle," "nanofluid," "fin," "composite PCM," "battery thermal management," "lithium-ion battery," "solar still," and "building." These keywords have been extensively studied and published on, with an average publication year of 2019. The current trends in PCM research indicate a focus on enhancing the thermal performance of PCMs by incorporating thermal conductivity enhancers like metal foam, fins, nanoparticles, and other similar materials. Additionally, there is a growing interest in exploring the diverse applications of PCMs, such as their utilisation in thermal management systems for buildings and batteries.

1.4 CONCLUSIONS

This chapter provides a bibliometric study of the research trends in the subject of PCMs, examining the data from the Web of Science database from 1948 through December 2022. Research papers were chosen using a conditional screener (TI = "Phase Change Material" OR AK = "Phase Change Material" OR TI = "Latent Heat Storage" OR AK = "Phase Change Material" OR TI = "PCM" OR TI = "Phase Change Materials" OR AK = "Phase Change Materials"). An investigation was conducted to analyse the data in terms of the number of publications per year. Additionally, the study examined the prominent nations and journals in which the

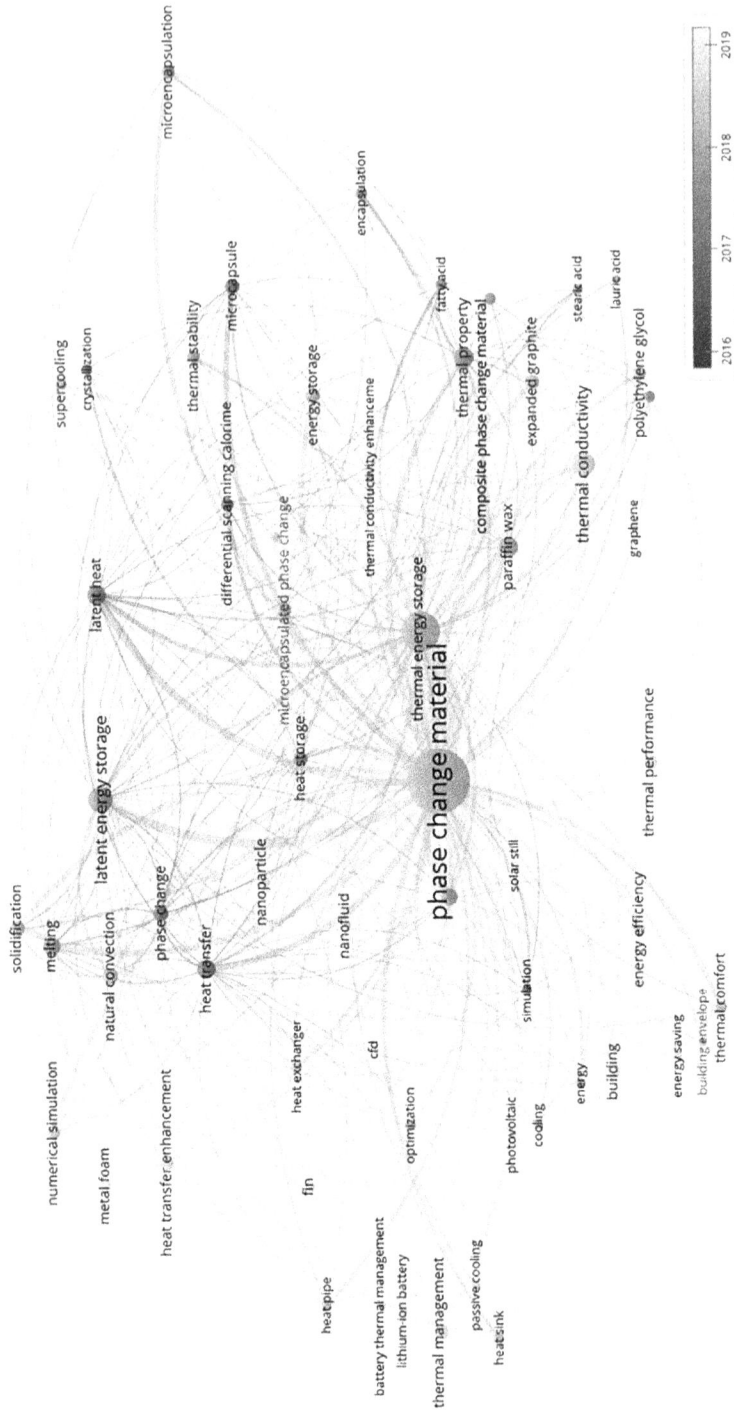

FIGURE 1.18   Bibliometric year-wise analysis based on author keywords.

research was published. Furthermore, a keyword network analysis was employed to gain further insights. The major conclusions from the analysis are summarised as follows.

- The statistics demonstrate a steady increase in research from year to year, indicating that PCMs have gained increasing attention. While Germany and Australia rank highest in the "citations per article" and "articles per million people" indices, China was shown to be the leader in PCM research, accounting for nearly 35% of all research publications. It was also observed that the *Journal of Energy Storage* has the highest number of publications.
- Furthermore, the research was found to revolve around five domains, i.e., "research into heat transfer enhancement," "research into material characterisation," "research into battery thermal management," "applications in solar and building temperature regulation," and "research into material and thermal enhancement of its properties."
- From the analysis, it is found that PCMs are most commonly used in solar energy applications such as PV/T systems, solar stills, building cooling/passive systems, and battery thermal management systems.
- It is interesting to see PCM microencapsulation become more popular in research. Paraffin wax has been extensively used by researchers. Additionally, extended graphene has been widely employed as a thermal performance enhancer.
- Nowadays, PCM research is focused on enhancing thermal performance by adding thermal conductivity enhancers such as metal foam, fins, nanoparticles, etc., and on expanding its use in applications like thermal management systems for buildings and batteries.

In conclusion, PCM has shown itself to be a potential solution across a variety of applications, and it is generating considerable excitement amongst researchers. However, the low thermal conductivity and thermal stability of PCMs represent a particular area of concern, and as a result, new techniques should be introduced with critical interactions between knowledge domains to accomplish innovative and environmental solutions.

REFERENCES

[1] British Petroleum, 2016. *BP Energy Outlook*, p. 98. http://doi.org/10.1017/CBO9781107415324.004
[2] International Energy Agency, 2015. *World Energy Outlook 2015*. Executive Summary. International Energy Agency Books Online, pp. 1–9.
[3] Yataganbaba, A., Ozkahraman, B. and Kurtbas, I., 2017. Worldwide trends on encapsulation of phase change materials: A bibliometric analysis (1990–2015). *Applied Energy*, *185*, pp. 720–731.
[4] Rathod, M.K. and Banerjee, J., 2013. Thermal stability of phase change materials used in latent heat energy storage systems: A review. *Renewable and Sustainable Energy Reviews*, *18*, pp. 246–258.

[5] Ali, H.M., Touseef, R., Arici, M., Said, Z., Durakovic, B., Mohhamed, H.I., Kumar, R., Rathod, M., Buyukdagli, O. and Teggar, M., 2024. Advances in thermal energy storage: Fundamentals and applications. *Progress in Energy and Combustion Science, 100*, p. 101109.

[6] Shukla, A., Kant, K., Sharma, A. and Biwole, P.H., 2017. Cooling methodologies of photovoltaic module for enhancing electrical efficiency: A review. *Solar Energy Materials and Solar Cells, 160*, pp. 275–286.

[7] Wong, H.S.P., Raoux, S., Kim, S., Liang, J., Reifenberg, J.P., Rajendran, B., Asheghi, M. and Goodson, K.E., 2010. Phase change memory. *Proceedings of the IEEE, 98*(12), pp. 2201–2227.

[8] Mselle, B.D., Zsembinszki, G., Borri, E., Vérez, D. and Cabeza, L.F., 2021. Trends and future perspectives on the integration of phase change materials in heat exchangers. *Journal of Energy Storage, 38*, p. 102544.

[9] Mongeon, P. and Paul-Hus, A., 2016. The journal coverage of Web of Science and Scopus: A comparative analysis. *Scientometrics, 106*(1), pp. 213–228.

[10] Van Eck, N. and Waltman, L., 2010. Software survey: VOSviewer, a computer program for bibliometric mapping. *Scientometrics, 84*(2), pp. 523–538.

[11] Cabeza, L.F., Chàfer, M. and Mata, É., 2020. Comparative analysis of web of science and scopus on the energy efficiency and climate impact of buildings. *Energies, 13*(2), p. 409.

[12] Yu, D. and He, X., 2020. A bibliometric study for DEA applied to energy efficiency: Trends and future challenges. *Applied Energy, 268*, p. 115048.

[13] Borri, E., Tafone, A., Zsembinszki, G., Comodi, G., Romagnoli, A. and Cabeza, L.F., 2020. Recent trends on liquid air energy storage: A bibliometric analysis. *Applied Sciences, 10*(8), p. 2773.

[14] Khor, K.A. and Yu, L.G., 2016. Influence of international co-authorship on the research citation impact of young universities. *Scientometrics, 107*(3), pp. 1095–1110.

2 Systematic Classification and Preparation Methods for Phase Change Materials

H. M. Teamah

HIGHLIGHTS

- Energy storage systems can be sensible, latent, or thermochemical. Latent storage has high storage capacity in different applications.
- When the system operating temperature range is narrow, latent storage works the best. This is attributed to the high latent heat of phase change materials (PCMs). They can be organics, inorganics, and composites.
- Organic PCMs include fatty acids and paraffins. They are of high stability, non-toxic, and non-corrosive.
- Integration of organic PCMs is hindered by their low thermal conductivity. Different enhancement techniques needs to be implemented to address it.
- Inorganic PCMs include hydrate salts and metals.
- Inorganic PCMs are well suited for relatively high-temperature applications.
- Different methods of preparation and handling of PCMs are explained in detail.
- Chemical methods of PCM preparation include interfacial polymerization, emulsion, and in situ suspension.
- Physical methods of PCMs preparation include centrifuge extrusion, fluidized beds, and spray drying.
- Physico-chemical methods of PCM preparation include the sol–gel method, complex coacervation, and solution casting.

NOMENCLATURE

C_p Material specific heat capacity
E_{stored} Energy stored in the system
K Thermal conductivity
k_{eff} Effective thermal conductivity

DOI: 10.1201/9781003331957-2

L_{pcm}	PCM latent heat of phase change
m_{pcm}	mass of PCM in the system
m_w	Water mass flow rate
N_c	Number of PCM cylinders
ΔT_m	Melting temperature range of the considered PCM
ΔT_w	System operating temperature range
T	Temperature of system operation
T_{st}	Initial temperature
V_w	Water volume in storage system

LATIN SYMBOLS

| ρ | Density |
| ρ_w | Water density |

ABBREVIATIONS

HTF	Heat transfer fluid
PCM	Phase change material
PMMA	Polymethyl methacrylate
PVA	Polyvinyl alcohol
SBES	Styrene-b-(ethylene-co-butylene)-b-styrene
SDHW	Solar domestic hot water system

2.1   BACKGROUND AND CLASSIFICATION OF PCMS

Today's era has witnessed a great deal of human revolution. This has been accompanied by a strong reliance on fossil fuels as traditional energy resources. It has increased the greenhouse gas emissions, which is a serious global problem. To remedy this, a shift to renewable resources has prevailed in recent decades. Due to the intermittent nature of renewable resources, thermal energy storage is crucial. It is a pivotal part of efficient energy management to ensure that renewable resources can meet the demand.

There are different forms of thermal energy storage systems (Figure 2.1). Sensible heat storage is when energy is stored in the form of increased temperature. There is no variation in the phase. The common abundant forms of common sensible storage systems include rocks and water. Due to their high availability, they are considered to be less costly. Latent storage is another form. As the name implies, the heat is stored through a change in phase. Phase change materials (PCMs) undergo a change in phase, which releases a large amount of heat. Thermochemical storage is the third storage type. In such storage, heat is stored in chemical reactions. Sensible and latent storage systems are prevalent in the majority of applications. A hybrid

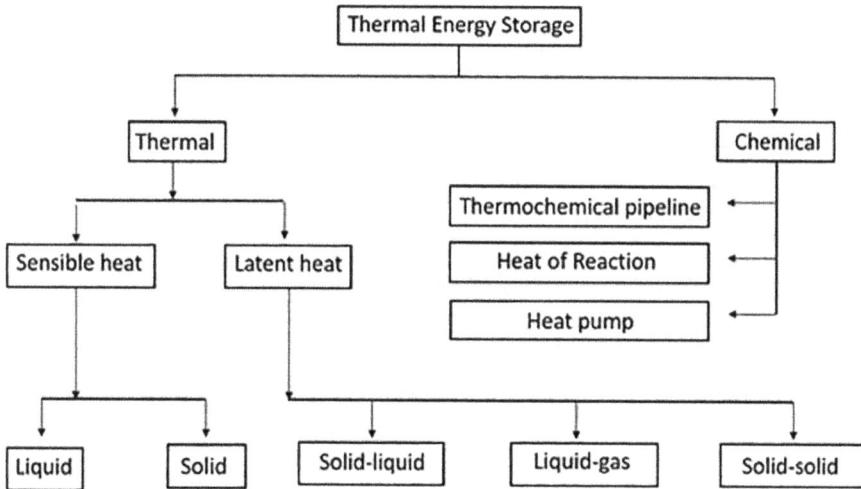

FIGURE 2.1 Thermal energy storage system forms [1].

storage contains both storage forms together. It maximizes the advantages of both and minimizes the disadvantages.

Materials are typically classified as solids, liquids, or gases. PCMs show different methods of transformation of phase. They can transform from solid to liquid, from liquid to gas, etc. Latent heat is released when PCMs undergo this phase transformation. Latent heat is significantly higher than sensible heat. The phase change heat of water is 2441 kJ/kg at 25°C. This is more than 50 times the water sensible storage capacity at 10°C of operating range. This feature makes PCMs a compact energy storage solution for a variety of applications. Those applications can be residential, commercial, institutional, or industrial ones.

PCMs have favorable thermal properties; however, their history emerged in the early 1900s. Although the majority of PCMs were discovered at that time, their applications were very limited. Since the 1980s, there has been a lot of research reported on PCMs and their applications [1]. The number of research works on the subject of PCMs has been growing rapidly in an exponential manner over the past few decades.

Different phase transformations are reported in PCMs. Figure 2.2 summarizes those transformations. They can be solid–solid, solid–liquid, liquid–gas, and solid–gas. The melting process is the most common, where transformation happens from solid to liquid. The inverse happens during solidification.

Two main categories of organic PCMs are fatty acids and paraffins. Their prices are higher than the corresponding prices for inorganic PCMs. Paraffins are prepared as a product for the refining of different oils. This makes them abundant; however, they are still pricy. Their price can be as high as USD $2 per kilogram. Table 2.1 includes the properties of common paraffins. Paraffin types are composed of chains of carbon atoms. Melting temperature gets higher with longer carbon

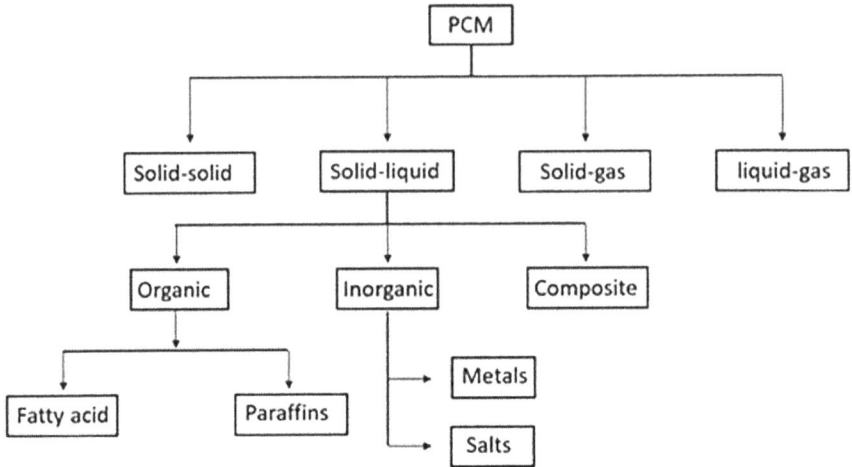

FIGURE 2.2   Common PCM classifications.

TABLE 2.1
Thermal Properties of Common Paraffins [2–4]

Paraffin	T_m (°C)	Density of solid phase (kg/m³)	Heat of change in phase (kJ/kg)	Conductivity (W/mK)
n-Hexadecane	18	770	237	0.2
n-Heptadecane	22	760	213	0.145
n-Octadecane	28	865	245	0.148
n-Nanodecane	32	830	222	0.22
n-Docosane	44.5	880	249	0.2

chains. However, the latent heat of phase change does not follow this correlation. Fatty acids are the other major category of organic PCMs. They comprise a family of acids whose melting temperatures range from 16°C to 65°C. Their prices are also high; however, they are not as high as those for paraffins. A kilogram of stearic acid costs about USD $1.5 per kilogram. Table 2.2 includes common fatty acid thermal properties.

Organic PCMs have numerous advantages. They have a high degree of chemical stability. They are also non-toxic, non-corrosive, and fire resistant. This facilitates the process of their encapsulation and handling. When a phase change occurs in organics, there is a marginal change in their volume that makes the sizing of their encapsulation straightforward. Fatty acids come from raw materials. Mixing two or more fatty acids can change the desired melting temperature. As they have a natural origin, this mitigates their environmental potential, as it lowers carbon dioxide emissions.

TABLE 2.2
Thermophysical Properties of Common Fatty Acids [5]

Fatty acid	T_m (°C)	Density (kg/m³)	Specific heat capacity (kJ/kgK)	Latent heat (kJ/kg)	Solid conductivity (W/mK)
Capric acid	31.5	886	NA*	153	0.149
Lauric acid	42	870	1.6 (solid), 2.3 (liquid)	178	0.147
Myristic acid	54	844	1.6 (solid), 2.7 (liquid)	187	NA*
Palmitic acid	63	847	NA*	187	0.165

NA*: value is not available in literature.

Organic PCMs have a major disadvantage of low thermal conductivity relative to that of inorganics. In the majority of applications, charging or discharging periods are limited. The low thermal conductivity risks that a partial melting/solidification will occur to the PCM. This allows for only partial utilization of phase change. The extent of this can be reduced by adding some nanoparticles to a plain PCM. However, this might alter the original properties of the plain PCM.

The second main category of PCMs is the inorganics. They are either hydrate salts or metals. Hydrate salts are a mixture of water and inorganic salts. When phase transition happens, the hydrate salt crystals divide into water and less aqueous salts. Table 2.3 includes the thermophysical properties of common salt hydrates. Metals are the second major category of inorganic PCMs. They have favorable thermal properties, including high thermal conductivity. Their phase change temperatures span a wide range. This makes them suitable for high-temperature applications. Metals such as gallium, indium, and cesium are applicable to low-temperature applications. Magnesium, aluminum, and zinc are suitable for high-temperature applications. Metal alloys with high temperatures (400–1000°C) are used for extremely high-temperature systems.

Inorganic PCMs generally have greater phase change temperatures than organics. This makes them better candidates for high-temperature applications. In addition, they have a lot of other advantages. They have high storage potential, as their latent heat is higher. They are available in the markets at lower prices compared to the organic PCMs. In addition, they do not have the drawback of low thermal conductivity like the organics. On the other hand, they have numerous disadvantages as well. These include the large volume change when phase transformation happens. This makes the sizing of the encapsulation very difficult. They also show high degree of phase segregation and supercooling. They also exhibit more corrosive behavior than organics. Table 2.3 includes thermal properties of common salt hydrates. Table 2.4 summarizes the main differences between organic and inorganic PCMs.

The last main classification is the composite phase change materials. These are the products of the addition of nanometals or graphite to a plain PCM. They have been receiving attention recently to overcome the hurdle of low thermal conductance in common PCMs. They represent modern ways of enhancing PCM conductivity.

TABLE 2.3
Thermal Properties of Common Salt Hydrates [6–10]

Hydrate salt	Chemical formula	T_m (°C)	Heat of fusion (kJ/kg)
Calcium chloride hexahydrate	$CaCl_2.6H_2O$	30	170
Sodium sulfate decahydrate	$Na_2SO_4.10H_2O$	32.4	239
Sodium hydrogen phosphate dodecahydrate	$Na_2HPO_4.12H_2O$	36.5	279
Zinc nitrate hexahydrate	$Zn(NO_3)_2.6H_2O$	36.4	147
Iron (III) chloride hexahydrate	$FeCl_3.6H_2O$	37	186.2
Calcium chloride tetrahydrate	$CaCl_2.4H_2O$	44.2	99.6
Calcium nitrate tetrahydrate	$Ca(NO_3)_2.4H_2O$	47	142
Sodium thiosulfate pentahydrate	$Na_2S_2O_3.5H_2O$	48	206
Sodium acetate trihydrate	$C_2H_3NaO_2.3H_2O$	58	252

TABLE 2.4
Comparison Summary between Organic and Inorganic PCMs

	Organic PCMs	Inorganic PCMs
Major categories	Fatty acids, paraffins	Salt hydrates
Melt temperature	Lower	Higher
Latent heat	Lower	Higher
Thermal conductivity	Lower	Higher
Cost	Higher	Lower
Chemical stability	Higher	Lower
Corrosiveness	Lower	Higher
Volume change after altering the phase	Lower	Higher
Subcooling	Lower	Higher
Phase segregation	Lower	Higher

However, there is not much recent work on the subject. Organics and inorganics have the highest market share.

Latent heat storage offers a compact option for storing energy. Its storage capacity is very high compared to that of sensible storage under certain conditions. Those are mainly a narrow range of temperature difference and a sufficient time for melting/solidifying the PCM. Figure 2.3 shows a plot comparing the energy that is

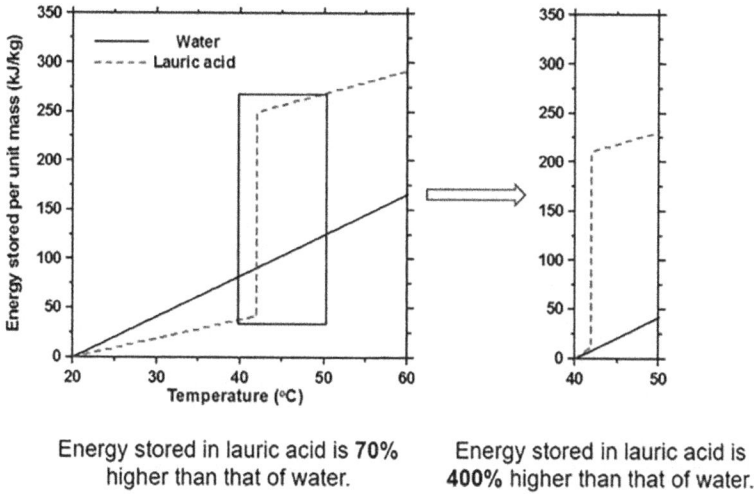

Energy stored in lauric acid is **70%** higher than that of water.

Energy stored in lauric acid is **400%** higher than that of water.

FIGURE 2.3 Energy stored in lauric acid compared to water.

stored in lauric acid (fatty acid) as a PCM relative to that in water. The melting temperature of lauric acid is 42°C. The plot on the left shows the comparison of energy stored when the system operating temperature range is 40°C. There is q 70% increase in storage with lauric acid compared to water. The heat that is released due to melting is the source of enhancement, as the specific heat capacity of lauric acid is less than that of water. The right plot shows the energy storage enhancement in narrow a operating temperature range (10°C). The enhancement is more than four times for lauric acid compared to water. Latent heat storage is definitely a better candidate for systems that have narrow system temperature ranges.

The majority of PCM research concentrates on utilizing organics and inorganics. To assess their suitability to different applications, their thermal properties are compared relative to those of water. Organic PCM density is approximately 80% that of water. The specific heat capacity is half the corresponding capacity of water. The phase transformation heat is ~150 MJ/m^3. Inorganic PCMs have higher densities that can be a bit less than double that of water. The specific heat capacity is not much different from that of the organic PCMs. It is around half of the corresponding capacity of water. Their latent heat of fusion is high and can reach two times that of the organic PCMs. This highlights that the feasibility of PCM integration mainly for narrow temperature ranges of operation. Under this condition, the high latent heat reflects the high storage capacity. However, when a higher operating temperature range is considered, the benefit of PCM integration fades, as water sensible properties are far better than those of the corresponding of PCM.

The high latent storage of PCMs and better sensible storage in water have motivated research to consider hybrid storage. Hybrid storage both maximizes storage media advantages and reduces the extent of disadvantages. A proper heat exchanger configuration is crucial for hybrid storage design. A common type is the shell-and-tube heat exchanger. The placement of PCMs in the tubes is more common. Water flows through the shell in parallel, counter, or cross flow patterns.

The energy stored in hybrid storage is the sum of energy for both sensible and latent storage. The following equations calculate the stored energy for water-based, latent, and hybrid storage configurations.

For water-based thermal energy storage:

$$E_{stored} = m_w \, C_{pw} \, \Delta T_w = \rho_w \, V_w \, C_{pw} \, \Delta T_w \tag{2.1}$$

where m_w is water mass, C_{pw} is water specific heat capacity, and ΔT_w is the system operating temperature range.

For latent thermal energy storage:

$$E_{stored} = m_{pcm} \, C_{pcm} \, \Delta T + m_{pcm} \, L_{pcm} \tag{2.2}$$

where m_{pcm} is PCM mass, C_{pcm} is specific heat capacity of the PCM, and ΔT is the system operating temperature range. L_{pcm} is the phase transformation latent heat of the PCM.

For hybrid energy storage:

$$E_{stored} = m_w \, C_{pw} \, \Delta T + m_{pcm} \, C_{pcm} \, \Delta T + m_{pcm} \, L_{pcm} \tag{2.3}$$

Hybrid storage energy is the summation of stored energy in water and the PCM. The stored energy in water is proportional to temperature difference (first term on the right-hand side of Equation 2.3). The energy stored in the PCM is both sensible and latent. The sensible term is correlated to the temperature difference (second term on the right-hand side of Equation 2.3). The latent portion is a function of the latent heat of phase change (third term on the right-hand side of Equation 2.3).

The volume percent of PCM is the most important factor in the system energy stored. The energy stored is higher as the percent of PCM increases. Figure 2.4

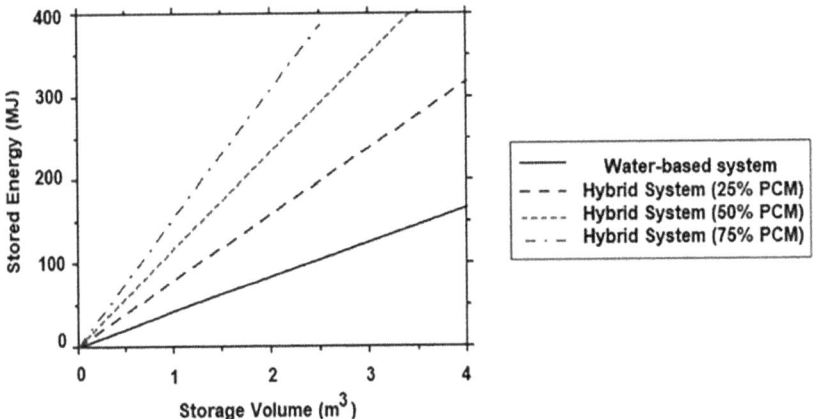

FIGURE 2.4   Integrated PCM volume fraction effect on stored energy in the system.

shows a graph for energy stored versus PCM volume percent. The graph is for an operating temperature range of 10°C. A 75% percent PCM hybrid system shows the highest storage potential. A volume of 3.7 m^3 is the needed water-based storage for 150 MJ. If the PCM volume percent is 25%, 1.9 m^3 is required. Only 1 m^3 of storage is required if the PCM volume percent is 75%. There is a significant storage volume reduction when PCM is integrated into the system. This is of the utmost importance in applications where the allocated storage space is of major concern. This can be applicable for supplying heat to multiresidential applications.

2.2   PREPARATION AND HANDLING OF PCMS

PCMs typically change their phase through melting. The handling process is crucial for realizing the benefit of PCM. Encapsulation is widely used to prevent liquid PCM leakage. Encapsulation involves two main parts: the core and the shell. PCM is main located in the core to be shielded. Polymer materials with favorable thermophysical properties are used in the shell. The shell needs to be well designed, particularly if the enclosed PCM has a degree of toxicity. Encapsulation can occur at different levels, depending on the application. These can be macroencapsulation (bulk), microencapsulation, and nanoencapsulation.

Macroencapsulation, or the bulk method, is the simplest handling method. It is also cheaper than the other methods. It is utilized in heat exchangers, solar thermal energy storage systems, and building applications [11,12]. Microencapsulation can take different forms: flat sheets, tubular, cylindrical, and spherical. Cylindrical encapsulations are the most common ones. They are integrated with solar thermal systems and with building applications. The effective heat transfer through such encapsulations is strongly dependent on size. The smaller the diameter of encapsulation, the better the heat transfer rate.

Microencapsulation of PCMs is the second way of handling PCM. It is higher in efficiency and cost compared to macroencapsulation. The sizes of such microencapsulations range from 1 μm to 1 mm. The process of microencapsulation is not a straightforward process. However, it results in a more efficient system. This is attributed to shorter charging and discharging times and a higher contact area for heat transfer. Microencapsulation involves two main methods: chemical and physical [13,14]. The chemical methods include different types of polymerization: interfacial polymerization, emulsion, and in situ suspension. The physical methods include: centrifuge extrusion, fluidized beds, and spray drying.

In interfacial polymerization, the microencapsulation wall is a result of the fast polymerization of lipophilic and hydrophilic monomers at the interface. The process involves two reaction monomers, continuous phase and dispersed phase. Those are not miscible during emulsion. The polymerization typically happens on the side of the organic phase of the interface. This method is relatively straightforward with no need for expensive equipment. However, the walls of microencapsulation are very thin, and thus they are easy to crack.

In in situ polymerization, there is mixing between dispersed-phase core material and water to form an emulsion. The catalyst and the shell monomer can be either outside or inside the core material droplet. Polymerization always takes place on the droplet surface. At first, the shell monomer starts polymerization, forming low weight prepolymer. As the size and the disposition increase, a solid capsule is finally formed.

In emulsion polymerization, high speed stirring is utilized. The PCM is suspended in an aqueous polar medium. To ensure more stabilization of particles, surfactants are used. After that, lipophilic monomers are added as a preparation for polymerization. A polymer is then formed and encapsulates the PCM. For better polymer properties, polymethyl methacrylate or polyvinyl alcohol can be added. They ensure a smooth surface of the formed microencapsulations [15,16].

Suspension polymerization involves the dispersion of a wall material monomer, core material, and catalyst for stirring in a suspension stabilizer. This will initiate the polymerization process. Disposition and continuous precipitation form the microencapsulation. Polymer wall materials such as vinylidene chloride and polyvinyl chloride are formed by this method. The product of such method is characterized by cleanliness and a high degree of purity. This makes it suitable for direct molding. Table 2.5 includes different microencapsulations from different research studies.

The physical methods include spray drying. Its principle relies on the net-like structure that is formed when preheating the shell material. This causes a separation between small-molecule core materials and large-molecule core materials. Evaporation causes the escape of small-molecule core materials. However, the large-molecule core materials remain in the film. The type of shell material is an influencing factor for the desired membrane pores [23].

The procedure of the spray drying method includes two different steps. The first step includes dissolving the wall material in a water solution. After that, there is the addition of an oil-soluble material to the core to form an emulsion. This emulsion goes through a liquefaction process and then through a high-temperature air flow. This causes the evaporation of the solvent of dissolved wall material. A microcapsule from a size as small as 5 μm to as big as 5000 μm is formed through the solidification process.

TABLE 2.5
Different Microencapsulations from Different Research Studies

Reference	PCM	Shell material	Particle size (μm)	Encapsulation method
Al-shannaq et al. [15]	RT21 paraffin	Polymethyl methacrylate with polyvinyl alcohol	15	Emulsion
Rahman et al. [16]	RT21 paraffin	Polymethyl methacrylate	20–40	Suspension
Sari et al. [17]	n-Nonadecane	Polymethyl methacrylate	8	Emulsion
Sari et al. [18]	n- Heptadecane	Polystyrene	<2	Emulsion
Jin et al. [19]	Paraffin	Urea–formaldehyde	20	In situ
Silakhori et al. [20]	Paraffin	Polyaniline	<1	Emulsion
Su and Huang [21]	Paraffin	Methanol–melamine–formaldehyde	10–30	In situ
Sanchez et al. [22]	Paraffin	Polystyrene-co- polymethyl methacrylate	20	Suspension

Spray drying method require a careful choice of wall and core materials. It also needs an optimum spray rate. The main advantages of such technology are its low cost and high efficiency [24,25]. It is considered an energy-conserving method widely used for the production of different materials. This method has been widely used in the pharmaceutical and food industries [26,27]. Studies have shown a high degree of stability and reversibility of the synthesized material. However, there are drawbacks that hinder the widespread application of such technology. This technology can not control the produced particle size effectively. In addition, the fast-flowing air sometimes can cause non-uniformity of the product material.

Solvent evaporation is a comparable method to spray drying. Its procedure is even easier and simpler. The first step involves the dissolution of the wall material. After that, when the solvent vaporizes, the droplet capsule wall is then formed [28,29]. Although the method is straightforward, it shows some limitations. A solvent must be selected as a volatile material. Solvent volatilization also takes a large period of time, which limits its applicability to large-scale industrial applications.

Centrifugal extrusion is an important physical method that dominates the pharmaceutical industry. The method involves the pumping of incompatible core and shell materials. This occurs at the same time in a nozzles that handle two fluids. The liquid droplets collide in the rotating ejection process [30]. A fluidized shell material film is formed around the core. This shell is solidified by the implementation of cooling system. It can be soaked in a gel bath, which results in the production of microcapsules.

There are also the physico-chemical methods like the sol–gel method. In this method, liquid compounds are mixed with high-activity chemical compounds. They are termed precursors. They can either be inorganic compounds or metal organic alkoxides. They mix uniformly with the liquid phase of the raw material. This helps in initiating hydrolysis and causing a condensation chemical reaction throughout the solution [31]. This yields a sol transparent system. The sol colloidal particles undergo polymerization to form a gel. This is composed of a network structure in three dimensions. The space in between the branches of the network is populated with a solvent, which gradually losses its fluidity. Polycondensation is then formed after sintering and heat treatment. This results in the enhancement of structural stability and mechanical properties.

The most important advantage of this method is the high toughness and strength of the prepared materials. In addition, the temperature of preparation is low. The solvent is easily removable during the process of treatment. The resulting material has a high purity and uniform components [32]. There are some disadvantages, as the precursors are pricy and toxic. Shrinkage can sometimes be produced during drying. The products are fragile, and cracking can be easy as well. The derived materials are applicable in energy, medicine, ceramics, electronics, and separation chromatography [33].

Another physico-chemical method is complex coacervation. This process includes the utilization of two types of polymers whose charges are opposite as wall materials. The pH value adjustment is crucial for the solution. It causes reduced solubility and neutralization. This results in phase separation followed by polymerization [34].

The main advantage of complex coacervation is its ability to produce water-insoluble solids. It can also result highly efficient liquid core materials. The microcapsules that are formed by this method protect the core material significantly. This is attributed to complex condensation and good moisture and thermal resistance. They does not show degradation in high humidity or high temperature environment [35,36].

Solution casting is a method which differs from the listed microencapsulation methods. It involves the wrapping of an organic PCM with a polymer material at the macro level. This acts to mitigate any leakage problems that might be encountered. If the mixture has a temperature greater than the melting temperature, the wrapped PCM is a solid. This reduces any possible increase in volume due to phase change. The produced mixture does not need additional packaging. This is considered a cost-effective method. Any size can be considered for the material, and is tailored relative to the actual demand [37–42].

Alkan et al. [42] studied the use of polymethyl methacrylate (PMMA) with fatty acids to enhance their thermophysical properties. PMMA was mixed with lauric acid, stearic acid, myristic acid, and palmitic acid. Different mass fractions of 50, 60, 70, 80, and 90% were used. They were able to find that the latent heat was enhanced up to 190 J/g. This method is best suited for applications with an operating temperature range from 40°C to 70°C.

Li et al. [43] studied the addition of polypropylene to a mix of liquid and solid paraffins. It acted as a supporting agent for preparing a flame-redundant material. After testing, the formed PCM had a high phase transition latent heat of 127 J/g. The phase change temperature was found to be lower than about 25°C.

Chen et al. [44] added paraffin powder to a styrene-b-(ethylene-co-butylene)-b-styrene network. It was used to coat high-density polyethylene. This resulted in the preparation of new shaped organic compound. Its latent heat reached 152 J/g, and melt temperature was 50.56°C. An accelerated degradation test was performed at 80°C. The material was proven to be stable after 50 cycles of heating or cooling.

As mentioned before, organic PCMs are a major category of PCMs used in a variety of applications. However, their thermal conductivity is low. This is critical for different applications like solar thermal systems, as solar resources are intermittent on a daily and seasonal basis. Insufficient charging and discharging periods may cause incomplete melting of the PCM. This leads to limiting the benefit of high phase change heat [45]. There has been a lot of research on innovative ways to enhance thermal conductivity using nanoparticles.

Different nanomaterials used to enhance PCM conductivity are shown in Figure 2.5. They are mainly organics, inorganics, and hybrids [46,47]. The organics are classified into graphite, carbon nanofibers, and nanotubes. Inorganics are several metal oxides and metalloid nanoparticles. The third type includes hybrids of both organics and inorganics.

Nanometals can be prepared in different ways. The most common techniques for nanoenhanced PCM preparation are the one-step method and the two-step method. The one-step method is more straightforward than the two-step method. In the one-step method, the synthesis and dispersion of nanoparticles are performed at the same time. This eliminates the need for storage or shipping. Small-scale manufacturing

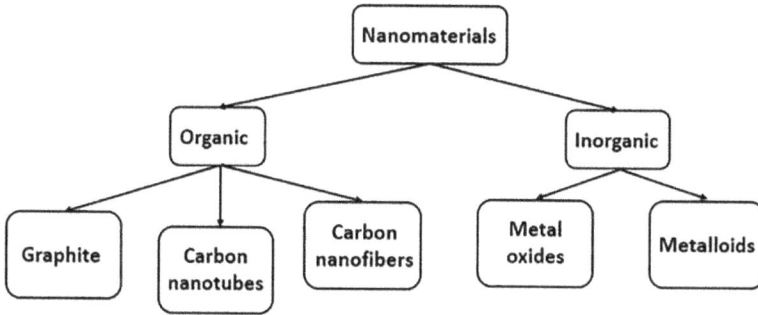

FIGURE 2.5   Nanometal types added to PCMs.

applications use the one-step method. The two-step method is suited for mass production. The first stage involves forming a powder from the synthetization process of nanoparticles. It is followed by the dissolution of nanoparticles in the PCM.

Punniakodi and Senthil [48] provided a recent comprehensive summary of the advances of nanoparticles integration with a PCM in solar thermal systems. The different nanomaterial types are metals, metal oxides, carbon-based, and non-carbon-based particles. The integration of nanometals to the PCM enhanced its thermal conductivity. However, the energy storage enhancement was not always guaranteed. Shoeibi et al. [49] presented a detailed analysis of that topic. They highlighted the favorable effect of nanomaterial integration on the PCM. However, they reported challenges from the economic and technical aspects.

Muzhanje et al. [50] summarized plenty of research that integrated nanomaterials in phase change materials. They highlighted the crucial importance of increasing thermal conductivity. They provided in-depth assessments of the challenges. The nanoenhanced material preparation was reported to require a special preparation procedure and equipment. However, they found out that although nanomaterials increase PCM thermal conductivity, they may result in lower phase change heat. Mixing nanoenhanced PCMs homogenously is a crucial cornerstone for a better system performance.

Zhang et al. [51] performed research on the characterization of carbon nanotube addition to PCMs. Thermophysical properties of microPCMs were assessed by in situ polymerization. They reported that nanotube weight increases, and the photo-thermal conversion performance and thermal conductivity were enhanced.

Huq et al. [52] analyzed the integration of aqueous graphene from different aspects of stability and heat transfer augmentation. Covalent functionalization was found to be the most effective. It enhanced the coefficient of heat transfer by 119% and the thermal conductivity by 30%.

Following the preparation of a PCM, it can be used in a variety of applications. These span the residential, commercial, and industrial sectors. PCMs were found to be promising when integrated with the building envelope and systems [53]. They stabilize the indoor temperature, which guarantees better thermal comfort for the occupants. Their integration with the wall system offers better thermal storage, as the heat that is absorbed during daylight time can be used during the night [54]. Although PCMs improve the system efficiency, they add a cost of integration. This needs more in-depth life-cycle analysis and performance optimization to find out the most favorable integration conditions from performance and economic aspects [55]. A recent summary of related research found that the integration of PCMs into building systems can have a feasible payback time as low as two years [56]. It can also result in very high payback that can exceed 15 years. Different systems need tailored models to ensure that PCMs are the right fit for the application. In addition to the financial aspect, the environmental aspect needs to be considered [57]. The PCMs that have bio-based origins tend to be more environmentally friendly than those that come from hydrocarbons. The life-cycle analysis needs to consider the magnitude of the emissions from the start of the preparation process until the end [58]. This will support the wider integration of PCMs with sustainable systems and development.

PCMs have a promising potential for being integrated in the industrial and institutional sectors as well. One area in the industrial sector is the efficient cooling of nuclear reactors [59]. PCMs act as passive coolants for reactor containment buildings. They have shown higher effectiveness in accident initiation than the traditional spray systems [60]. In the institutional field, PCMs show great feasibility in medicine. As they show a high degree of biocompatibility and non-toxicity, they were used efficiently in tumor therapy [61]. In addition, fatty acids and eutectics are effective in the field of temperature-controlled drug delivery, as they offer a high level of temperature modulation to the process [62].

In addition to solar systems, PCMs can be integrated with other renewable technologies. The high latent heat of PCMs offers improved performance of hydrogen storage systems [63]. An optimum PCM amount can enhance the release/absorption rates of a system [64]. PCM integration with either fins, nanometals, or both has shown better performance improvement as well. PCMs can be integrated with the grout material of the geothermal system [65]. This reduces the need for drilling deeper boreholes and offers a compact storage system at a lower drilling cost [66].

2.3 CONCLUSIONS

This introductory chapter presented an overview on PCM classification, preparation, and handling. Thermal energy storage has been a crucial state of the art, especially with the wide integration of renewable resources. Latent energy storage systems are an alluring option for different applications. They are of special importance in applications with a narrow operating temperature range. PCMs are broad classified as organics, inorganics, or composites. Organics and inorganics are more widely used. Organics mainly include fatty acids and paraffins. Inorganics include salt hydrates and metals. The following key points summarize the research outcomes on PCM classifications:

- Organic PCMs have a high degree of chemical stability. They are non-corrosive and non-toxic, which makes their handling process easier.
- Organic PCMs do not exhibit high volume changes when they are liquefied.

- The main drawbacks of organic PCMs are that they are of a higher cost relative to inorganics, they have lower latent heat than inorganics, and their thermal conductivity is low.
- Inorganic PCMs have a great phase change heat. They also have higher melting temperatures, especially for metals. This makes them a better fit for applications involving high temperatures, like concentrated solar power systems.
- Inorganics are of a lower cost compared to organics. Their thermal conductivity is higher. This makes them suitable for applications with limited charging and discharging periods.
- The main drawbacks of inorganics is that they are more toxic and corrosive than organics. They also show a high volume change when liquefied. This makes their handling process less straightforward than that of organics.

This chapter included the different methods of preparation and handling of PCMs. It provided an overview of macroencapsulation (bulk), microencapsulation, and nano-encapsulation. The macroencapsulation occurs on a large scale that differs from microencapsulation and nanoencapsulation. There are different methods for preparing microencapsulations: chemical, physical, and physico-chemical methods. The following are the main research highlights on chemical methods:

- Interfacial polymerization is where the wall of the microencapsulation is a result of the rapid polymerization of lipophilic and hydrophilic monomers at the interface. This method is simple, with no need for expensive equipment. However, the walls of microencapsulation are very thin, and thus they are easy to crack.
- In situ polymerization is where there is mixing between a dispersed-phase core material and water to form an emulsion. The catalyst and the shell monomer can be either outside or inside the core material droplet.
- Emulsion polymerization is where high speed stirring is utilized. The PCM is suspended in an aqueous polar medium. To ensure more stabilization of particles, surfactants are used. For better polymer properties, PMMA or polyvinyl alcohol can be added.
- Suspension polymerization involves the dispersion of a wall material monomer, core material, and catalyst for stirring in a suspension stabilizer. Polymer wall materials such as vinylidene chloride and polyvinyl chloride are formed by this method. The product of this method is characterized by cleanliness and a high degree of purity. This makes it suitable for direct molding.

There are physical methods of microencapsulation, including:

- Spray drying. Its principle relies on the net-like structure that is formed when preheating the shell material. A microcapsule of a size as small as 5 μm to as big as 5000 μm is formed through the process. It is considered an energy-conserving method widely used for different material production.

However, there are drawbacks that hinder the widespread application of such technology. This technology can not control the produced particle size effectively. In addition, the fast-flowing air sometimes can cause non-uniformity of the product material.

- Solvent evaporation is easier and simpler than spray drying. It is composed of two steps. At first, the dissolution of the wall material occurs in a continuous phase. After, the formation of a droplet capsule wall happens as the solvent vaporizes. Solvent volatilization usually takes a large period of time, which limits its applicability to large-scale industrial applications.
- Centrifugal extrusion involves the pumping of incompatible core and shell materials. This occurs at the same time in a nozzle that handles two fluids. It has proven its effectiveness in the pharmaceutical industry.

There are physico-chemical methods of microencapsulation, including:

- The sol–gel method. In this method, liquid compounds are mixed with high-activity chemical compounds. They are termed precursors. They mix uniformly with the liquid phase of the raw material. This helps in initiating hydrolysis and creating a condensation chemical reaction throughout the solution. This results in high toughness, purity, and strength of the prepared materials. There are some disadvantages, as the precursors are pricy and toxic. The products are fragile, and cracking can be easy as well.
- Complex coacervation. This process includes the utilization of two types of polymers, whose charges are opposite, as materials for the wall. Adjusting the solution pH value in the process is crucial. It causes reduced solubility and neutralization. The microcapsules that are formed by this method significantly protect the core material.

To overcome the drawback of low PCM thermal conductivity, nanomaterials are added. They are classified into organics, inorganics, and hybrids. The organics are graphite, carbon-based, and non-carbon-based nanomaterials. Inorganics are different oxides of metals. Nanometals can be prepared by different ways. The most common techniques for nanoenhanced PCM preparation are the one-step method and the two-step method. Small-scale manufacturing uses the one-step method, as it is simple. The synthesis and dispersion of nanoparticles is done simultaneously. This is well suited to small-scale manufacturing applications. The two-step method suits large-scale production, as it includes two stages. Nanoparticle synthetization in a powder form occurs first. After that, the dissolution of nanoparticles in the PCM happens.

REFERENCES

[1] Abhat A., "Low temperature latent thermal energy storage system: Heat storage materials", *Solar Energy*. 1983;30:313–332.
[2] Alieva R.V., Vakhshouri A.R., et al., "Heat storage materials based on polyolefins and low molecular weight waxes", *Plastics (Russian Language)*. 2012;10:42–46.

[3] Haynes W.M., *CRC Handbook of Chemistry and Physics*. 91st ed. Boca Raton, FL: CRC Press Inc.; 2010–2011.

[4] Khan Z., Khan Z., Ghafoor A., "A review of performance enhancement of PCM based latent heat storage system within the context of materials, thermal stability and compatibility", *Energy Conversion and Management*. 2016;115:132–158.

[5] Teamah H.M., Lightstone M.F., Cotton J.S., "Numerical investigation and nondimensional analysis of the dynamic performance of a thermal energy storage system containing phase change materials and liquid water", *ASME Journal of Solar Energy Engineering*. 2017;139:021004-021004-14.

[6] Guion J., Sauzade J.D., Laugt M., "Critical examination and experiemental determination of melting enthalpies and entropies of salt hydrates", *Thermochimica Acta*. 1983;67:167–179.

[7] Bajnóczy G., Gagyi Pálffy E., Prépostffy E., Zöld A., "Thermal properties of a heat storage device containing sodium acetate trihydrate", *Periodica Polytechnica Chemical Engineering*. 1995;39(2):129–135.

[8] Zhang Y., Jiang Y., Jiang Y., "A simple method, the T-history method, of determining the heat of fusion, specific heat and thermal conductivity of phase-change materials", *Measurement Science and Technology*. 1999;10(3):201.

[9] Tyagi V.V., Buddhi D., "Thermal cycle testing of calcium chloride hexahydrate as a possible PCM for latent heat storage", *Solar Energy Materials and Solar Cells*. 2008;92(8):891–899.

[10] Ushak S., Suárez M., Véliz S., Fernández A.G., Flores E., Galleguillos H.R., "Characterization of calcium chloride tetrahydrate as a phase change material and thermodynamic analysis of the results", *Renewable Energy*. 2016;95:213–224.

[11] Almadhoni K., "A review—An optimization of macro-encapsulated paraffin used in solar latent heat storage unit", *International Journal of Engineering Research and Technology*. 2016;5(1):729–736.

[12] Calvet N., Py X., et al., "Enhanced performances of macro-encapsulated phase change materials (PCMs) by intensification of the internal effective thermal conductivity", *Energy*. 2013;55;956–964.

[13] Gulfam R., Zhang P., Meng Z., "Advanced thermal systems driven by paraffin-based phase change materials—A review", *Applied Energy*. 2019;238:582–611.

[14] Nazir H., Batool M., et al., "Recent developments in phase change materials for energy storage applications: A review", *International Journal of Heat and Mass Transfer*. 2019;129:491–523.

[15] Al-shannaq R., Farid M., Dickinson M., Behzadi S., "Microencapsulation of phase change materials for thermal energy storage in building application", in: *Chemeca 2010: Quality of Life through Chemical Engineering*, Wellington, New Zealand and Barton, ACT: Engineers Australia; September 2012, pp. 943–952.

[16] Rahman A., Dickinson M.E., Farid M.M., "Microencapsulation of a PCM through membrane emulsification and nanocompression-based determination of microcapsule strength", *Materials for Renewable and Sustainable Energy*. 2012;1(4).

[17] Sari A., Alkan C., et al., "Micro/nano-encapsulated n-nonadecane with poly(methyl methacrylate) shell for thermal energy storage", *Energy Conversion and Management*. 2014;86:614–621.

[18] Sari A., Alkan C., et al., "Micro/nano-encapsulated n-heptadecane with polystyrene shell for latent heat thermal energy storage", *Solar Energy Materials and Solar Cells*. 2014;126:42–50.

[19] Jin Z., Wang Y., Liu J., Yang Z., "Synthesis and properties of paraffin capsules as phase change materials", *Polymer*. 2008;49:2903–2910.

[20] Silakhori M., Metselaar H.S.C., Mahlia T.M.I., Fauzi H., "Preparation and characterisation of microencapsulated paraffin wax with polyaniline-based polymer shells for thermal energy storage", *Materials Research Innovations*. 2014;18(6):480–484.

[21] Su J.F., Huang Z., "Fabrication and properties of microencapsulated-paraffin/gypsum-matrix building materials for thermal energy storage", *Energy Conversion and Management*. 2012;55:101–107.

[22] Sanchez Silva L. et al., "Microencapsulation of PCMs with a styrene-methyl methacrylate copolymer shell by suspension-like polymerization", *Chemical Engineering Journal*. 2010;157:216–222.

[23] Magendran S.S., Khan F.S.A., Mubarak N.M., Walvekar M.V.R., et al., "Synthesis of organic phase change materials (PCM) for energy storage applications: A review", *Nano-Structures & Nano-Objects*. 2019;20:100399.

[24] Loi C.C., Eyres G.T., Pat S., Birch E.J., "Preparation and characterisation of a novel emulsifier system based on glycerol monooleate by spray drying", *Journal of Food Engineering*. 2020;285:110100.

[25] Tan S., Zhong C., Langrish T., "Encapsulation of caffeine in spray dried micro-eggs for controlled release: The effect of spray-drying (cooking) temperature", *Food Hydrocoll.* 2020;108:105979.

[26] McDonagh A.F., Duffff B., Brennan L., Tajber L., "The impact of the degree of intimate mixing on the compaction properties of materials produced by crystallo-co-spray drying", *European Journal of Pharmaceutical Sciences*. 2020;154:105505.

[27] Umaña M., Turchiuli C., Rosselló C., Simal S., "Addition of a mushroom by-product in oil-in-water emulsions for the microencapsulation of sunflower oil by spray drying", *Food Chemistry*. 2020;343:18429.

[28] Ekanem E., Ekanem Z.Z., Vladisavljevic E.T., "Facile microfluidic production of composite polymer core-shell microcapsules and crescent-shaped microparticles", *Journal of Colloid and Interface Science*. 2017;498:387–394.

[29] Wang X., Yin H., Chen Z., Xia L., "Epoxy resin/ethyl cellulose microcapsules prepared by solvent evaporation for repairing microcracks: Particle properties and slow-release performance", *Materials Today Communications*. 2020;22:100854.

[30] Zhu K., Wang S., Qi H., Liu H., Zhao Y., Yuan X., "Supercooling suppression of microencapsulated n-Alkanes by introducing an organic gelator", *Chemical Research in Chinese Universities*. 2012;28:539–541.

[31] John D.M., Bescher E.P., "Chemical routes in the synthesis of nanomaterials using the sol-gel process", *Accounts of Chemical Research*. 2007;40(9):810–817.

[32] Fang G., Li H., Xu L., "Preparation and properties of lauric/silicon composites as form-stable phase change materials for thermal energy storage", *Materials Chemistry and Physics*. 2010;122:533–536.

[33] Livage J., "Sol-gel processes", *Current Opinion in Solid State and Materials Science*. 1997;2:132–138.

[34] Jamekhorshid A., Sadrameli S.M., Farid M., "A review of microencapsulation methods of phase change materials (PCMs) as a thermal energy storage (TES) medium", *Renewable and Sustainable Energy Reviews*. 2014;31:531–542.

[35] Dai R., Gang W., Li W., Qiang Z., Li X., Chen H., "Gelatin/carboxymethyl cellulose/dioctyl sulfosuccinate sodium microcapsule by complex coacervation and its application for electrophoretic display", *Colloids and Surfaces A: Physicochemical and Engineering Aspects*. 2010;362:84–89.

[36] Fuquet E., Platerink C.V., Janssen H., "Analytical characterization of glutardialdehyde cross-linking products in gelatin-gum Arabic complex coacervates", *Analytica Chimica Acta*. 2007;604:45–53.

[37] Alkan C., Sari A., Uzun O., "Poly(ethylene glycol)/acrylic polymer blends for latent heat thermal energy storage [J]", *AICHE Journal*. 2006;52(9):3310–3314.

[38] Sari A., Alkan C., Karaipekli A., Onal A., "Preparation, characterization and thermal properties of styrene anhydrate copolymer (SMA)/fatty acids composites as form stable phase change materials", *Energy Conversion and Management*. 2008;49:373–380.

[39] Chen K., Yu X., Tian C., Wang J., "Preparation and characterization of form-stable paraffin/polyurethane composites as phase change materials for thermal energy storage", *Energy Conversion and Management*. 2014;77:13–21.

[40] Ehid R., Fleischer A.S., "Development and characterization of paraffin-based shape stabilized energy storage materials", *Energy Conversion and Management*. 2012;53:84–91.

[41] Sari A., "Form-stable paraffifin/high density polyethylebe composites as solid–liquid phase change material for thermal energy storage: Preparation and thermal properties", *Energy Conversion and Management*. 2004;45:2033–2042.

[42] Alkan C., Sari A., "Fatty acid/poly(methyl methacrylate) (PMMA) blends as form-stable phase change materials for latent heat thermal energy storage", *Solar Energy*. 2008;82:118–124.

[43] Li L., Wang G., Guo C., "Influence of intumescent flame retardant on thermal and flame retardancy of eutectic mixed paraffin/polypropylene formstable phase change materials", *Applied Energy*. 2016;162:428–434.

[44] Chen P., Gao X., Wang Y., Xu T., Fang Y., Zhang Z., "Metal foam embedded in SEBS/paraffin/HDPE form-stable PCMs for thermal energy Storage", *Solar Energy Materials and Solar Cells*. 2016;147:60–65.

[45] Fertahi S.D., Jamil A., Benbassou A., "Review on Solar Thermal Stratified Storage Tanks (STSST): Insight on stratification studies and efficiency indicators", *Solar Energy*. 2018;176:126–145.

[46] Sezer N., Atieh M.A., Koc M., "A comprehensive review on synthesis, stability, thermophysical properties, and characterization of nanofluids", *Powder Technology*. 2019;344:404–431.

[47] Mahian O., Kolsi L., Amani M., Estellé P., Ahmadi G., Kleinstreuer C., Marshall J.S., Taylor R.A., Abu-Nada E., Rashidi S., Niazmand H., Wongwises S., Hayat T., Kasaeian A., Pop I., "Recent advances in modeling and simulation of nanofluid flows-part II: Applications", *Physics Reports*. 2019;791:1–59.

[48] Punniakodi B.M.S., Senthil R., "Recent developments in nano-enhanced phase change materials for solar thermal storage", *Solar Energy Materials and Solar Cells*. 2022;238:111629.

[49] Shoeibi S., Kargarsharifabad H., Mirjalily S.A.A., Sadi M., Arabkoohsar A., "A comprehensive review of nano-enhanced phase change materials on solar energy applications", *Journal of Energy Storage*. 2022;50:104262.

[50] Muzhanje A.T., Hassan M.A., Ookawara S., Hassan H., "An overview of the preparation and characteristics of phase change materials with nanomaterials", *Journal of Energy Storage*. 2022;51:104353.

[51] Zhang X., Zhang Y., Yan Y., Chen Z., "Synthesis and characterization of hydroxylated carbon nanotubes modified microencapsulated phase change materials with high latent heat and thermal conductivity for solar energy storage", *Solar Energy Materials and Solar Cells*. 2022;236:111546.

[52] Huq T., Ong H.C., Chew B.T., Leong K.Y., Kazi S.N., "Review on aqueous graphene nanoplatelet Nanofluids: Preparation, Stability, thermophysical Properties, and applications in heat exchangers and solar thermal collectors", *Applied Thermal Engineering*. 2022;210:118342.

[53] Wang X., Li W., Luo Z., Wang K., Shah S.P., "A critical review on phase change materials (PCM) for sustainable and energy efficient building: Design, characteristic, performance and application", *Energy and Buildings*. 2022;260:111923.

[54] Kumar N., Rathore P.K.S., Sharma R.K., Gupta N.K., "Integration of lauric acid/zeolite/graphite as shape stabilized composite phase change material in gypsum for enhanced thermal energy storage in buildings", *Applied Thermal Engineering*. 2023;224:120088.

[55] Jelle B.P., Kalnæs S.E., "Chapter 3—phase change materials for application in energy-efficient buildings", in *Cost-Effective Energy Efficient Building Retrofitting Materials, Technologies, Optimization and Case Studies*, Amsterdam, Netherlands: Elsevier; 2017, pp. 57–118.

[56] Struhala K., Ostrý M., "Life-cycle assessment of phase-change materials in buildings: A review", *Journal of Cleaner Production*. 2022;336:130359.

[57] Dincer I., Bicer Y., "1.27 life cycle assessment of energy", in *Comprehensive Energy Systems*, Editor(s): Ibrahim Dincer, Amsterdam, Netherlands: Elsevier Inc.; 2018, pp. 1042–1084.

[58] Aridi R., Yehya A., "Review on the sustainability of phase-change materials used in buildings", *Energy Conversion and Management: X*. 2022;15:100237.

[59] Shin S.G., Cho J.O., Ko A., Jung H., Lee J.I., "Preliminary design of safety system using phase change material for passively cooling of nuclear reactor containment building", *Applied Thermal Engineering*. 2022;200:117672.

[60] Saeed R.M., Schlegel J.P., Sawafta R., "Characterization of high-temperature PCMs for enhancing passive safety and heat removal capabilities in nuclear reactor systems", *Energy*. 2019;189:116137.

[61] He M., Wang Y., Li D., Zhang M., Wang T., Zhi F., Ji X., Ding D., "Recent applications of phase-change materials in tumor therapy and theranostics", *Biomaterials Advances*. 2023;147:213309.

[62] Zhu C., Huo D., Chen Q., Xue J., Shen S., Xia Y., "A eutectic mixture of natural fatty acids can serve as the gating material for near-infrared-triggered drug release", *Advanced Materials*. 2017:201703702.

[63] El Mghari H., Huot J., Xiao J., "Analysis of hydrogen storage performance of metal hydride reactor with phase change materials", *International Journal of Hydrogen Energy*. 2019;44(54):28893–28908.

[64] Nguyen H.Q., Mourshed M., Paul B., Shabani B., "An experimental study of employing organic phase change material for thermal management of metal hydride hydrogen storage", *Journal of Energy Storage*. 2022;55(B):105457.

[65] Mahmoud M., Ramadan M., Naher S., Pullen K., Olabi A., "Advances in grout materials in borehole heat exchangers", *Reference Module in Materials Science and Materials Engineering*. 2021:9780128035818.

[66] Aljabr A., Chiasson A., Alhajjaji A., "Numerical modeling of the effects of micro-encapsulated phase change materials intermixed with grout in vertical borehole heat exchangers", *Geothermics*. 2021;96:102197.

3 Development of Microencapsulated Phase Change Materials

Mohammed Fareed Rahi, Pratyush Anand, Mohd Naqueeb Shaad Jagirdar, Hafiz Muhammad Ali, and Hakeem Niyas

HIGHLIGHTS

1. Microencapsulated phase change material development techniques are discussed.
2. Coating phase change material droplets with polymeric layers enhances performance.
3. Minimal thermal property impact is observed post-microencapsulation.
4. Applications in textile, food, and vital sectors are being actively explored.
5. Microencapsulation proves vital for efficient thermal energy storage.

NOMENCLATURE

Symbols

ΔH	Latent heat
A	Surface area of the sample
d	Diameter
h	Thickness of the sample
T	Temperature of the surface
θ	Heat through the sample
k	Thermal conductivity
η	Efficiency

Subscripts

c	Crystallization
core	Core material
ee	Encapsulation efficiency
m	Melting

DOI: 10.1201/9781003331957-3

mc   Microcapsules
p   Single PCM
s   Shell material

ABBREVIATIONS

AA/AG   Agar-agar and arabic gum
AFM   Atomic force microscopy
DSC   Differential scanning calorimetry
FT-IR   Fourier-transform infrared spectroscopy
mPCM   Microencapsulated phase change material
MFPCM   Multi-functional phase change material
PCM   Phase change material
PMMA   Polymethyl methacrylate
SiO_2 Silicon dioxide
SG/AG   Sterilized gelatin and arabic gum
TGA   Thermo-gravimetric analysis
TES   Thermal energy storage
XRD   X-ray diffraction

3.1 INTRODUCTION

Phase change materials (PCMs) have grown in prominence as a feasible method for storing thermal energy in recent years. When a PCM switches from one state to another, such as when it melts or solidifies, it can absorb or release a lot of heat. Latent heat storage has numerous benefits over sensible heat storage, including that it offers higher energy density and operates within a more precise temperature range [1].

Although utilizing thermal energy storage (TES) has many benefits, there are also some drawbacks to traditional applications. These include the need for specialized equipment or heat exchange surfaces, which can increase costs and decrease thermal performance [2]. Additionally, when the PCMs change their phase to liquid from solid, they can be difficult to handle, and when mixed with other elements like textiles and building materials, the salt hydrates can float and sweat out or wash out in humid environments. Furthermore, the crystal water content of the PCMs may change in a wide range of humidity conditions [3]. When paraffin wax is melted, it can quickly diffuse through other components and even evaporate into the air, potentially increasing its volatile organic content [4]. As a result, it is generally not advised to apply PCMs without encapsulation [3]. An appropriate packaging method is necessary to ensure that PCMs maintain a stable form, whether liquid or solid, and are kept isolated from surrounding materials. This will address the issues that arise when using bulk PCMs in applications [5]. Encapsulating PCMs within inert materials, as shown in Figure 3.1, is considered the most effective and practical way to create form-stable PCMs. This technique involves trapping tiny solid or liquid particles within a sturdy solid barrier; this limits environmental interaction with the

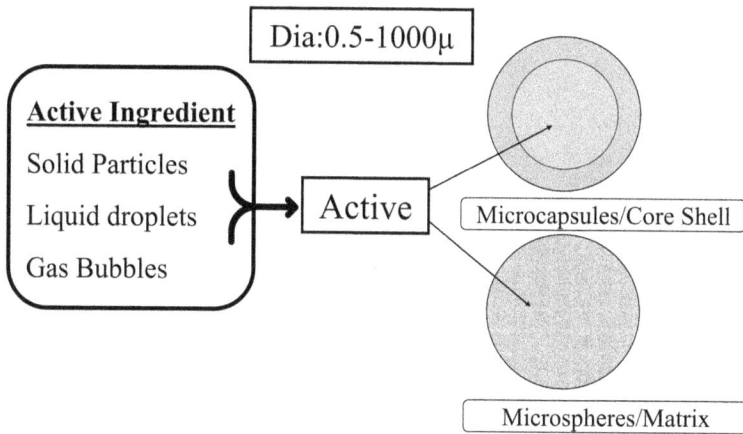

FIGURE 3.1 Structure of microencapsulated PCMs [3].

phase change behavior and prevents PCMs from escaping, expands the area of heat transfer, and facilitates the management of liquid PCMs during phase transition [6].

3.2 MICRO/NANOENCAPSULATED PCM DEVELOPMENT

The microencapsulation of PCMs is a method where tiny particles or drops are added a protective material or shell, making small capsules with unique properties [7]. The inner part is called the core, and the outer covering is called the shell or capsule material. Encapsulation is divided into microcapsules with diameters greater than 1 mm, microcapsules with diameters between 1 mm and 1 μm, and nanocapsules with diameters less than 1 μm [8].

Physical methods of microencapsulation employ processes such as adhesion, dehydration, and drying, which are utilized as physical techniques of microencapsulation to create microcapsule shells. Solvent evaporation and spray drying are two of the most popular physical techniques for encapsulating PCMs. Figure 3.2 provides a thorough classification of all the techniques.

3.2.1 PHYSICAL PROCESSES

3.2.1.1 Spray Drying

Spray drying is a multi-step process that involves mixing PCMs and shell materials in an oil–water emulsion, atomizing the emulsion, spraying it into a drying chamber where it is dried by a stream of hot gas, and finally sorting out the solid particles using a cyclone filter [10]. Borreguero et al. [11] developed microcapsules by using a combination of a Rubitherm® RT27 paraffin core and an ethylene vinyl acetate outer layer. They also experimented with incorporating carbon nanofibers into the capsules. The findings demonstrated that the carbon nanofiber-infused capsules retained their capacity for heat storage while having stronger mechanical characteristics and

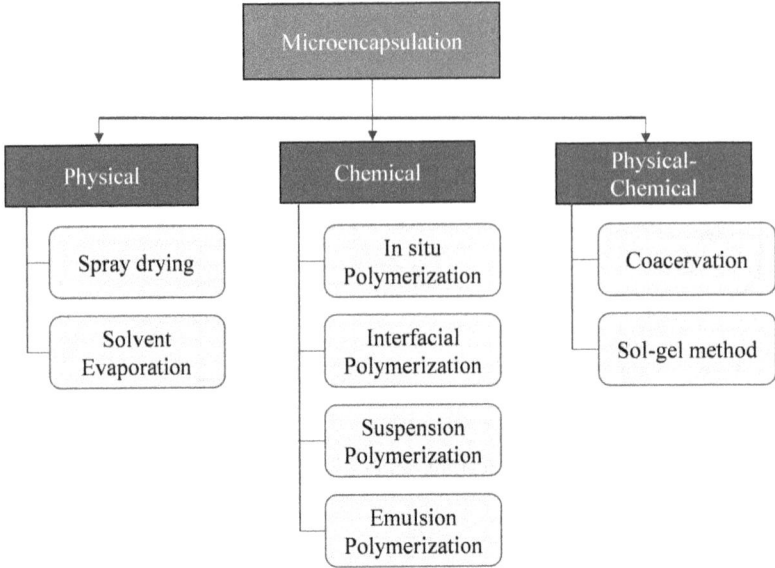

FIGURE 3.2 Categorization of microencapsulation techniques [9].

improved thermal conductivity. Furthermore, experiments based on differential scanning calorimetry (DSC) showed that the microcapsules' thermal characteristics remained robust even after several charge and discharge cycles. This approach was used by Wu et al. [12] to improve heat transmission. Encapsulating paraffin in polystyrene results in microcapsules having spherical forms. They discovered that when there is a 28% paraffin nanoparticle volume percentage in slurry, the spray approach provides an increase of 70% in the average heat transfer coefficient compared to that of water.

3.2.1.2 Solvent Evaporation

The procedure of solvent evaporation consists of four phases. The shell material is first dissolved in a volatile solvent to create a polymer solution. Then, PCMs are added to form an oil-in-water mixture. Third, the solvent is evaporated to form shells around the droplets, and finally, drying and filtration are used to create microcapsules. Fashandi et al. [13] employed solvent evaporation to produce microcapsules from palmitic and polylactic acids. They used hydrolyzed polyvinyl alcohol and sodium dodecyl sulfate as emulsifying agents. It was found that the quantity of emulsifier had to be increased with any emulsifier until the critical micelle concentration was achieved. During this process, the size of the microcapsule decreased because the surface energy at the oil–water interface decreased. This method was used by Wang et al. [14] to create microcapsules with a polymethyl methacrylate (PMMA) shell and a sodium phosphate dodecahydrate core. The microcapsules had melting/solidifying temperatures of 53.32/44.44 °C and melting/solidifying enthalpies of 122.61/104.24 J/g. Through systematic analysis, optimal parameters were determined for creating high-performance microcapsules during the microencapsulation process. These parameters were a synthesis temperature of 80–90 °C, a reaction time of 240 mins, and a stirring rate of 90 rpm. The microcapsules created under these conditions had an energy storage value of approximately 143 J/g at a maximum temperature of 52 °C.

3.2.2 Chemical Processes

3.2.2.1 In Situ Polymerization

Chemical microencapsulation techniques involve a continuous phase reaction between two immiscible liquids, similar to emulsion/mini-emulsion, suspension, and interfacial polycondensation [15]. Chemical microencapsulation typically includes four stages: forming an oil-in-water emulsion, preparing a prepolymer mixture, adding this mixture to the emulsion to coat the core particles, and washing and drying the resulting microencapsulated phase change material (mPCM). Tetradecane capsules were made by Choi et al. [16] using a melamine formaldehyde outer layer and an emulsifying agent called styrene-maleic anhydride-monomethyl in a 5% solution. The researchers discovered that when the emulsion speed was set at 8000 rpm, this procedure decreased the size of the capsules and increased their homogeneity. In order to create an mPCM paraffin, Jin et al. [17] used urea–formaldehyde as the material for the outer layer and hydrolyzed styrene-alt-maleic anhydride as the emulsifying agent. They boosted the shell material's content to 28% before putting it through multiple heat cycles. As a result of the findings, the capsules were demonstrated to have higher thermal stability than bulk paraffin. Using melamine formaldehyde as the outside layer and different amounts of emulsifier and cyclohexane as the interior materials, Zhang et al. [18] produced capsules that contained n-octadecane. They discovered that shrinking the microcapsule size was caused by raising the stirring rate from 3000 to 9000 rpm. In addition to using aminoplast as the outer layer, in situ polymerization was employed by Song et al. [19] to produce microencapsulated bromohexadecane utilizing nano-silver particles. The study showed that adding nano-silver particles to the process improved thermal stability and mechanical stability without producing any particle aggregation.

3.2.2.2 Interfacial Polycondensation

By generating walls at the boundary between two phases, each of which contains an appropriate reaction monomer, interfacial polymerization is a technique for in situ microcapsule formation. Using PCMs and a hydrophobic monomer to create an emulsion, a hydrophilic monomer is added to start polymerization under the right circumstances. The microcapsules are then separated by filtration, washed, and dried. Polyurea and polyurethane are commonly used as organic shell materials for microencapsulation [10]. For example, Zhang et al. [20] used polyurea to microencapsulate n-octadecane. Tolylene 2,4-diisocyanate and three distinct amines (ethylene diamine, diethylene triamine, and Jeffamine T403) were used to synthesize the shells. They created an oil solution by combining tolylene 2,4-diisocyanate and n-octadecane; to produce an oil-in-water emulsion, they added water to the mixture and stirred it at 3000 rpm. The interfacial polymerization procedure involved adding an ethylene diamine solution containing 0.1 weight percent styrene-maleic anhydride-monomethyl to the emulsion and stirring at 600 rpm. Jeffamine was used as the amine monomer to create the mPCM with the greatest anti-osmosis characteristics. According to Li et al. [21], the core material used was paraffin, while the outside layer was silicon dioxide (SiO_2). Their evaluation findings demonstrated that even after several repeated cycles of melting and freezing, the capsules could sustain a steady phase transition without any leakage. Through interfacial polymerization, Lu et al. [22] encased a butyl stearate core in a crosslinked network shell made of polyurethane.

3.2.2.3 Suspension Polymerization

Following a set of conventional processes, an mPCM is created using the suspension polymerization method in the following steps: (1) in the organic phase (core materials), the polymer monomer is dissolved; (2) an oil/water (O/W) emulsion is formed; (3) monomer molecules are separated and precipitated from the base materials, resulting in a solid shell. Li et al. [23] used various shell materials, including polydivinylbenzene, styrene-divinylbenzene copolymer, styrene-divinylbenzene-1,4-butylene glycol diacrylate copolymer, and styrene-divinylbenzene copolymer, to microencapsulate n-octadecane. Later, they used scanning electron microscopy (SEM) images to study the morphology of the microencapsulated particles and discovered that the most effective particles had styrene-divinylbenzene-1,4-butylene glycol diacrylate copolymer shells. The microsuspension polymerization method has also been used to create much smaller mPCM/nPCM capsules. As seen in Figure 3.3 [24], n-octadecane was enclosed in a polydivinyl benzene shell to produce mPCMs with an average size of 1.5 µm. Sánchez et al. [25] attracted significant attention by using the suspension polymerization approach to encapsulate nonpolar

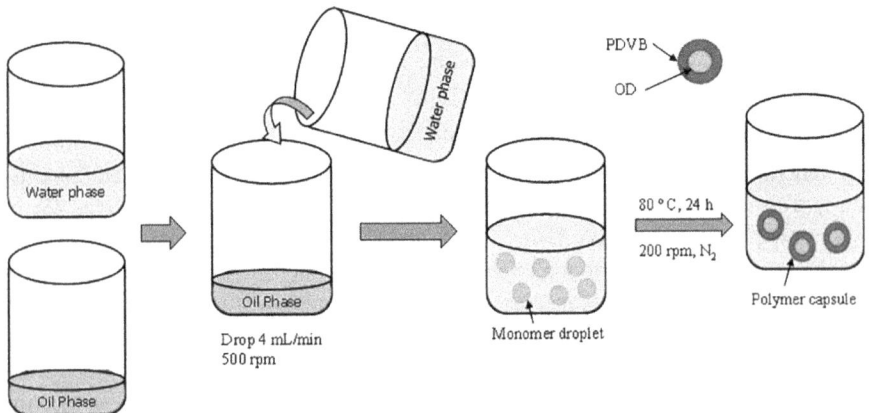

FIGURE 3.3 Description of suspension emulsion [24].

PCMs with a polystyrene shell. Nonpolar PCMs like tetradecane, Rubitherm RT27, and others may be encapsulated with over 50% core content using this method, whereas polar PCMs like PEG cannot be encapsulated due to their hydrophilic nature.

3.2.2.4 Emulsion Polymerization

Multiple chemical and physical processes occur simultaneously with the production and growth of particles during emulsion polymerization. Three primary processes for the formation of particles have been postulated by researchers. These mechanisms include: (1) an aqueous-phase free radical entering an emulsifier micelle that has been inflated by a monomer and continuing to develop there; (2) an aqueous-phase free radical that has been polymerized beyond its solubility limit; or (3) aqueous-phase free radicals expanding above their solubility limit before precipitating to form particle nuclei [26]. In the emulsion polymerization technique, the common material used for shells to create mPCMs is PMMA. Small-scale capsules with a typical size of 150 nm were developed by Chen et al. [27] utilizing PMMA as the shell and n-dodecanol as the core. The most significant values of 98.8 J/g and 82.2% for energy storage and encapsulation efficiency, respectively, were obtained by using a 3% proportion of a polymerizable emulsifier to the inner material and a 2% proportion of a co-emulsifier (hexadecane) to the inner material. Using emulsion polymerization and a polystyrene covering, Sari et al. [28] created microcapsules that contained a combination of capric, lauric, and myristic acids. The melting and freezing points of these microcapsules ranged from 22 to 48 °C and 19 to 49 °C, respectively. The range of the latent heat of melting and freezing was from 84 to 96 J/g and from 87 to 98 J/g, respectively. The researchers discovered that these microcapsules' thermal characteristics remained constant after 5000 thermal cycle tests, proving their thermal stability.

3.2.3 PHYSICOCHEMICAL PROCESSES

3.2.3.1 Complex Coacervation

This process involves creating copolymer shells by combining crosslinks with two or more different types of polymers that have opposite charges. The overall process, as illustrated in Figure 3.4, is divided into three stages and is carried out under continuous agitation. (1) In the first process, the coating substance is dissolved in water. The coating material is dispersed in water in the first step, a phase separation catalyst is created, and the core material is added to create an O/W emulsion. (2) In the subsequent stage, the O/W emulsion is introduced to a colloid solution with an opposing electric charge, and the solution's pH is modified as necessary. (3) Cooling the mixture, encasing the core material, and harvesting the mPCM bring the third stage to a close. The biggest issue with this strategy is that it might be challenging to scale up the procedure [29].

Despite the challenges, many advancements have been made in encapsulation using the complex coacervation method in recent years. To successfully create RT-27

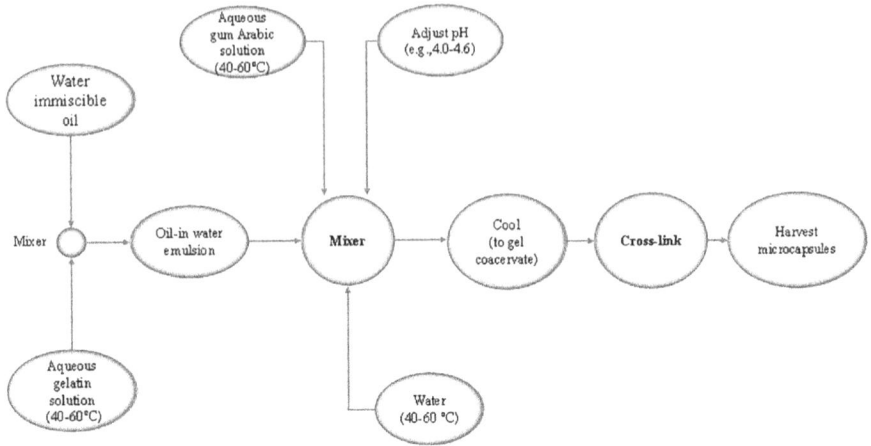

FIGURE 3.4   Illustration of complex coacervation [29].

microcapsules, Bayes-Garcia et al. [30] used two distinct coacervates: sterilized gelatin and arabic gum (SG/AG technique), and agar-agar and arabic gum (AA/AG method). The typical size of the capsules created by the SG/AG technique was 12 μm, according to the particle size analysis, whereas the AA/AG method produced capsules with an average diameter of 104 nm. Hawlader et al. [31] employed complex coacervation techniques to produce paraffin particles with encapsulation. For coacervation, a homogenization duration of 10 mins is advised. A volume of 6–8 ml of crosslinking agent should be employed. Coacervated microcapsules have a TES or release capability of about 145–240 J/g. The DSC results showed that the created microcapsules are ideal for use as a storage medium of energy.

3.2.3.2 Sol–Gel Method

Solution–gelling refers to a method of synthesizing mainly inorganic materials. The process involves a solution transitioning to a gel, which is distinguished by a large three-dimensional network structure that is evenly dispersed across the liquid medium [15]. Wang et al. [32] used silicon to encapsulate PCMs and were the first to do so. It was discovered that using cationic surfactants such as cetyltrimethylammonium chloride, dodecyl tri methyl ammonium chloride, and dodecyl tri methyl ammonium bromide as emulsifiers was effective in producing mPCMs. Chen et al. [33] developed mPCM capsules using SiO_2 as the shell material and stearic acid (90.7%) as the core material. Tests revealed that the pills were thermally stable and could store 162–171 J/g of energy for phase shift at temperatures between 52.6 and 53.5 °C. Sara et al. [34] created nPCMs with palmitic acid and SiO_2 as the core and shell, respectively. At a higher pH level, the nanocapsules' size and energy storage capacity was seen to be enhanced. For example, once the pH level rose from 11 to 12, the particle diameter increased by 183.7–722.5 nm, and energy storage capacity increased by 168–181 J/g.

3.3 CHARACTERIZATION TECHNIQUES OF MPCMS

The materials and methods employed during microencapsulation significantly affect the resulting microcapsules' physical, thermal, chemical, and mechanical characteristics. Therefore, it is vital to assess these properties with precision to understand and use PCM microcapsules in various applications fully.

3.3.1 PHYSICAL CHARACTERIZATION

The size, shape, and dispersion of mPCMs are commonly measured during physical characterization. Two methods are generally used to evaluate the distribution of microcapsule sizes: particle size analysis and observation statistics. Microcapsules must first be placed in proper media before performing particle size analysis. The microencapsulation process of PCMs, which involves encapsulating the materials in a shell, does not apply to all PCMs. Measuring the productive ratio, which assesses the ratio of the mass of microcapsules to the overall mass of the components used to create the core and shell, is one technique to assess the efficiency of the process. This ratio, however, can be affected by several factors, such as the proportion of materials in the core to those in the shell, the concentration of emulsifiers used, and the agitation rate. Therefore, these factors must be considered and controlled to evaluate the microencapsulation process accurately [35,36].

Encapsulation efficiency, which describes the proportion in the ratio of shell to core material in a microcapsule, is a measure of how much of the core is encapsulated within the shell. However, determining the exact density of the shell and core in individual microcapsules can be difficult. As a result, it is common to compare the latent heat of PCM microcapsules to that of pure PCMs to assess encapsulation efficiency, and Equations 3.1 and 3.2 are used for determining efficiency [20,37].

$$\eta_{ee} = \frac{\Delta H_{mc}}{\Delta H_{PCMs}} \times 100\% \tag{3.1}$$

$$\eta_{ee} = \frac{\Delta H_{m,mc} + \Delta H_{c,mc}}{\Delta H_{m,PCMs} + \Delta H_{c,PCMs}} \times 100\% \tag{3.2}$$

where ΔH_{mc} stands for the latent heat of the PCM, and ΔH_{PCMs} stands for the latent heat of the microcapsules encapsulating the PCM. The latent heat of the microcapsules' melting and crystallization is indicated by $\Delta H_{m,mc}$ and $\Delta H_{c,mc}$, respectively; the related latent heats of crystallization and melting for pure PCMs are denoted by the symbols $\Delta H_{m,PCMs}$ and $\Delta H_{c,PCMs}$ respectively. When supercooling conditions occur, there is a small discrepancy between Equations (3.1) and (3.2).

Electron microscopy-based methods are most frequently utilized to investigate the microcapsules' structure. To assess the composite materials' compatibility and the PCM's capacity to be absorbed by the supporting substance during the manufacturing of these compounds, SEM is usually used as a practical tool. [38]. When the mPCM has a small size, transmission electron microscopy can be used to study the particle shape and form [39]. Microcapsules can differ in terms of smoothness, size, and form because the microencapsulation process for PCMs is affected by several variables, such as the materials and synthesis circumstances. An SEM picture of microcapsules with flat, uniform surfaces is represented in Figure 3.5 (a). In contrast, Figure 3.5 (b) displays the thickness of the shell.

3.3.2 Thermal Characterization

Thermal characterization of an mPCM is essential for understanding the material's behavior and performance in TES systems and for optimizing the design of those systems. It includes assessing its thermal characteristics, including its thermal stability, heat capacity, and thermal conductivity. The material's effectiveness in applications for TES can be impacted by the mPCM's thermal stability, which is a crucial feature. mPCMs possess a substantial latent heat capacity and quick heat transmission rates. Different standard instruments and methods, such as thermal conductivity apparatus, thermo-gravimetric analysis (TGA), and DSC, are employed to assess the thermal characteristics of mPCMs. The thermal conductivity apparatus is used to

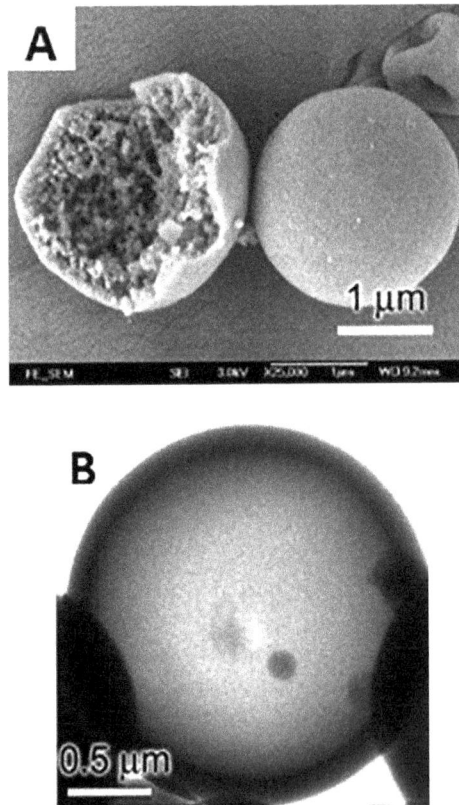

FIGURE 3.5   SEM images of flat and uniform microcapsules with defect (a), (b) thickness of shell [40].

measure the thermal conductivity of mPCMs. Thermal stability is measured using TGA, and heat capacity is assessed using DSC. The heat capacity of PCM is highly dependent on temperature, making accurate characterization essential. While commercial DSC equipment gauges the amount of heat a material absorbs as its temperature fluctuates, which evaluates the heat capacity of storage materials, a common method is to apply a constant heating or cooling rate and then integrate the thermal gradient signal to obtain the enthalpy curve [41].

DSC is a technique that can be utilized to evaluate the thermal characteristics of PCMs and mPCMs. DSC measures a material's heat flow through a controlled temperature shift. The technique can determine characteristics like composition, purity, specific heat capacity, glass transition temperature, and melting/solidifying point. Many different materials are often analyzed using DSC, including coatings, composites, metals, plastics, and polymers. However, DSC does have some drawbacks, such as difficulty in interpreting the results, sensitivity to changes in the sample, and inability to optimize resolution and sensitivity during measurement [42]. By using DSC to examine n-octadecane with a silica shell and microencapsulated methyl palmitate, Methaapanon et al. [43] discovered that the melting point of the mPCMs was marginally higher than that of the PCMs.

PCM microcapsules are frequently exposed to difficult conditions, particularly hot temperatures, that could result in serious thermal damage. The ability of PCM microcapsules to withstand thermal decomposition is known as thermal stability. By evaluating the sample's weight loss when the temperature is steadily increased at regular intervals, TGA is used to assess thermal stability. However, this approach has certain drawbacks, including that it only functions with solid materials and that temperature measurements are not very precise. Thermal reliability is the ability of a material or device to maintain its performance under thermal stress. One way to evaluate thermal reliability is to subject the material or device to many thermal cycles, simulating the conditions it may experience in real-world use. For example, the material or device may be repeatedly heated and cooled to simulate solidification and liquefaction. The material or gadget is regarded as thermally dependable if its thermal characteristics and chemical composition do not appreciably alter as a result of this testing. This type of testing may involve exposing the material or device to up to 5000 thermal cycles and then using analytical techniques such as X-ray diffraction (XRD), TGA, Fourier-transform infrared spectroscopy (FT-IR), and DSC to assess changes in thermal characteristics and chemical composition. Using the DSC technique, Xuan et al. [44] determined an mPCM suspension's heat capacity. The results demonstrated that as the volume proportion of mPCM particles increased, the heat capacity of the mPCM slurry at the same temperature dropped. Naikwadi et al. [45] developed an mPCM the with core material as n-nonadecane and PMMA-co-butyl acrylate-co-methacrylic acid material for the shell. The microcapsule's thermal curves were obtained using DSC after 1, 50, and 100 cycles of phase change. The researchers noticed that the DSC curves had hardly changed since the initial cycle, indicating that the microcapsules were stable during thermal cycling. They also compared the two spectra using FT-IR and discovered that there was no change in the resonant frequencies of the distinctive peaks, demonstrating that the temperature cycling did not affect the microcapsules' chemical composition.

Under steady-state conditions, a material's thermal conductivity describes how much heat travels through a certain region when there is a particular temperature gradient. Faster heat transfer from materials with enhanced thermal conductivity is important in a variety of applications. Energy release and storage both can be enhanced by increasing thermal conductivity. However, given their small size, it might be challenging to determine the heat conductivity of individual microcapsules, so a theoretical method has been proposed to calculate this conductivity [46].

$$\frac{1}{k_p d_p} = \frac{1}{k_c d_c} + \frac{d_p - d_c}{k_s d_p d_c} \tag{3.3}$$

Here, k_c and k_s are the core and shell materials' thermal conductivity, respectively; d_c and d_p represent the mPCM core material and single mPCM particle diameters, respectively; and lastly, the thermal conductivity of an individual mPCM particle is denoted by k_p. The thermal conductivity of a large number of microcapsules is essential in practical applications. As a result, it is a common practice to analyzing bulk microcapsules' thermal conductivity. This can be done by setting up an axial temperature gradient, placing the capsules sandwiched between two plates, and measuring the sample's heat flow. The temperature difference can then be noted using heat produced by the heat flow transducer, for which Equation 3.4 is utilized [47].

$$\theta = A \frac{k}{h} (T_1 - T_2) \tag{3.4}$$

k, A, h, and θ are the thermal conductivity, surface area, thickness, and heat flowing through the sample, respectively, and T represents the temperature of the surface/plate. Microcapsules with inorganic shells often have better thermal conductivities than those with PCM cores because inorganic materials have a higher thermal conductivity. Calcium carbonate was used by Yu et al. [48] as a shell material to encase n-octadecane, eventually leading to an 8.3´ increase in thermal conductivity. Similar to this, by using silica as a shell for n-octadecane cores, Zhang et al. [49] achieved increased thermal conductivity. According to the research, the thermal conductivity improves along with the fraction of inorganic components in the shell, demonstrating that these materials are crucial for improving thermal conductivity, and it has also been discovered that thicker shells and smaller pores are better for heat transfer. With the addition of high-thermal conductivity core or shell materials like graphite, carbon nanotubes, and nanometals, the thermal conductivity of mPCM can be improved, especially when it is utilized to create microcapsules with organic shells. Microcapsules made of carbon nanotubes exhibit increased thermal conductivity and compatibility by 79.2%, as determined by Li et al. [50], who covered PCMs in melamine resin and grafted stearyl alcohol onto carbon nanotubes.

3.3.3 CHEMICAL CHARACTERIZATION

FT-IR can be utilized to examine the structure and chemical properties of PCMs, shells, partial additives, and microcapsules. This method allows for the evaluation of the material of the microcapsule shell and investigates the probability of microsphere external degradation. The absorption, emission, photoconductivity, or Raman scattering of a solid, liquid, or gas can be captured in an infrared spectrum using the FT-IR technique. Raman spectroscopy is a powerful tool for learning more about lower frequency modes, crystal lattice vibrations, and molecular structures. However, it has a few limitations, such as its inapplicability to alloys or metals and the fact that the material might get damaged if exposed to intense laser radiation. Additionally, it is not frequently used to study the characterization of encapsulated mPCMs. Raman spectroscopy was employed by researchers like Yuan et al. [51] to examine mPCMs with silica/graphene oxide as the shell and paraffin as the PCM. The Raman spectra of the paraffin@silica/graphene oxide mPCMs exhibited two peaks, whereas paraffin@silica showed no peaks, indicating that the graphene has been successfully incorporated into the silica shell.

Another method for accurately identifying the individual components in samples is XRD. When the shell material is inorganic, XRD is a better tool for determining its structure [52]. The technique is based on the atoms in a sample of scattering X-rays, and it is used to figure out the size and orientation of the crystals as well as the spacing between layers or rows of atoms [53]. Lin et al. [54] applied XRD to study an mPCM, in which the PCM used was myristic acid, while the shell was made up of ethyl cellulose. According to the results of the XRD study, even after being enclosed, the PCM retained a stable crystal state because of the shell's strong crystalline structure.

3.3.4 MECHANICAL CHARACTERIZATION

mPCMs can undergo deformation and fracture when exposed to external forces, such as during mixing or transportation, or internal forces, like osmotic pressure or phase conversion. The strength and durability of the shell material can affect how much deformation and cracking occurs. Additionally, the properties of the core material and its temperature can also contribute to deciding these characteristics.

The evaluation of the mechanical properties of mPCMs can be classified into two approaches: the analysis of individual microcapsules and the examination of multiple microcapsules simultaneously. The mechanical characteristics of individual microcapsules can be evaluated using techniques such as micro/nanoindentation detection, atomic force microscopy (AFM), and microcompression [55]. These techniques involve employing a minor amount of force on an individual microcapsule and observing its deformation or failure characteristics in real time. Based on the outcomes, the samples' mechanical characteristics can then be researched. Comparing the rupture loads of different microcapsules can reveal information about their strengths. Examining the surface morphology of solid materials is performed using the AFM technique. During an AFM test, a microcantilever is set up with one end fixed and the other lightly touching the sample's surface with a tiny tip. As a result of the needle tip's weak repelling force in opposition to an atom on the surface of the sample, the needle-tipped microcantilever will move perpendicularly. Based on this idea, microcantilevers can be used to load test microcapsules. The roughness of mPCMs with n-octadecane as the core and melamine formaldehyde as the shell was measured using AFM by Huang et al. [56]. Roughness measurements showed that the root-mean-square roughness was 0.017 m and the average roughness was 0.021 m.

Microcapsule mechanical characteristics can be easily determined using AFM, although it has several drawbacks. Microcantilever displacement calculations may contain errors. The micro/nanoindentation method employs a variety of loads and indenter forms to assess the mechanical characteristics of microcapsules. It is more precise than AFM analysis, notably while evaluating each microcapsule's mechanical characteristics. Nanoindentation is thought to be the most practical and common method [57]. The single microcapsule typically deforms elastically and plastically; however, if the pressure is released prior to the yield point, it can revert to its previous shape. When the pressure exceeds the yield threshold, irreversible deformation will cause continued expansion of the mPCM until the microcapsules burst [58]. Figure 3.6 presents a schematic diagram of the micro/nanoindentation technique. This method can be used to determine Young's modulus and the hardness of microcapsules.

3.4 APPLICATIONS

PCM microcapsules offer new possibilities for utilizing PCMs due to their outstanding qualities, which include a large energy storage capacity, resilience in chemical and heat environments, and an ideal solid-to-liquid phase transition. In addition,

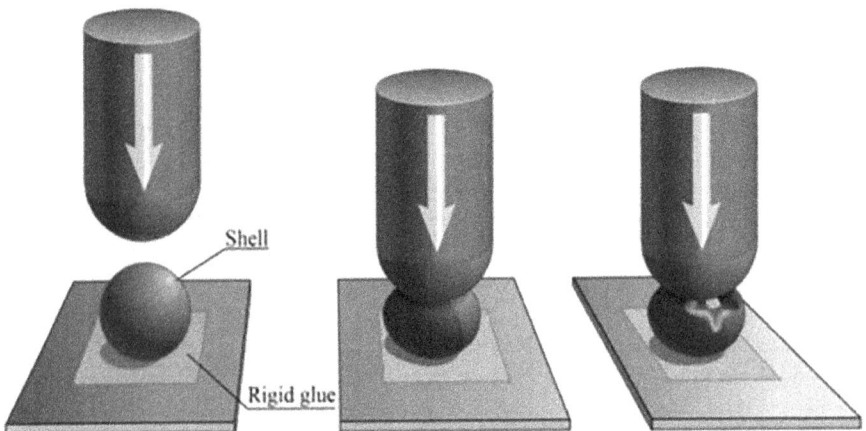

FIGURE 3.6 Nanoindentation technique [47].

incorporating these microcapsules into polymers and other materials can enhance the performance of PCMs, lower toxicity, and safeguard the PCM's core components from external factors. As depicted in Figure 3.7, PCM microcapsules have a broad variety of possible applications, which will be covered in more detail in the next section.

3.4.1 Textiles

mPCMs are used to provide additional protection against severe climatic conditions in a variety of outdoor wear, including gloves, ear warmers, trousers, boots, and snowsuits. To prolong the longevity of perfumes, they can be incorporated into the fiber or coated on the fabric's surface. They are also used in insect repellents and employed for their thermal storage capabilities [59]. Gao et al. [60] looked at the impact of wearables combined with mPCMs. The human skin temperature was monitored after a mPCM was integrated into cooling vests, and the results

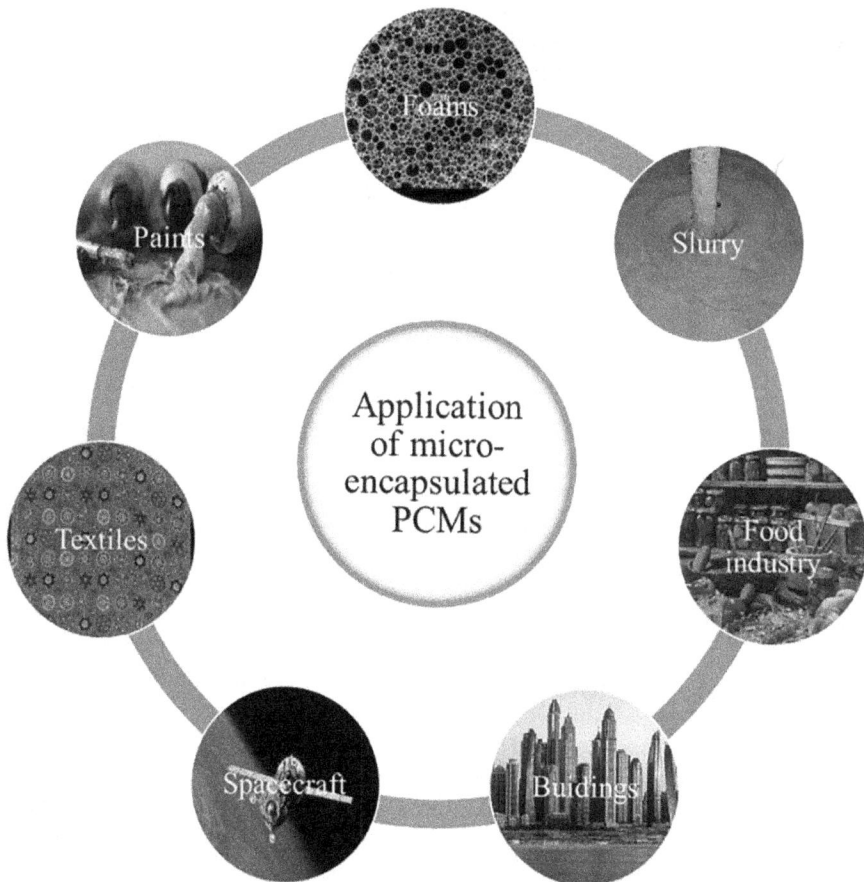

FIGURE 3.7   Applications of microencapsulated PCMs in different sectors [71].

showed that the temperature was reduced by 2–3 °C and stayed at 33 °C. The entire body's thermal comfort was improved. Currently, thermoregulating textiles are created using a variety of methods, including spinning fibers with various coating PCMs, which are infused into organic fabrics and foams, as well as adding PCMs to foams, fabrics, and fibers [61]. Using PCMs as core–shell substrates opens a world of possibilities because it limits the PCM's reactivity with external components like water and oxygen and controls the core material evaporation. Textiles that have been modified with PCMs can regulate temperature by releasing or absorbing heat as the temperature differs, resulting in thermal comfort. The polymer fiber matrix can incorporate PCMs directly. By integrating n-octadecane PCMs, Nejman et al. [62] tried to increase a fabric's thermal characteristics and air permeability. They used three methods, namely padding, printing, and coating, to add the PCMs to the fabric. According to the investigation, the fabric treated by printing had the highest enthalpy value, while the fabric modified through padding had the lowest enthalpy value, with the opposite trend observed for air permeability. The coated fabric had moderate air permeability and enthalpy. Scacchetti et al. [63] studied the impact of silver zeolites and chitosan zeolite composites, as well as mPCMs, on the thermal and antimicrobial properties of cotton. They found that using chitosan zeolite in textile production improved thermoregulating and antibacterial properties and that incorporating PCM microcapsules enhanced thermal comfort and flame retardancy. Additionally, the research team observed that these PCM microcapsules were evenly distributed on textile surfaces and remained effective after multiple washings. They also presented data from thermal history measurements to support the thermoregulating effects of the PCM-enhanced fabrics [64].

3.4.2 SLURRY

PCM microcapsules have a plethora of applications. PCM microcapsules, which possess a high latent heat, are particularly noteworthy for their use in the slurry industry. Because of their outstanding heat transfer capabilities, they are extensively used for heating and cooling, enhancing heat transfer fluid performance and serving as a TES medium. Using PCM microcapsules in a slurry also resolves the issue of droplets merging, as the PCM particles are encased in protective shells [65]. Using experimental and numerical methods, in a circular tube, Zeng et al. [66] evaluated the convective heat transfer coefficients of mPCM slurries. As a sine function of temperature, this study examined heat capacity. Utilizing the dimensionless wall temperature and the Nusselt number, heat transfer was examined. The addition of microcapsules increased effective thermal conductivity.

Under identical mass flows and thermal conditions, Rao et al. [67] evaluated the cooling behavior of water and microencapsulated n-octadecane slurries in rectangle-shaped microchannels. At low mass flows, the mPCM slurry was more effective in cooling than water, with better results seen at higher PCM concentrations. Due to factors like shorter residence time, higher thermal conductivity, and particle movement, the cooling effect of the mPCM slurry diminished as the mass flow increased. With the mass flow and PCM concentration, the Nusselt number and heat transfer coefficient rose, although the rise was barely detectable at higher concentrations. A multi-functional phase change material (MFPCM) slurry is a combination of a liquid and dispersed PCM particles. It is a very effective thermal storage and transfer medium because of its high latent heat capacity. An enhanced heat transfer fluid has been used extensively in heating and cooling systems because of its tremendous capacity to transmit heat [68]. The thermal properties of MFPCM slurry and its potential use in cooling systems have been discussed extensively. One such application is its use in a cooling ceiling system that is paired with an MFPCM slurry storage tank, which can be used to predict thermal conditions in office spaces and calculate energy consumption. The cooling ceiling system with MFPCM slurry uses very little energy. The cooling ceiling and MFPCM work together to deliver a cooling-shifting and energy-saving solution. The effects of utilizing MFPCM slurry with a metallic coating in a stainless steel tube microchannel heat exchanger as opposed to MFPCM with a non-metallic coating having the same PCM concentration were compared in this study. According to the findings, the non-metallic coating equivalent with a similar PCM concentration had a heat transfer coefficient that was 10% lower than that of the metal-coated MFPCM slurry. In their investigation, Yuan et al. [51] discovered that a 10% mPCM slurry was created by dispersing paraffin@SiO_2/graphene oxide and paraffin@SiO_2 into water. The paraffin@SiO_2 mPCM slurry was found to have a 6.5% greater thermal conductivity than water; the heat conductivity of the paraffin@SiO_2/graphene oxide mPCM slurry, however, was 11% greater than that of water.

3.4.3 BUILDINGS

The addition of PCM microcapsules to a variety of construction materials, including concrete, wall boards, and plaster, can greatly enhance energy efficiency in heating, cooling, and lighting systems [69]. The application of an mPCM slurry to a ceiling panel was studied by Griffiths et al. [70]. In a test chamber (with a working temperature of 16–18 °C), an mPCM slurry with 2–8 μm microcapsules and a 40% concentration was utilized as a heat transfer fluid. Concrete, being the primary construction material in buildings, has a higher thermal and acoustic insulation when PCM microcapsules are embedded into it. Konuklu et al. [71] developed a sandwich panel made of mPCMs and studied the results; they demonstrated that the use of microPCMs decreased the heating and cooling loads. On a typical summer day in a crowded laboratory with microencapsulated paraffin-enhanced wallboards, Kuznik and Virgone [72] conducted an evaluation using experimental data and found that the inclusion of a mPCM might enhance the air's free convection mixing and, as a result, efficiently lower the surface temperature of the walls to avoid an excessively hot daytime environment in the space.

Lecompte et al. [73] thoroughly investigated the properties of concrete and cement mortar containing mPCMs by testing compressive strength and thermal capacity, and the findings showed that while the heat capacity of concrete and cement mortar rose, the compressive strength of cement and concrete remained same. Advanced gypsum composites with mPCM were developed by Borreguero et al. [74] to increase TES capability in building materials for more comfortable construction. They discovered that increasing the mPCM content in gypsum composites reduces thermal conductivity and density while raising the heat capacity and accumulated thermal power. Additionally, while mPCM diminishes the compressive strength of gypsum, it does not restrict its usage in construction. In another study, the thermal characteristics of three different types of gypsum boards, one without mPCM, one with 4.7% mPCM, and one with 7.5% mPCM, were evaluated by Borreguero et al. [75]. They found that as the mPCM content in the gypsum boards increases, the temperature curve of the outer wall becomes more consistent, and the time it takes to reach a steady-state temperature is prolonged.

3.4.4   FOOD INDUSTRY

To maintain the safety and freshness of food products, they must be stored and transported under freezing or refrigerated conditions. To increase the shelf life of food products and avoid spoilage, the integrity of the cold chain must be preserved throughout distribution and sales. PCMs have become extensively employed in the food industry for storage and energy management. They are used in heat storage and transit systems, including chilled storage and heat treatment sections, as well as packaging solutions. Devahastin and Pitaksuriyarat [76] investigated experimentally to evaluate the potential of using paraffin wax as a PCM to store latent heat in the drying process of sweet potatoes. The results showed that an inlet surrounding air velocity of 1 m/s could save 40% of the energy. Similarly, Lu and Tassou [77] evaluated the effect of using PCMs in heat pipe prototypes. One prototype included PCMs, while the other did not. The study found that incorporating PCMs into the cabinet structure enhanced the distribution of food temperature. Furthermore, research has also shown that using a composite PCM made of nano-structured calcium silicate and paraffin can improve thermal cushioning in food packaging. The energy efficiency and thermal performance of chillers and other cold storage applications are improved in the food industry by using PCMs. However, subcooling is a typical issue when employing PCMs in cooling processes. This happens if a substance is cold without the crystallization process starting below the freezing point. To solve this issue, studies have shown that utilizing PCM emulsions can give greater heat transfer rates than using bigger PCM modules.

3.4.5   FOAMS

The thermal performance of foams can be considerably improved by using mPCMs, particularly in terms of insulation. Research has shown that incorporating PCM microcapsules into rigid polyurethane foams can increase the TES capacity [78]. The addition of metal foam decreased the surface temperature of mPCM foam composites by 47% and centralized the inside temperatures, according to Li et al. [79] in an evaluation of their thermal performance. Additionally, incorporating PCM microcapsules into cellular metal foam has been found to improve thermal management by reducing surface temperature and eliminating issues of leakage and low thermal conductivity.

3.4.6 PAINTS

Paints are colored emulsions that dry into a solid covering when applied to a surface. They are commonly used to protect surfaces from damage and to enhance their appearance with color and texture. Various specialized types of paint, such as those that resist corrosion, UV rays, fire, and fouling, are available on the market to suit different applications. One promising area of research is the development of thermo-regulating paints that use mPCMs, which have the potential to benefit a wide range of structures such as roofs, warehouses, cold storage facilities, vehicles, and more. Additionally, the use of these paints in equipment and electronics that generate heat as a byproduct could also be beneficial [80].

3.4.7 SPACECRAFT APPLICATIONS

Future spacecraft systems can benefit from recent developments in PCM research and development. Materials with high thermal conductivity and cutting-edge heat-transporting techniques, including capillary heat pipes, loop heat pipes, passive coatings, and fixed conductance heat pipes with low absorption and high emissivity, are all feasible solutions for managing heat in spacecraft. PCMs constructed from salt hydrates are especially well-suited for spaceship transportation.

3.4.8 OTHER APPLICATIONS

Microcapsules containing PCMs have potential uses outside of TES, including electrical energy storage, medical imaging, drug delivery, biomedicine, and solar energy storage. Recent research has demonstrated that using mPCMs for photovoltaic (PV) cell passive temperature control can increase the PV system's overall efficiency, and the coupling of a PV cell and mPCMs in a single module can significantly increase the reliability of the PV system. Research showed that using the mPCM slurry at high flow rates decreased the PV system's module temperature and enhanced both thermal and electrical efficiency [81]. Other studies have proposed using PCM microcapsules in Joule heating systems to improve the convective heat transfer of electrothermal systems and have found that even at lower voltages and surrounding temperatures, with just 5% PCM microcapsules, the working temperature can be increased by 30% [82]. Additionally, PCM microcapsules have been developed for use in sterilization and have been found to have high antibacterial activity against a range of bacteria, with up to 99.1% inhibition upon two-hour contact [83]. Tomizwa et al. [84] experimentally studied the effects of mPCMs on the thermal management of mobile phones, and the results revealed that using the mPCM sheets has many disadvantages for the thermal management of mobile phones.

3.5 KEY CHALLENGES

Microencapsulation of PCMs is a complex process that involves encapsulating liquid PCM within small particles or capsules, typically with diameters in the range of micrometers or millimeters. This process can be challenging to optimize, as it involves several steps and can be affected by several factors. Here are some of the key challenges related to the development of microencapsulated PCMs:

1. Encapsulation methods: Microencapsulation of PCMs can be achieved using various methods, such as spray drying, coacervation, and in situ polymerization. There are benefits and drawbacks to each technique. For example, spray drying is a widely used method that can produce micro-capsules with a high surface-area-to-volume ratio, but it can also lead to high rates of PCM loss and the formation of irregularly shaped particles. Coacervation is a method that uses the differences in solubility between two immiscible liquids to form the capsules, but this method can be sensi-tive to pH and ionic strength, which can make it difficult to control. In situ polymerization is a method that involves the formation of the capsule shell through the polymerization of a monomer within the droplets of the liquid core, but this method can be time consuming and complex.
2. PCM leakage: mPCMs can experience leakage over time, which can nega-tively impact their thermal performance. Leakage can occur through defects in the capsule shell, poor adhesion between the capsule shell and the PCM, or the gradual dissolution of the PCM in the host matrix. This is especially a concern for high-temperature applications where the melting point of the PCM is close to the process temperature.
3. Mechanical properties: The encapsulation process can also affect the mechanical properties of the microcapsules, such as their strength and flex-ibility. These properties are important because they affect the handling and storage of the capsules, as well as the release of the PCM. Capsules that are too brittle may break easily, while capsules that are too flexible may not be able to retain the PCM.
4. Cost: The production of mPCMs can be costly, especially when large quan-tities are needed. The cost of the raw materials, energy, and equipment used to produce them are factors that need to be considered for any commercial application.
5. Compatibility: mPCMs must be chemically compatible with the host matrix in which they are integrated. This can be challenging to ensure, as different PCMs and host matrices may have different chemical properties.
6. Incompatibility can lead to poor adhesion between the capsule and the host matrix, which can increase the risk of leakage.
7. Thermal performance: Microencapsulation can change the thermal char-acteristics of the PCM, leading to reduced thermal performance. Also, the encapsulation shell can insulate the PCM and prevent effective heat trans-fer. This can be a concern for applications where thermal conductivity is an important factor, such as thermal management in electronics. Therefore, the PCM and shell thermal properties must be well selected and tailored.

Despite these challenges, microencapsulated PCMs have the potential to be useful in a wide range of applications, including energy storage, heating and cooling, and thermal management in electronics. Therefore, ongoing research efforts are focused on developing new and improved methods for microencapsulation that can overcome these challenges and improve the thermal performance of mPCMs.

3.6   CONCLUSIONS

The chapter highlights the significance of PCMs in a constantly evolving industry; importantly, mPCMs are discussed. mPCMs have been tested on a range of building materials, including concrete, floor tiles, walls, gypsum plaster, bricks, ceilings, roofs, cement mortar, windows, shutters, glasses, insulation panels, etc. Further research is required to improve these methods and evaluate the long-term performance of microencapsulated PCMs. The major conclusions reached are listed here:

- PCMs offer significant energy storage capacity within a smaller temperature range when combined with conventional sensible heat storage materials.
- Various physical, physicochemical, and chemical fabrication techniques for mPCM development, including in situ, emulsion, interfacial, coacervation, solvent evaporation, and spray drying methods, have been evaluated.
- Different fabrication techniques have their advantages and disadvantages, such as solvent evaporation (simple and cost effective, but may result in porous capsules), melt encapsulation (produces mechanically strong capsules, but scaling up can be challenging), and spray drying (versatile but can be costly).
- Additional techniques like interfacial polymerization, emulsion, and in situ polymerization are also available for mPCM fabrication.
- The choice of fabrication technique depends on the specific application and desired properties of the microencapsulated PCM.
- The characterization of mPCMs involves multiple techniques, including FT-IR for chemical structure and configuration, and DSC for thermal property assessment.
- mPCMs' versatility allows for their use in various applications such as construction materials, foams, heat transfer fluid, TES, and textiles.
- The combination of mPCMs with building materials for passive TES has the potential to significantly reduce energy usage in buildings.
- mPCM slurry serves as an efficient heat transfer fluid and heat storage medium.

REFERENCES

[1] E.R.G. Eckert, R.J. Goldstein, W.E. Ibele, S.V. Patankar, T.W. Simon, T.H. Kuehn, P.J. Strykowski, K.K. Tamma, A. Bar-Cohen, J.V.R. Heberlein, J.H. Davidson, J. Bischof, F.A. Kulacki, U. Kortshagen, S. Garrick, Heat transfer—a review of 1997 literature, *Int J Heat Mass Transf.* (2000) 98. https://doi.org/10.1016/S0017-9310(99)00196-9.

[2] M.J.H. Rawa, N.H. Abu-Hamdeh, A. Karimipour, O.K. Nusier, F. Ghaemi, D. Baleanu, Phase change material dependency on solar power plant building through examination of energy-saving, *J Energy Storage.* 45 (2022) 103718. https://doi.org/10.1016/j. est.2021.103718.

[3] C.Y. Zhao, G.H. Zhang, Review on microencapsulated phase change materials (MEPCMs): Fabrication, characterization and applications, *Renew Sustain Energy Rev.* 15 (2011) 3813–3832. https://doi.org/10.1016/j.rser.2011.07.019.

[4] S. Himran, A. Suwono, G.A. Mansoori, Characterization of alkanes and paraffin waxes for application as phase change energy storage medium, *Energy Sources.* 16 (1994) 117–128. https://doi.org/10.1080/00908319408909065.

[5] K. Cho, S.H. Choi, Thermal characteristics of paraffin in a spherical capsule during freezing and melting processes, *Int J Heat Mass Transf.* (2000) 14. https://doi.org/10.1016/S0017-9310(99)00329-4.

[6] A. Sarı, C. Alkan, A. Karaipekli, Preparation, characterization and thermal properties of PMMA/n-heptadecane microcapsules as novel solid–liquid microPCM for thermal energy storage, *Appl Energy.* 87 (2010) 1529–1534. https://doi.org/10.1016/j.apenergy.2009.10.011.

[7] A. Poshadri, K. Aparna, Microencapsulation technology: A review, *J Res Angrau.* 38 (2010) 86–102.

[8] R. Sinaga, J. Darkwa, S.A. Omer, M. Worall, The microencapsulation, thermal enhancement, and applications of medium and high-melting temperature phase change materials: A review, *Int J Energy Res.* 46 (2022) 10259–10300. https://doi.org/10.1002/er.7860.

[9] M. Teggar, M. Arıcı, M.S. Mert, S.S. Mousavi Ajarostaghi, H. Niyas, E. Tunçbilek, K.A.R. Ismail, Z. Younsi, A.T. Benhouia, E.H. Mezaache, A comprehensive review of micro/nano enhanced phase change materials, *J Therm Anal Calorim.* 147 (2022) 3989–4016. https://doi.org/10.1007/s10973-021-10808-0.

[10] G. Alva, Y. Lin, L. Liu, G. Fang, Synthesis, characterization and applications of microencapsulated phase change materials in thermal energy storage: A review, *Energy Build.* 144 (2017) 276–294. https://doi.org/10.1016/j.enbuild.2017.03.063.

[11] A.M. Borreguero, J.L. Valverde, J.F. Rodríguez, A.H. Barber, J.J. Cubillo, M. Carmona, Synthesis and characterization of microcapsules containing Rubitherm®RT27 obtained by spray drying, *Chem Eng J.* 166 (2011) 384–390. https://doi.org/10.1016/j.cej.2010.10.055.

[12] W. Wu, H. Bostanci, L.C. Chow, S.J. Ding, Y. Hong, M. Su, J.P. Kizito, L. Gschwender, C.E. Snyder, Jet impingement and spray cooling using slurry of nanoencapsulated phase change materials, *Int J Heat Mass Transf.* 54 (2011) 2715–2723. https://doi.org/10.1016/j.ijheatmasstransfer.2011.03.022.

[13] M. Fashandi, S.N. Leung, Preparation and characterization of 100% bio-based polylactic acid/palmitic acid microcapsules for thermal energy storage, *Mater Renew Sustain Energy.* 6 (2017) 14. https://doi.org/10.1007/s40243-017-0098-0.

[14] T.Y. Wang, J. Huang, Synthesis and characterization of microencapsulated sodium phosphate dodecahydrate, *J Appl Polym Sci.* 130 (2013) 1516–1523. https://doi.org/10.1002/app.39249.

[15] S.K. Ghosh, Functional coatings and microencapsulation: A general perspective, in: *Functional Coatings*, John Wiley & Sons, Ltd, 2006, pp. 1–28. https://doi.org/10.1002/3527608478.ch1.

[16] J.K. Choi, J.G. Lee, J.H. Kim, H.S. Yang, Preparation of microcapsules containing phase change materials as heat transfer media by in-situ polymerization, *J Ind Eng Chem.* 7 (2001) 358–362.

[17] Z. Jin, Y. Wang, J. Liu, Z. Yang, Synthesis and properties of paraffin capsules as phase change materials, *Polymer.* 49 (2008) 2903–2910. https://doi.org/10.1016/j.polymer.2008.04.030.

[18] X.X. Zhang, Y.F. Fan, X.M. Tao, K.L. Yick, Fabrication and properties of microcapsules and nanocapsules containing n-octadecane, *Mater Chem Phys.* 88 (2004) 300–307. https://doi.org/10.1016/j.matchemphys.2004.06.043.

[19] Q. Song, Y. Li, J. Xing, J.Y. Hu, Y. Marcus, Thermal stability of composite phase change material microcapsules incorporated with silver nano-particles, *Polymer.* 48 (2007) 3317–3323. https://doi.org/10.1016/j.polymer.2007.03.045.

[20] H. Zhang, X. Wang, Synthesis and properties of microencapsulated n-octadecane with polyurea shells containing different soft segments for heat energy storage and thermal regulation, *Sol Energy Mater Sol Cells.* 93 (2009) 1366–1376. https://doi.org/10.1016/j.solmat.2009.02.021.

[21] B. Li, T. Liu, L. Hu, Y. Wang, L. Gao, Fabrication and properties of microencapsulated paraffin@SiO2 phase change composite for thermal energy storage, *ACS Sustain Chem Eng.* 1 (2013) 374–380. https://doi.org/10.1021/sc300082m.

[22] S. Lu, T. Shen, J. Xing, Q. Song, J. Shao, J. Zhang, C. Xin, Preparation and characterization of cross-linked polyurethane shell microencapsulated phase change materials by interfacial polymerization, *Mater Lett.* 211 (2018) 36–39. https://doi.org/10.1016/j.matlet.2017.09.074.

[23] W. Li, G. Song, G. Tang, X. Chu, S. Ma, C. Liu, Morphology, structure and thermal stability of microencapsulated phase change material with copolymer shell, *Energy.* 36 (2011) 785–791. https://doi.org/10.1016/j.energy.2010.12.041.

[24] P. Chaiyasat, Md. Z. Islam, A. Chaiyasat, Preparation of poly(divinylbenzene) microencapsulated octadecane by microsuspension polymerization: Oil droplets generated by phase inversion emulsification, *RSC Adv.* 3 (2013) 10202. https://doi.org/10.1039/c3ra40802g.

[25] L. Sánchez, P. Sánchez, A. de Lucas, M. Carmona, J.F. Rodríguez, Microencapsulation of PCMs with a polystyrene shell, *Colloid Polym Sci.* 285 (2007) 1377–1385. https://doi.org/10.1007/s00396-007-1696-7.

[26] M. Nomura, H. Tobita, K. Suzuki, Emulsion polymerization: Kinetic and mechanistic aspects, in: M. Okubo (Ed.), *Polymer Part*, Springer, Berlin, Heidelberg, 2005, pp. 1–128. https://doi.org/10.1007/b100116.

[27] Z.-H. Chen, F. Yu, X.-R. Zeng, Z.-G. Zhang, Preparation, characterization and thermal properties of nanocapsules containing phase change material n-dodecanol by miniemulsion polymerization with polymerizable emulsifier, *Appl Energy.* 91 (2012) 7–12. https://doi.org/10.1016/j.apenergy.2011.08.041.

[28] A. Sarı, C. Alkan, A. Altıntaş, Preparation, characterization and latent heat thermal energy storage properties of micro-nanoencapsulated fatty acids by polystyrene shell, *Appl Therm Eng.* 73 (2014) 1160–1168. https://doi.org/10.1016/j.applthermaleng.2014.09.005.

[29] Y.P. Timilsena, T.O. Akanbi, N. Khalid, B. Adhikari, C.J. Barrow, Complex coacervation: Principles, mechanisms and applications in microencapsulation, *Int J Biol Macromol.* 121 (2019) 1276–1286. https://doi.org/10.1016/j.ijbiomac.2018.10.144.

[30] L. Bayés-García, L. Ventolà, R. Cordobilla, R. Benages, T. Calvet, M.A. Cuevas-Diarte, Phase change materials (PCM) microcapsules with different shell compositions: Preparation, characterization and thermal stability, *Sol Energy Mater Sol Cells.* 94 (2010) 1235–1240. https://doi.org/10.1016/j.solmat.2010.03.014.

[31] M.N.A. Hawlader, M.S. Uddin, M.M. Khin, Microencapsulated PCM thermal-energy storage system, *Appl Energy.* 74 (2003) 195–202. https://doi.org/10.1016/S0306-2619(02)00146-0.

[32] L.-Y. Wang, P.-S. Tsai, Y.-M. Yang, Preparation of silica microspheres encapsulating phase-change material by sol-gel method in O/W emulsion, *J Microencapsul.* 23 (2006) 3–14. https://doi.org/10.1080/02652040500286045.

[33] Z. Chen, L. Cao, F. Shan, G. Fang, Preparation and characteristics of microencapsulated stearic acid as composite thermal energy storage material in buildings, *Energy Build.* 62 (2013) 469–474. https://doi.org/10.1016/j.enbuild.2013.03.025.

[34] S. Tahan Latibari, M. Mehrali, M. Mehrali, T.M. Indra Mahlia, H.S. Cornelis Metselaar, Synthesis, characterization and thermal properties of nanoencapsulated phase change materials via sol–gel method, *Energy.* 61 (2013) 664–672. https://doi.org/10.1016/j.energy.2013.09.012.

[35] J. Huo, Z. Peng, Q. Feng, Y. Zheng, X. Liu, Controlling the heat evaluation of cement slurry system used in natural gas hydrate layer by micro-encapsulated phase change materials, *Sol Energy.* 169 (2018) 84–93. https://doi.org/10.1016/j.solener.2018.04.035.

[36] Z. Zhao, X. Zhou, Q. Tian, X. Wang, W. Li, D. Liu, Microencapsulation of triglycidyl isocyanurate by solvent evaporation method for UV and thermal dual-cured coatings, *J Appl Polym Sci.* 131 (2014). https://doi.org/10.1002/app.41008.

[37] G. Fang, Z. Chen, H. Li, Synthesis and properties of microencapsulated paraffin composites with SiO2 shell as thermal energy storage materials, *Chem Eng J.* 163 (2010) 154–159. https://doi.org/10.1016/j.cej.2010.07.054.

[38] L. Wang, D. Meng, Fatty acid eutectic/polymethyl methacrylate composite as form-stable phase change material for thermal energy storage, *Appl Energy.* 87 (2010) 2660–2665. https://doi.org/10.1016/j.apenergy.2010.01.010.

[39] F. Wang, Y. Zhang, X. Li, B. Wang, X. Feng, H. Xu, Z. Mao, X. Sui, Cellulose nanocrystals-composited poly (methyl methacrylate) encapsulated n-eicosane via a Pickering emulsion-templating approach for energy storage, *Carbohydr Polym.* 234 (2020) 115934. https://doi.org/10.1016/j.carbpol.2020.115934.

[40] B. Fei, H. Lu, K. Qi, H. Shi, T. Liu, X. Li, J.H. Xin, Multi-functional microcapsules produced by aerosol reaction, *J Aerosol Sci.* 39 (2008) 1089–1098. https://doi.org/10.1016/j.jaerosci.2008.07.007.

[41] G.W.H. Höhne, W.F. Hemminger, H.-J. Flammersheim, Theoretical fundamentals of differential scanning calorimeters, in: *Differential Scanning Calorimetry*, Springer Berlin Heidelberg, Berlin, Heidelberg, 2003, pp. 31–63. https://doi.org/10.1007/978-3-662-06710-9_3.

[42] O. Koshy, L. Subramanian, S. Thomas, Differential scanning calorimetry in nanoscience and nanotechnology, in: *Thermal and Rheological Measurement Techniques for Nanomaterials Characterization*, Elsevier, 2017, pp. 109–122. https://doi.org/10.1016/B978-0-323-46139-9.00005-0.

[43] R. Methaapanon, S. Kornbongkotmas, C. Ataboonwongse, A. Soottitantawat, Microencapsulation of n-octadecane and methyl palmitate phase change materials in silica by spray drying process, *Powder Technol.* 361 (2020) 910–916. https://doi.org/10.1016/j.powtec.2019.10.114.

[44] Y. Xuan, Y. Huang, Q. Li, Experimental investigation on thermal conductivity and specific heat capacity of magnetic microencapsulated phase change material suspension, *Chem Phys Lett.* 479 (2009) 264–269. https://doi.org/10.1016/j.cplett.2009.08.033.

[45] A.T. Naikwadi, A.B. Samui, P. Mahanwar, Experimental investigation of nano/microencapsulated phase change material emulsion based building wall paint for solar thermal energy storage, *J Polym Res.* 28 (2021) 438. https://doi.org/10.1007/s10965-021-02808-3.

[46] L. Chen, T. Wang, Y. Zhao, X.-R. Zhang, Characterization of thermal and hydrodynamic properties for microencapsulated phase change slurry (MPCS), *Energy Convers Manag.* 79 (2014) 317–333. https://doi.org/10.1016/j.enconman.2013.12.026.

[47] Y.-D. Guo, J.-F. Su, R. Mu, X.-Y. Wang, X.-L. Zhang, X.-M. Xie, Y.-Y. Wang, Y.-Q. Tan, Microstructure and properties of self-assembly graphene microcapsules: Effect of the pH value, *Nanomaterials.* 9 (2019) 587. https://doi.org/10.3390/nano9040587.

[48] S. Yu, X. Wang, D. Wu, Microencapsulation of n-octadecane phase change material with calcium carbonate shell for enhancement of thermal conductivity and serving

durability: Synthesis, microstructure, and performance evaluation, *Appl Energy.* 114 (2014) 632–643. https://doi.org/10.1016/j.apenergy.2013.10.029.

[49] H. Zhang, X. Wang, D. Wu, Silica encapsulation of n-octadecane via sol–gel process: A novel microencapsulated phase-change material with enhanced thermal conductivity and performance, *J Colloid Interface Sci.* 343 (2010) 246–255. https://doi.org/10.1016/j.jcis.2009.11.036.

[50] M. Li, M. Chen, Z. Wu, Enhancement in thermal property and mechanical property of phase change microcapsule with modified carbon nanotube, *Appl Energy.* 127 (2014) 166–171. https://doi.org/10.1016/j.apenergy.2014.04.029.

[51] K. Yuan, H. Wang, J. Liu, X. Fang, Z. Zhang, Novel slurry containing graphene oxide-grafted microencapsulated phase change material with enhanced thermo-physical properties and photo-thermal performance, *Sol Energy Mater Sol Cells.* 143 (2015) 29–37. https://doi.org/10.1016/j.solmat.2015.06.034.

[52] S. Liang, Q. Li, Y. Zhu, K. Chen, C. Tian, J. Wang, R. Bai, Nanoencapsulation of n-octadecane phase change material with silica shell through interfacial hydrolysis and polycondensation in miniemulsion, *Energy.* 93 (2015) 1684–1692. https://doi.org/10.1016/j.energy.2015.10.024.

[53] M. Kaliva, M. Vamvakaki, Nanomaterials characterization, in: *Polymer Science and Nanotechnology*, Elsevier, 2020, pp. 401–433. https://doi.org/10.1016/B978-0-12-816806-6.00017-0.

[54] Y. Lin, C. Zhu, G. Alva, G. Fang, Microencapsulation and thermal properties of myristic acid with ethyl cellulose shell for thermal energy storage, *Appl Energy.* 231 (2018) 494–501. https://doi.org/10.1016/j.apenergy.2018.09.154.

[55] S. Magonov, Visualization of polymers at surfaces and interfaces with atomic force microscopy, in: *Handbook of Surfaces and Interfaces of Materials*, Elsevier, 2001, pp. 393–430. https://doi.org/10.1016/B978-012513910-6/50029-3.

[56] Y.-T. Huang, H. Zhang, X.-J. Wan, D.-Z. Chen, X.-F. Chen, X. Ye, X. Ouyang, S.-Y. Qin, H.-X. Wen, J.-N. Tang, Carbon nanotube-enhanced double-walled phase-change microcapsules for thermal energy storage, *J Mater Chem A.* 5 (2017) 7482–7493. https://doi.org/10.1039/C6TA09712J.

[57] J.-F. Su, X.-Y. Wang, S. Han, X.-L. Zhang, Y.-D. Guo, Y.-Y. Wang, Y.-Q. Tan, N.-X. Han, W. Li, Preparation and physicochemical properties of microcapsules containing phase-change material with graphene/organic hybrid structure shells, *J Mater Chem A.* 5 (2017) 23937–23951. https://doi.org/10.1039/C7TA06980D.

[58] X.-L. Zhang, Y.-D. Guo, J.-F. Su, S. Han, Y.-Y. Wang, Y.-Q. Tan, Investigating the electrothermal self-healing bituminous composite material using microcapsules containing rejuvenator with graphene/organic hybrid structure shells, *Constr Build Mater.* 187 (2018) 1158–1176. https://doi.org/10.1016/j.conbuildmat.2018.08.071.

[59] G. Nelson, Application of microencapsulation in textiles, *Int J Pharm.* 242 (2002) 55–62. https://doi.org/10.1016/S0378-5173(02)00141-2.

[60] C. Gao, K. Kuklane, F. Wang, I. Holmér, Personal cooling with phase change materials to improve thermal comfort from a heat wave perspective: Phase change materials to improve thermal comfort, *Indoor Air.* 22 (2012) 523–530. https://doi.org/10.1111/j.1600-0668.2012.00778.x.

[61] Z. Qiu, X. Ma, P. Li, X. Zhao, A. Wright, Micro-encapsulated phase change material (MPCM) slurries: Characterization and building applications, *Renew Sustain Energy Rev.* 77 (2017) 246–262. https://doi.org/10.1016/j.rser.2017.04.001.

[62] A. Nejman, E. Gromadzińska, I. Kamińska, M. Cieślak, Assessment of thermal performance of textile materials modified with PCM microcapsules using combination of DSC and infrared thermography methods, *Molecules.* 25 (2019) 122. https://doi.org/10.3390/molecules25010122.

[63] F.A.P. Scacchetti, E. Pinto, G.M.B. Soares, Thermal and antimicrobial evaluation of cotton functionalized with a chitosan-zeolite composite and microcapsules of phase-change materials, *J Appl Polym Sci.* 135 (2018) 46135. https://doi.org/10.1002/app.46135.

[64] S. Alay Aksoy, C. Alkan, M.S. Tözüm, S. Demirbağ, R. Altun Anayurt, Y. Ulcay, Preparation and textile application of poly(methyl methacrylate-co-methacrylic acid)/n-octadecane and n-eicosane microcapsules, *J Text Inst.* 108 (2017) 30–41. https://doi.org/10.1080/00405000.2015.1133128.

[65] R. Yang, H. Xu, Y. Zhang, Preparation, physical property and thermal physical property of phase change microcapsule slurry and phase change emulsion, *Sol Energy Mater Sol Cells.* 80 (2003) 405–416. https://doi.org/10.1016/j.solmat.2003.08.005.

[66] X. Wang, J. Niu, Y. Li, X. Wang, B. Chen, R. Zeng, Q. Song, Y. Zhang, Flow and heat transfer behaviors of phase change material slurries in a horizontal circular tube, *Int J Heat Mass Transf.* 50 (2007) 2480–2491. https://doi.org/10.1016/j.ijheatmasstransfer.2006.12.024.

[67] Y. Rao, F. Dammel, P. Stephan, G. Lin, Convective heat transfer characteristics of microencapsulated phase change material suspensions in minichannels, *Heat Mass Transf.* 44 (2007) 175–186. https://doi.org/10.1007/s00231-007-0232-0.

[68] X. Wang, J. Niu, Performance of cooled-ceiling operating with MPCM slurry, *Energy Convers Manag.* 50 (2009) 583–591. https://doi.org/10.1016/j.enconman.2008.10.021.

[69] W. Su, T. Zhou, Y. Li, Y. Lv, Development of microencapsulated phase change material with poly (methyl methacrylate) shell for thermal energy storage, *Energy Procedia.* 158 (2019) 4483–4488. https://doi.org/10.1016/j.egypro.2019.01.764.

[70] P.W. Griffiths, P.C. Eames, Performance of chilled ceiling panels using phase change material slurries as the heat transport medium, *Appl Therm Eng.* 27 (2007) 1756–1760. https://doi.org/10.1016/j.applthermaleng.2006.07.009.

[71] G. Peng, G. Dou, Y. Hu, Y. Sun, Z. Chen, Phase Change Material (PCM) microcapsules for thermal energy storage, *Adv Polym Technol.* 2020 (2020) e9490873. https://doi.org/10.1155/2020/9490873.

[72] F. Kuznik, J. Virgone, Experimental assessment of a phase change material for wall building use, *Appl Energy.* 86 (2009) 2038–2046. https://doi.org/10.1016/j.apenergy.2009.01.004.

[73] T. Lecompte, P. Le Bideau, P. Glouannec, D. Nortershauser, S. Le Masson, Mechanical and thermo-physical behaviour of concretes and mortars containing phase change material, *Energy Build.* 94 (2015) 52–60. https://doi.org/10.1016/j.enbuild.2015.02.044.

[74] A.M. Borreguero, I. Garrido, J.L. Valverde, J.F. Rodríguez, M. Carmona, Development of smart gypsum composites by incorporating thermoregulating microcapsules, *Energy Build.* 76 (2014) 631–639. https://doi.org/10.1016/j.enbuild.2014.03.005.

[75] A.M. Borreguero, M. Carmona, M.L. Sanchez, J.L. Valverde, J.F. Rodriguez, Improvement of the thermal behaviour of gypsum blocks by the incorporation of microcapsules containing PCMS obtained by suspension polymerization with an optimal core/coating mass ratio, *Appl Therm Eng.* 30 (2010) 1164–1169. https://doi.org/10.1016/j.applthermaleng.2010.01.032.

[76] S. Devahastin, S. Pitaksuriyarat, Use of latent heat storage to conserve energy during drying and its effect on drying kinetics of a food product, *Appl Therm Eng.* 26 (2006) 1705–1713. https://doi.org/10.1016/j.applthermaleng.2005.11.007.

[77] W. Lu, S.A. Tassou, Characterization and experimental investigation of phase change materials for chilled food refrigerated cabinet applications, *Appl Energy.* 112 (2013) 1376–1382. https://doi.org/10.1016/j.apenergy.2013.01.071.

[78] A.M. Borreguero, J.F. Rodríguez, J.L. Valverde, T. Peijs, M. Carmona, Characterization of rigid polyurethane foams containing microencapsulted phase change materials: Microcapsules type effect, *J Appl Polym Sci.* 128 (2013) 582–590. https://doi.org/10.1002/app.38226.

[79] W. Li, H. Wan, H. Lou, Y. Fu, F. Qin, G. He, Enhanced thermal management with microencapsulated phase change material particles infiltrated in cellular metal foam, *Energy.* 127 (2017) 671–679. https://doi.org/10.1016/j.energy.2017.03.145.

[80] T.P. Tumolva, N.S. Sabarillo, Department of chemical engineering, University of the Philippines, characterization of MEPCM-incorporated paint as latent heat storage system, *Int J Chem Eng Appl.* 8 (2017) 203–209. https://doi.org/10.18178/ijcea.2017.8.3.657.

[81] Z. Qiu, X. Zhao, P. Li, X. Zhang, S. Ali, J. Tan, Theoretical investigation of the energy performance of a novel MPCM (Microencapsulated Phase Change Material) slurry based PV/T module, *Energy.* 87 (2015) 686–698. https://doi.org/10.1016/j.energy.2015.05.040.

[82] Z. Zheng, J. Jin, G.-K. Xu, J. Zou, U. Wais, A. Beckett, T. Heil, S. Higgins, L. Guan, Y. Wang, D. Shchukin, Highly stable and conductive microcapsules for enhancement of joule heating performance, *ACS Nano.* 10 (2016) 4695–4703. https://doi.org/10.1021/acsnano.6b01104.

[83] X. Zhang, X. Wang, D. Wu, Design and synthesis of multifunctional microencapsulated phase change materials with silver/silica double-layered shell for thermal energy storage, electrical conduction and antimicrobial effectiveness, *Energy.* 111 (2016) 498–512. https://doi.org/10.1016/j.energy.2016.06.017.

[84] Y. Tomizawa, K. Sasaki, A. Kuroda, R. Takeda, Y. Kaito, Experimental and numerical study on phase change material (PCM) for thermal management of mobile devices, *Appl Therm Eng.* 98 (2016) 320–329. https://doi.org/10.1016/j.applthermaleng.2015.12.056.

4 Physical and Thermal Properties with Measurement Methods for Phase Change Materials

S. Shankara Narayanan, Apurv Yadav, and Venkata Reddy Poluru

FEATURES

- Describes the importance of thermal energy storage and different types of thermal energy storage systems.
- Provides classification of phase change materials into different types.
- Describes the thermal and physical properties of phase change materials.
- Explains the techniques involved in the characterization of phase change materials.

COMPLETE DETAILS OF NOMENCLATURE AND ABBREVIATIONS

Abbreviation	Nomenclature
DSC	Differential scanning calorimetry
DTA	Differential thermal analysis
LFA	Laser flash analysis
LH	Latent heat
LHTES	Latent heat thermal energy storage
PCM	Phase change material
SH	Sensible heat
STES	Sensible thermal energy storage
TES	Thermal energy storage
THS	Transient hot strip
THW	Transient hot wire
TPS	Transient plane source

DOI: 10.1201/9781003331957-4

4.1 INTRODUCTION

There is a crucial need for an uninterrupted energy supply globally to maintain the momentum of economic and technological advancement [1]. The perpetual depletion faced by conventional energy resources and the surge in global warming has shifted the emphasis on the exploitation of sustainable energy resources [2]. Renewable energy sources provide an undeniable edge over the utilization of conventional energy resources due to their ease of access, eco-friendliness, abundance and unending supply [3]. However, these sources have an intermittent nature that poses a major limitation in their widespread commercial applications, and additional arrangements are required to increase their adaptability [4]. The reliance on carbon-based energy sources can be tremendously reduced by applying thermal energy storage (TES) systems, which will make renewable energy sources more dependable and cut down on global warming [5]. Also, if the energy-production-to-energy-demand ratio becomes greater than one, the excess energy can be stored and reused during the demand for peak power [6]. There are many integration methods of TES systems with non-conventional energy generation systems like solar, wind, hydrogen, hydropower, geothermal power, or even recovery of waste heat [7]. TES is performed by either heating or cooling the storage media, which can be later used for the desired applications.

Chemical and thermal methods are the two fundamental thermal energy storage methods, as presented in Figure 4.1 [8]. The thermal methods are further classified into sensible heat and latent heat thermal energy storage (LHTES). In a chemical TES

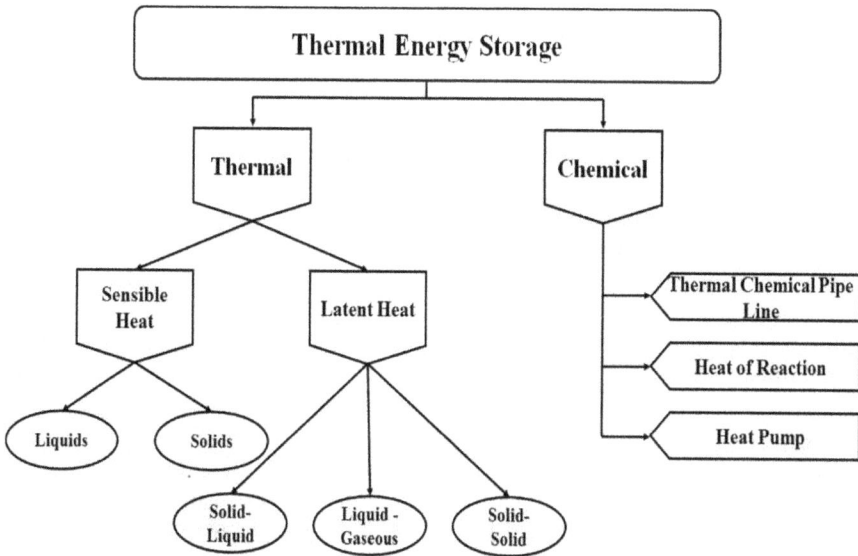

FIGURE 4.1 Types of thermal energy storage systems [8].

system, the storage and release of energy occur during a reversible chemical reaction [9, 10]. The thermal cycle of the chemical TES system comprises the charging process, storing process and discharging process, and it stores energy as a result of endothermic dissociation and retrieves energy during the reversible reaction [11, 12].

Sensible thermal energy storage (STES) and LHTES are two types of TES methods. In STES, the thermal energy is stored or released by either increasing or decreasing the temperature, respectively, of the storage material [13]. This increase in temperature can be 'sensed', and the heat stored is thus known as sensible heat. STES, in most of the cases, uses storage materials either in the form of solids (e.g. stone, brick) or liquids (e.g. water). On the other hand, in LHTES, thermal energy is stored or released in the form of latent heat when the material undergoes a phase change under isothermal condition [14]. Such materials are known as phase change materials (PCMs).

The phase change process can be of different types: solid to solid, solid to gas, solid to liquid, liquid to gas and vice-versa. Solid-to-gas and liquid-to-gas phase transitions in TES systems are not practical due to high system pressure, more volumetric changes and larger volume requirements. In solid–solid PCMs, energy absorbed or released during the change from one crystalline phase to another is responsible for TES, but these transitions involve low phase transition enthalpies. Solid-to-liquid phase transition is the most widely used type because of its huge heat storage density and relatively small changes in volume compared to solid-to-gas or liquid-to-gas transitions and solid-to-solid transitions, which possess a small phase transition energy [15].

During a solid-to-liquid phase change process (e.g. melting), heat is transferred to the heat storage material while the material remains in the isothermal condition at the melting point. When the thermal energy has been transferred to the storage material, the melting is completed, and the heat thus stored is known as latent heat. Materials with a solid-to-liquid phase transition, which are fit for energy storage in the form of latent heat, are commonly referred to as latent heat storage materials or simply phase change materials (PCMs). These materials absorb or release heat during their phase transitions, making them ideal for hot or cold storage applications. While STES systems are heavier and more space-consuming, LHTES systems are lighter and less volume-occupying and can store 10 to 100 times more thermal energy than STES systems [16–18].

Figure 4.2 presents an overview of the density, temperature range, durability and commercial feasibility of TES systems employing STES, LHTES and chemical means [14]. The highest heat storage density is obtained in the case of chemical heat storage, followed by LHTES and then STES. Although LHTES systems have high availability, compared to STES systems, their progress is slower due to material characteristic limitations. The durability of STES materials is four times that of LHTES materials and 10 times that of chemical energy storage [15]. However, both chemical energy storage and STES systems have significantly lower ranges of operating temperatures than LHTES, which offers a very flexible range of operating temperatures [19].

The classification of PCMs based on chemical composition is presented in Figure 4.3. In general, PCMs can be broadly classified into two types due to their different

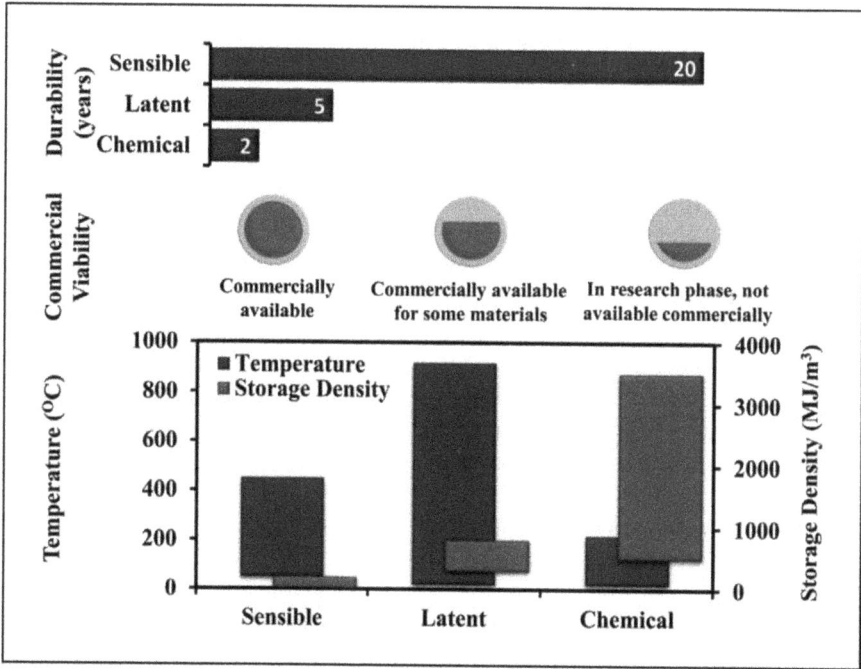

FIGURE 4.2 Overview of properties and feasibility of TES systems [14].

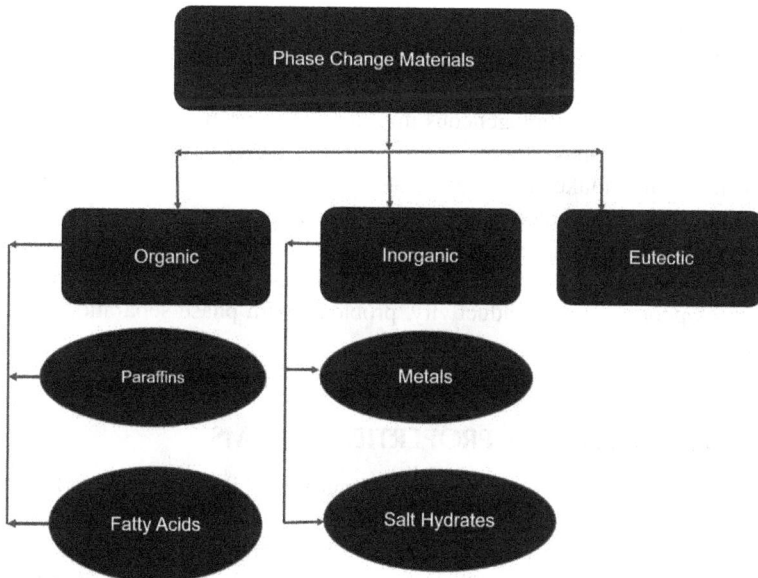

FIGURE 4.3 Classification of PCMs [22].

chemical structure: organic and inorganic PCMs. Organic PCMs are further subdivided into two categories: paraffins and non-paraffins. Similarly, inorganic PCMs are subdivided into two categories: salt hydrates and metals. Organic PCMs cover a range of melting points varying from 0 to 200 °C. It is important to note that most of the organic PCMs are unstable at high temperatures owing to their high levels of carbon and hydrogen. Organic PCMs usually have lower heats of fusion per volume than their inorganic counterparts. Paraffins are compounds which generally refer to paraffin waxes with a general chemical formula C_nH_{2n+2}. These are chemically inert and stable and have large value of latent heat of fusion. Paraffin waxes display several undesirable properties such as incompatibility with plastic containers, low thermal conductivity and moderate flammability. On the other hand, non-paraffin PCMs, which include a number of esters, fatty acids and fatty alcohols, exhibit both desirable (such as large heats of fusion and no or limited supercooling for fatty acids) and undesirable (such as poor thermal conductivity, non-flammability and instability at high temperatures) characteristics.

Inorganic PCMs are classified as either salt hydrates or metals. Inorganic salts containing few moles of H_2O that form a typical crystalline solid are known as salt hydrates. The most striking properties of salt hydrates include high latent heat, relatively high thermal conductivity (almost twice as that of paraffins) and negligible changes in volume upon melting. Apart from the salt hydrates, which are currently the most prominent group of PCMs which have been extensively studied and used in LHTES systems, the other broad category of inorganic PCMs is metals. The characteristics of the metallic PCMs include a large value for latent heat, higher thermal conductivity, lower specific heat and relatively low vapor pressure. On account of the advancement in technology for the fabrication of nanomaterials, metal PCMs have turned more viable. Their physical properties such as high thermal conductivity and sharp, well-defined melting points have made them attractive for myriad applications.

A eutectic PCM is a homogeneous mixture of two or more PCMs [20]. Eutectic PCMs have a fixed phase transition point that is lower than that of any of their constituents, which makes them ideal for targeting ranges of phase transition temperatures that single-component PCMs are unable to reach. Although the eutectic PCMs are suitable as a TES material, they are obtained by blending a variety of single-component PCMs, and hence they still face certain disadvantageous properties such as poor thermal conductivity, problems with phase separation, corrosion and supercooling.

4.2   THERMOPHYSICAL PROPERTIES OF PCMS

To choose a suitable PCM for TES, a PCM must exhibit certain desirable thermal, physical, kinetic and chemical properties. Additionally, its cost and easy availability has to be taken into consideration. For an excellent thermal storage system, the PCM to be used should possess desirable thermophysical, chemical and kinetic properties, which are as described herein [21, 22]:

Thermophysical properties

   (i)   Appropriate phase transition temperature.
  (ii) High latent heat.
 (iii) High thermal conductivity for better heat transfer.
 (iv) Phase stability.
  (v) High density.
 (vi) Small volume change.

In order to select a PCM for a particular application, the operating temperature of the heating or cooling system must match with the transition temperature of the PCM. Also, the latent heat per unit volume of the PCM should be high to minimize the physical size of the storage system. The high thermal conductivity of a PCM is an important requirement which enables an improved heat transfer rate, thus decreasing the rate of charging (melting) or discharging (solidification) of the PCM. Phase stability during the melting/freezing of PCMs aids in setting heat storage, and high density of PCMs is desirable to allow a smaller storage container. Small volume changes during phase transformations at melting/freezing temperatures help in reducing the containment problem.

Kinetic properties

   (i)   Negligible supercooling effect.
  (ii) Ample crystallization rate.

In supercooling, a PCM cools below the freezing point without solidification or crystallization. It is one of the most undesirable properties for PCMs, especially salt hydrates, since it interferes with proper heat extraction from the PCMs. It is seen that 5–10 °C supercooling can prevent the entire heat extraction from the PCMs.

Chemical properties

   (i)   Long-term chemical stability.
  (ii) Compatibility with the material of the container.
 (iii) Low or no toxicity.
 (iv) Non-flammability.

PCMs can suffer in their TES capacity by chemical decomposition or incompatibility with materials of the container. A suitable PCM must be non-toxic and non-flammable for safety. Besides these properties, a PCM must possess good thermal and cycling stability, which implies that no change in the PCM properties (such as latent heat, phase transition temperature, supercooling degree, etc.) should occur after it undergoes several melting–solidification cycles. The cost-effectiveness and large-scale availability of PCMs are also important parameters to be considered for their large-scale use.

    There is an abundance of organic and inorganic PCMs having different properties. However, no single PCM can possess all the required properties for an ideal

TES system, and therefore, we have to choose the best suitable available PCM and try to make up for any of its poor thermophysical properties by an adequate system design.

In general, the favourable characteristics of organic PCMs for thermal energy storage are: high heat of fusion, uniform melting and better nucleating properties. However, they exhibit certain undesirable characteristics such as: (i) poor thermal conductivity, (ii) moderate flammability, (iii) non-compatibility with the plastic container and (iv) high temperature instability.

On the other hand, inorganic PCMs are more advantageous for energy storage applications since they are endowed with attractive properties such as: (i) high latent heat, (ii) relatively high thermal conductivity, (iii) small volume changes during phase transition and (iv) non-flammability. The main drawbacks encountered in using inorganic PCMs (such as salt hydrates) are: incongruent melting, segregation, supercooling and corrosiveness. On the other hand, eutectic PCMs have gathered much attention due to their ability to regulate the required temperature and other properties by tuning the proportion of added components. However, their commercialization is restricted by difficulties such as low thermal conductivity, supercooling, phase separation and corrosion. High thermal conductivity and large values for latent heat are the two basic requirements for a PCM to be applied in TES. Other requirements include low cost, chemical stability, minimum subcooling, congruent melting, non-corrosivity, non-toxicity and a practical melting temperature range. Materials studied in the previous five decades are fatty acids, paraffin wax, hydrated salts, metals and eutectics of organic–organic, inorganic–inorganic or organic–inorganic compounds [23].

For example, Liu et al. [24] focussed on the selection criteria of PCMs and various techniques for enhancing the thermal conductivity and improving the latent heat of fusion of PCMs. They covered over 200 PCMs as candidates with an operating temperature range of 0–100 °C and compared them in terms of different selection criteria. Lu et al. [25] synthesized a kind of magnetically tightened, form-stable phase change material using neodymium iron boron (NdFeB) and paraffin as the input materials for their system. NdFeB particles were surface modified to NdFeB@Ag through in situ silver plating to improve the thermal conductivity of the form-stable phase change materials. The resultant form-stable phase change materials exhibited superior thermal conductivity (increase of 1400–1600%) and electrical (>104 S/m) conductivity, as well as prominent compressive strength. Kalombe [26] et al. investigated the application of mixtures of 50 wt% ratios of coconut oil and soybean oil (COSO) and of paraffin wax and soybean oil (PWSO) as blended organic–organic PCMs. The results from analysis of the as-prepared PW or prepared PWSO, and of COSO before or after multiple thermal cycles, showed they possessed high latent heat of fusion, which makes them suitable to store a large quantity of heat. Table 4.1 showcases the thermophysical properties of selected phase change materials in the range of 0–250 °C. In another study, Li et al. [27] reported a novel strategy in which carbon dots (CDs) evenly distributed into polyethylene glycol (PEG) acted as local self-heating agents to shorten thermal transfer distances, which resulted in the enhancement of the phase change enthalpies of PEG-related PCMs by their interface interaction. The obtained PCM composite showed a long heat storage duration and achieved 82.57% of the solar-to-thermal storage efficiency as well as 92.27% of the heat release efficiency.

TABLE 4.1
Thermophysical Properties of Some Common Phase Change Materials

PCM	Melting point (°C)	Latent heat (kJ/Kg)	Thermal conductivity W/(m-K)	Density Kg/m³	Reference
Water	0	333	1.6	920	[28]
Formic acid	7.8	277	0.3	1226	[28]
Acetic acid	16.7	192	0.26	1214	[28]
66.6% CaCl₂·6H₂O + 33.3% MgCl₂·6H₂O	25	127	0.93	1661	[28]
Calcium chloride hexahydrate	30	125	1.09	1710	[28]
Urea CH₃COONa·3H₂O	30	200	0.63	1370	[31]
Sodium sulphate decahydrate	32	180	0.56	1485	[28]
Lauric acid	44	212	0.22	1007	[28]
Sodium thiosulphate pentahydrate	46	210	0.76	1666	[28]
61.5% Mg(NO₃)₂·6H₂O + 38.5% NH₄NO₃	52	125	0.59	1672	[31]
Urea–acetamide	53	224	0.51	1216	[32]
Stearic acid	54	157	0.29	940	[29]
Sodium acetate trihydrate	58	265	0.73	1450	[28]
58.7% Mg(NO₃)₂·6(H₂O) + 48.3% MgCl₂·6(H₂O)	59	132	0.67	1610	[28]
Palmitic acid	61	222	0.21	989	[28]
Stearic acid–acetamide	65	213	0.3	972	[33]
86% (MgNO₃)₂·6(H₂O) + 14% LiNO₃.3H₂O	72	180	0.7	1713	[33]
Urea–LiNO₃	76	218	0.85	1438	[34]
Barium hydroxide octahydrate	78	280	1.26	2180	[28]
Acetamide	82	260	0.4	1160	[28]
Urea–NaNO₃	83	187	0.75	1502	[33]
Magnesium nitrate hexahydrate	89	140	0.65	1640	[28]
Urea–NH₄Cl	102	214	0.76	1348	[33]
Urea–K₂CO₃	102	206	0.78	1415	[33]
Oxalic acid dihydrate	105	264	0.9	1653	[29]
Urea–KNO₃	109	195	0.81	1416	[33]
Urea–NaCl	112	230	0.82	1372	[33]
Urea–KCl	115	227	0.83	1370	[33]
Erythritol	117	340	0.73	1450	[29]
Magnesium chloride hexahydrate	117	150	0.7	1570	[28]
High density poly ethylene (HDPE)	130	255	0.48	952	[28]
Urea	134	250	0.8	1320	[29]
Maleic acid	141	285	.	1590	[29]
KNO₃ + NaNO₂	141	97	0.73	1994	[35]
53% KNO₃+ 40% NaNO₃+ 7%NaNO₂	142	110	0.72	2006	[35]

(Continued)

TABLE 4.1

Thermophysical Properties of Some Common Phase Change Materials
(Continued)

PCM	Melting point (°C)	Latent heat (kJ/Kg)	Thermal conductivity W/(m-K)	Density Kg/m³	Reference
KNO$_3$ + NaNO$_3$	149	124	0.58	2080	[35]
LiNO$_3$ + NaNO$_2$	156	233	1.12	2296	[33]
LiNO$_3$ + KCl	160	272	1.31	2196	[28]
LiNO$_3$ + NaNO$_3$ + KCl	160	266	0.88	2297	[33]
d-Mannitol	165	300	0.19	1490	[30]
Hydroquinone	172	258	.	1300	[28]
LiOH + LiNO$_3$	183	352	1.33	2124	[35]
LiNO$_3$ + NaNO$_3$	194	262	0.87	2317	[35]
LiNO$_3$ + NaCl	208	369	1.35	2350	[28]
KNO$_3$ + KOH	214	83	0.88	1905	[28]
KNO$_3$ + NaNO$_3$	222	110	0.73	2028	[33]
LiBr + LiNO$_3$	228	279	1.14	2603	[35]
LiOH + NaNO$_3$ + NaOH	230	184	0.78	2154	[33]
NaNO$_2$ + NaNO$_3$	233	163	0.59	2210	[33]
CaCl$_2$ + LiNO$_3$	238	317	1.37	2362	[33]
LiCl + LiNO$_3$	244	342	1.37	2351	[33]
NaNO$_3$ + NaOH	250	160	0.66	2241	[28]

4.3 TECHNIQUES FOR MEASURING DIFFERENT THERMOPHYSICAL PROPERTIES OF PCMS

4.3.1 MELTING POINT, LATENT HEAT AND SPECIFIC HEAT

The thermophysical properties of PCMs, namely phase transition temperatures, latent heat and specific heat capacity, are the major considerations for the suitable selection of PCMs. Therefore, it is critical to obtain an effective and accurate determination of these properties. Currently, the thermophysical properties of PCMs are most effectively determined by either of the most widely used techniques: differential scanning calorimetry (DSC) and differential thermal analysis (DTA).

DSC works by measuring the difference in heat flow which is required to keep a sample and reference material at the same temperature as it is heated and cooled. DSC is used to measure the heat absorbed (endothermic reaction) and released (exothermic reaction) when a material is subjected to thermal energy. The exothermic peaks on a DSC curve represent heat release, while the endothermic curve represents heat absorption. On the other hand, DTA is a technique which is used to measure the temperature difference between the sample and a reference material when the applied heat is held constant. Materials respond to temperatures in different ways, resulting in differential temperatures. Because of this, DTA can measure material conductivity and phase changes unrelated to enthalpy. The range of maximum temperature that can be detected in DTA can reach up to 1500–1700 °C, while in DSC, it is only up to 1000 °C, but this limit also suits most PCM requirements. However, DTA has a somewhat lower precision and sensitivity than DSC. Therefore, DSC is utilized more in the experimental measurement of these properties.

Tables 4.2 and 4.3 present the data on DSC and DTA techniques which are used to measure the thermophysical properties of PCMs. In the case of DSC, some features which may affect the test outcomes, like oxidation, leakage and corrosion, should be considered initially to ensure the test accuracy.

TABLE 4.2

List of Thermophysical Property-Measuring Instruments for PCMs Using the DSC Technique

Materials	Instruments	Manufacturer	Reference(s)
Al, Mg, Zn	Q10-V5.1-Build191, STA449C	TA Instruments, NETZSCH	[36, 39, 40, 66, 67]
$C_6H_6O_2$, $C_6H_{14}O_6$	DSC822e	Mettler-Toledo	[61]
Fe_3O_4, Al_2O_3	Diamond	Perkin Elmer	[55, 92]
Li_2CO_3, K_2CO_3	–	–	[39, 49]
$LiNO_3$	DSC200PC, Q100	Mettler-Toledo, TA Instruments	[39, 42, 111, 72]
NaCl, KCl, $MgCl_2$, $CaCl_2$	SDT Q600	TA Instruments	[43, 44]
$NaNO_2$, $Ca(NO_3)_2$	STA-409PC	TA Instruments	[37, 38]
$NaNO_3$, KNO_3	STA-409PC, DSC 8000, DSC111, Q100	TA Instruments, Perkin Elmer, Vaisala, TA Instruments	[37, 38, 41, 42, 111, 69, 72, 75]
Si, Cu	STA449C	TA Instruments	[36, 39, 40, 66]
Ti, Ag	Q2000, DSC4000	TA Instruments, Perkin Elmer	[56, 64]
Carbon fiber, carbon nanofiber, carbon nanotube	Q1000	TA Instruments	[51, 53, 54]
Expanded fly ash, expanded clay, expanded perlite	Q100	TA Instruments	[76]
Graphene nanoplatelets, exfoliated graphite nanoplatelets	Q20, 2920	TA Instruments	[48, 50, 53]
HDPE	DSC8500, DSC204F1	Perkin Elmer, NETZSCH	[46, 62]
Neopentyl glycol	DSC200F3	NETZSCH	[68]
PEG, expanded graphite	Q20, Jade	TA Instruments, Perkin Elmer	[48, 69]
Single walled carbon nanotube, multi-walled carbon nanotube, C60	–	–	[49]
RT100, RT40, RT20	DSC204F1	NETZSCH	[45, 62, 70]

(Continued)

TABLE 4.2

List of Thermophysical Property-Measuring Instruments for PCMs Using the DSC Technique *(Continued)*

Materials	Instruments	Manufacturer	Reference(s)
Capric acid, myristic acid, stearic acid	DSC-131, STA449C, DSC200F3	SETARAM Instrumentation, NETZSCH	[59, 74, 73]
Cement mortar, concrete mortar	Q200, Micro DS3, DSC200F3	TA Instruments, Mettler-Toledo, NETZSCH	[57, 58, 74]
Docosane, polymethyl methacrylate	DSC131	SETARAM Instrumentation	[60]
Erythritol	Q2000	TA Instruments	[51, 56]
Graphene	Q100	TA Instruments	[49, 111]
Graphite	DSC8500, DSC111, DSC131, DSC200F3, DSC8000, Q100, Q200, Jade, STA449C	Perkin Elmer, SETARAM Instrumentation, TA Instruments	[46, 57, 63, 110, 111, 68, 69, 71, 73–75]
Halloysite nanotube	STA449C	TA Instruments	[73]
Montmorillonite, vermiculite, diatomite	DSC-131, Jade	SETARAM Instrumentation, Perkin Elmer	[110, 70, 71]
n-Hexadecane/n-octadecane	DSC4000	Perkin Elmer	[64]
Paraffin	DSC8500, 2920, Q1000, Diamond	Perkin Elmer, TA Instruments	[46, 50, 52, 54, 55, 92]
Polyaniline	DSC200F3	NETZSCH	[59]
Paraffin wax	Q1000	TA Instruments	[54]

TABLE 4.3

List of Thermophysical Property-Measuring Instruments for PCMs Using the DTA Technique

Materials	Instrument	Reference(s)
Ag	Du Pont 900 DTA, SEIKO 6200	[40, 64]
Al, Ca, Cu, Mg, P, Si, Zn	Du Pont 900 DTA	[40, 65]
$NaNO_3$, KNO_3, NaOH, KOH, $ZnCl_2$, NaCl, KCl	TG SETSYS 7000	[47]
n-Hexadecane, n-octadecane	SEIKO 6200	[64]

4.3.2 THERMAL CONDUCTIVITY MEASUREMENT METHODS

Thermal conductivity signifies the heat transfer capacity of a PCM and therefore is the most crucial thermophysical property. Thermal conductivity is usually obtained by experimental methods. The techniques for thermal conductivity measurement can be classified into two types: steady-state methods and transient methods [77, 78]. Steady-state methods are based on Fourier's law of heat conduction in a one-dimensional steady state. However, it takes a longer duration to achieve a stabilized state in actual experiments. An example of a typical steady-state method is the guarded hot plate method [79, 80], which is the standard method suitable for low-thermal-conductivity sample measurement.

The unsteady heat conduction equation is the basis of transient methods. The thermal conductivity in these methods is determined by transient temperature variation detection during sample heating. Due to high accuracy and quick measurement results, transient methods are more widely used. Laser flash analysis (LFA), transient hot strip (THS), transient hot wire (THW) and transient plane source (TPS) are commonly used transient methods for thermal conductivity measurements. The TPS method determines a sample's thermal conductivity by detecting the temperature change during heating by an infinite medium-embedded disc-type heating source [81–87]. In the THW method, the temperature rise is detected in an infinite sample-embedded long, thin wire heater. Then, this data is used to measure the sample's thermal conductivity [88–97]. It is widely used for the measurement of the thermal conductivity of fluids and powdered samples. THS is similar in principle to THW, but instead of a thin wire, heat is supplied by a thin metal strip [98–101]. On the other hand, the LFA method directly measures the thermal diffusivity of the sample, through which thermal conductivity is derived [102, 103]. The advantages of LFA include measurements in a wide temperature range and quick measurement results. Apart from metals, LFA can also measure the thermal diffusivity of liquids [104], molten compounds [105], thin-film materials [106], multilayer composites [107, 108] and translucent materials [109].

Table 4.4 lists some experiments and techniques to measure the thermal conductivity of PCMs. In addition to these methods, thermal conductivity has also been

---

TABLE 4.4
List of Some Common Methods for Measuring the Thermal Conductivity of PCMs

Method	Thermal conductivity range [W/mK]	Materials	Reference(s)
Steady state	0.005–2	Low conductivity materials	[46, 94, 95, 112–116, 124, 125, 127, 131, 133, 136]
Transient plane source	0.005–500	Homogeneous solid materials, heterogeneous materials and porous materials	[46, 48, 55, 57, 64, 81–87, 117–121, 123, 126, 129, 130, 132, 135,137]
Transient hot wire/ transient hot strip	0.02–20	Insulation materials, loose or powder nonmetal materials	[53, 54, 73, 75, 88–101, 122, 124, 127, 128]
Laser flash analysis	0.02–2000	Metal, conductor, semiconductor, building materials, film materials, multilayer materials and liquid materials	[49, 51, 102–109, 134]

measured by estimating the component material thermal conductivity and observing the duration of temperature change [110, 111]. All these transient and steady-state methods have achieved remarkable progress after years of research, practice and development.

Although there are myriad techniques to measure thermal conductivity, a single method is not suitable for all materials. If quick results are expected, then transient methods are preferred. For steady-state methods, especially at high temperatures, the measuring time is too long, as stabilizing the surroundings requires a longer time. THW, THS and TPS are more suitable for low-thermal-conductivity samples, and LFA is adaptable to electrically conductive and high-thermal-conductivity materials like metal alloys.

4.4   CONCLUSIONS

- Among several thermal energy storage methods, latent heat storage is more preferable than both thermochemical and sensible heat storage, as it has several advantages such as high energy storage density and a flexible range of operating temperatures. LHTES materials, which are also known as PCMs, absorb or release thermal energy during their phase transitions at constant temperature. Out of all the different types of phase transitions possible in PCMs, the solid-to-liquid transition is preferred, as the phase transition energy in solid–liquid PCMs is significantly higher than in other kinds of PCMs.

- Out of the two general categories of PCMs, viz. organic and inorganic PCMs, the former category fulfils the requirements of good chemical stability and non-corrosiveness. However, organic PCMs are flammable and have generally low thermal conductivity, which makes them unsuitable for wide applications. On the other hand, inorganic PCMs possess high latent heat, higher thermal conductivity, small volume change and narrow phase change temperature ranges as compared to organic PCMs, which make them potential materials in the TES field, especially in the medium-temperature to high-temperature applications where organic PCMs are not a feasible option.

- Accurate measurement of thermophysical properties of PCMs such as melting and freezing point, latent heat and thermal conductivity is a prerequisite for the development and application of PCM-based TES systems since the aforementioned properties determine the performance of TES systems. Therefore, it is imperative to gain knowledge about the techniques such as DSC and DTA. DSC helps us with an understanding of a material's endothermic and exothermic phase transition, which thereby helps in the determination of thermophysical properties such as melting temperature, specific heat capacity and latent heat of fusion or crystallization. DTA is mainly used for determining the temperature of the phase transitions of PCMs, such as melting point and solidification point. The methods for the measurement of thermal conductivity are classified as steady-state methods and transient methods. Thermal conductivity is another important thermal property which determines the heat transfer rate in and out of the PCMs and the use of its total TES capacity. Techniques like LFA, THS, THW and TPS can be easily employed for carrying out thermal conductivity measurements of PCMs.

REFERENCES

1. Yadav, A., Pal, N., Patra, J., & Yadav, M. (2020). Strategic planning and challenges to the deployment of renewable energy technologies in the world scenario: Its impact on global sustainable development. *Environment, Development and Sustainability, 22*(1), 297–315.
2. Da Cunha, J. P., & Eames, P. (2016). Thermal energy storage for low and medium temperature applications using phase change materials—a review. *Applied Energy, 177*, 227–238.
3. Hussain, A., Arif, S. M., & Aslam, M. (2017). Emerging renewable and sustainable energy technologies: State of the art. *Renewable and Sustainable Energy Reviews, 71*, 12–28.
4. Satish, M., Santhosh, S., Yadav, A., Kalluri, S., & Madhavan, A. A. (2021). Optimization and thermal analysis of Fe_2O_3 nanoparticles embedded myristic acid-lauric acid phase change material. *Journal of Electronic Materials, 50*, 1608–1614.
5. Anisur, M. R., Mahfuz, M. H., Kibria, M. A., Saidur, R., Metselaar, I. H. S. C., & Mahlia, T. M. I. (2013). Curbing global warming with phase change materials for energy storage. *Renewable and Sustainable Energy Reviews, 18*, 23–30.
6. Flueckiger, S. M., Iverson, B. D., & Garimella, S. V. (2014). Economic optimization of a concentrating solar power plant with molten-salt thermocline storage. *Journal of Solar Energy Engineering, 136*(1), 011015.
7. Gil, A., Medrano, M., Martorell, I., Lázaro, A., Dolado, P., Zalba, B., & Cabeza, L. F. (2010). State of the art on high temperature thermal energy storage for power generation. Part 1—Concepts, materials and modellization. *Renewable and Sustainable Energy Reviews, 14*(1), 31–55.
8. Yadav, S. (2018). Application of combined materials for baby incubator. *Procedia Manufacturing, 20*, 24–34.
9. Socaciu, L. G. (2012). Thermal energy storage with phase change material. *Leonardo Electronic Journal of Practices and Technologies, 20*, 75–98.
10. Gugulothu, R., Somanchi, N. S., Reddy, K. V. K., & Gantha, D. (2015). A review on solar water distillation using sensible and latent heat. *Procedia Earth and Planetary Science, 11*, 354–360.
11. Dinker, A., Agarwal, M., & Agarwal, G. D. (2017). Heat storage materials, geometry and applications: A review. *Journal of the Energy Institute, 90*(1), 1–11.
12. Pardo, P., Deydier, A., Anxionnaz-Minvielle, Z., Rougé, S., Cabassud, M., & Cognet, P. (2014). A review on high temperature thermochemical heat energy storage. *Renewable and Sustainable Energy Reviews, 32*, 591–610.
13. Mehling, H. (2015, May 19–21). Investigation of the options for thermal energy storage from the viewpoint of the energy form [Paper presentation]. *13th International Conference on Energy Storage GREENSTOCK 2015*, Beijing, China.
14. Nazir, H., Batool, M., Osorio, F. J. B., Isaza-Ruiz, M., Xu, X., Vignarooban, K., Phelan, P., Inamuddin, & Kannan, A. M. (2019). Recent developments in phase change materials for energy storage applications: A review. *International Journal of Heat and Mass Transfer, 129*, 491–523.

15. Abedin, A. H., & Rosen, M. A. (2011). A critical review of thermochemical energy storage systems. *The Open Renewable Energy Journal, 4*, 42–46.

16. Nazir, H., Batool, M., Osorio, F. J. B., Isaza-Ruiz, M., Xu, X., Vignarooban, K., Phelan, P., Inamuddin, & Kannan, A. M. (2019). Recent developments in phase change materials for energy storage applications: A review. *International Journal of Heat and Mass Transfer, 129*, 491–523.

17. Sarbu, I., & Sebarchievici, C. (2018). A comprehensive review of thermal energy storage. *Sustainability, 10*(1), 191.

18. Sharshir, S. W., Joseph, A., Elsharkawy, M., Hamad, M. A., Kandeal, A. W., Elkadeem, M. R., & Arıcı, M. (2023). Thermal energy storage using phase change materials in building applications: A review of the recent development. *Energy and Buildings*, 112908.

19. Yadav, A., & Shivhare, M. K. (2020). Nanoparticle enhanced PCM for solar thermal energy storage. *Advances in Science and Engineering Technology International Conferences (ASET)* (pp. 1–3). Dubai, United Arab Emirates.

20. Sun, M., Liu, T., Sha, H., Li, M., Liu, T., Wang, X., & Jiang, D. (2023). A review on thermal energy storage with eutectic phase change materials: Fundamentals and applications. *Journal of Energy Storage, 68*, 107713.

21. Samykano, M. (2022). Role of phase change materials in thermal energy storage: Potential, recent progress and technical challenges. *Sustainable Energy Technologies and Assessments, 52*, 102234.

22. Yadav, A., Barman, B., Kumar, V., Kardam, A., Narayanan, S. S., Verma, A., Madhwal. D., Shukla, P., & Jain, V. K. (2017). A review on thermophysical properties of nanoparticle-enhanced phase change materials for thermal energy storage. In *Recent Trends in Materials and Devices* (pp. 37–47). Springer, Cham.

23. Lawag, R. A., & Ali, H. M. (2022). Phase change materials for thermal management and energy storage: A review. *Journal of Energy Storage, 55*, 105602.

24. Liu, Y., Zheng, R., & Li, J. (2022). High latent heat phase change materials (PCMs) with low melting temperature for thermal management and storage of electronic devices and power batteries: Critical review. *Renewable and Sustainable Energy Reviews, 168*, 112783.

25. Lu, Y., Yu, D., Dong, H., Lv, J., Wang, L., Zhou, H., & He, Z. (2022). Magnetically tightened form-stable phase change materials with modular assembly and geometric conformality features. *Nature Communications, 13*(1), 1397.

26. Kalombe, R. M., Sobhansarbandi, S., & Kevern, J. (2023). Assessment of low-cost organic phase change materials for improving infrastructure thermal performance. *Construction and Building Materials, 369*, 130285.

27. Li, J., Chang, Q., Xue, C., Yang, J., & Hu, S. (2023). Carbon dots efficiently enhance photothermal conversion and storage of organic phase change materials through interfacial interaction. *Carbon, 203*, 21–28.

28. Jankowski, N. R., & McCluskey, F. P. (2014). A review of phase change materials for vehicle component thermal buffering. *Applied Energy, 113*, 1525–1561.

29. Haillot, D., Bauer, T., Kröner, U., & Tamme, R. (2011). Thermal analysis of phase change materials in the temperature range 120–150 C. *Thermochimica Acta, 513*(1–2), 49–59.

30. Solé, A., Neumann, H., Niedermaier, S., Martorell, I., Schossig, P., & Cabeza, L. F. (2014). Stability of sugar alcohols as PCM for thermal energy storage. *Solar Energy Materials and Solar Cells, 126*, 125–134.

31. Pielichowska, K., & Pielichowski, K. (2014). Phase change materials for thermal energy storage. *Progress in Materials Science, 65*, 67–123.

32. Zalba, B., Marın, J. M., Cabeza, L. F., & Mehling, H. (2003). Review on thermal energy storage with phase change: Materials, heat transfer analysis and applications. *Applied Thermal Engineering, 23*(3), 251–283.

33. Hewitt, N. J. (2012). Heat pumps and energy storage—The challenges of implementation. *Applied Energy, 89*(1), 37–44.

34. Sharma, A., Tyagi, V. V., Chen, C. R., & Buddhi, D. (2009). Review on thermal energy storage with phase change materials and applications. *Renewable and Sustainable Energy Reviews, 13*(2), 318–345.

35. Janz, G. J., & Tomkins, R. P. T. (1983). Molten salts: Volume 5, Part 2. Additional single and multi-component salt systems. Electrical conductance, density, viscosity and surface tension data. *Journal of Physical and Chemical Reference Data, 12*(3), 591–815.

36. Guocai, Z. H. A. N. G., Zhe, X. U., Yunfa, C. H. E. N., & Jianqiang, L. I. (2012). Progress in metal-based phase change materials for thermal energy storage applications. *Energy Storage Science and Technology, 1*(1), 74.

37. Wang, C., Ren, N., Wu, Y., et al. (2012). Development and thermal properties measurement of new low melting point mixed molten salt. In: *Proceedings of Chinese Society of engineering thermophysics heat and mass transfer conference 2012*. Dongguan: Chinese Society of Engineering Thermophysics; pp. 1–5.

38. Ren N, Wang C, Chen C, et al. (2012). Preparation and properties measurement of mixed molten salt. In: *Proceedings of Chinese Society of Engineering Thermophysics Heat and Mass Transfer Conference 2012*. Dongguan: Chinese Society of Engineering Thermophysics; pp. 1–8.

39. Shidong, L. I., Renyuan, Z. H. A. N. G., & Feng, L. I. (2010). Research progress in thermal storage materials applied in concentrating solar power. *Materials Review, 24*(21), 51–55.

40. Tyagi, V. V., Chopra, K., Sharma, R. K., Pandey, A. K., Tyagi, S. K., Ahmad, M. S., Sari, A., & Kothari, R. (2022). A comprehensive review on phase change materials for heat storage applications: Development, characterization, thermal and chemical stability. *Solar Energy Materials and Solar Cells, 234*, 111392.

41. Oh, J., Jung, H., & Yoh, J. J. (2022). Observation of gunpowder-like thermochemical responses of a thermal energy storage system based on KNO3/NaNO3/Graphite exposed to a heat transfer fluid. *Applied Thermal Engineering, 207*, 118215.

42. Kundu, R., Kar, S. P., & Sarangi, R. (2022). Performance enhancement with inorganic phase change materials for the application of thermal energy storage: A critical review. *Energy Storage, 4*(5), e320.

43. Zhang, C., Han, S., Wu, Y., Zhang, C., & Guo, H. (2022). Investigation on convection heat transfer performance of quaternary mixed molten salt based nanofluids in smooth tube. *International Journal of Thermal Sciences, 177*, 107534.

44. Baohua, H., Jing, D., & Xiaolan, W. (2010). Test of thermal physics and analysis on thermal stability of high temperature molten salt. *Inorganic Chemicals Industry, 42*(1), 22–24.

45. Agyenim, F., Eames, P., & Smyth, M. (2010). Heat transfer enhancement in medium temperature thermal energy storage system using a multitube heat transfer array. *Renewable Energy, 35*(1), 198–207.

46. AlMaadeed, M. A., Labidi, S., Krupa, I., & Karkri, M. (2015). Effect of expanded graphite on the phase change materials of high density polyethylene/wax blends. *Thermochimica Acta, 600*, 35–44.

47. Pincemin, S., Olives, R., Py, X., & Christ, M. (2008). Highly conductive composites made of phase change materials and graphite for thermal storage. *Solar Energy Materials and Solar Cells, 92*(6), 603–613.

48. Qi, G. Q., Yang, J., Bao, R. Y., Liu, Z. Y., Yang, W., Xie, B. H., & Yang, M. B. (2015). Enhanced comprehensive performance of polyethylene glycol based phase change material with hybrid graphene nanomaterials for thermal energy storage. *Carbon*, *88*, 196–205.

49. Tao, Y. B., Lin, C. H., & He, Y. L. (2015). Preparation and thermal properties characterization of carbonate salt/carbon nanomaterial composite phase change material. *Energy Conversion and Management*, *97*, 103–110.

50. Kim, S., & Drzal, L. T. (2009). High latent heat storage and high thermal conductive phase change materials using exfoliated graphite nanoplatelets. *Solar Energy Materials and Solar Cells*, *93*(1), 136–142.

51. Nomura, T., Tabuchi, K., Zhu, C., Sheng, N., Wang, S., & Akiyama, T. (2015). High thermal conductivity phase change composite with percolating carbon fiber network. *Applied Energy*, *154*, 678–685.

52. Li, S., Wang, H., Gao, X., Niu, Z., & Song, J. (2023). Design of corn straw/paraffin wax shape-stabilized phase change materials with excellent thermal buffering performance. *Journal of Energy Storage*, *57*, 106217.

53. Fan, L. W., Zhu, Z. Q., Zeng, Y., Xiao, Y. Q., Liu, X. L., Wu, Y. Y., . . . & Cen, K. F. (2015). Transient performance of a PCM-based heat sink with high aspect-ratio carbon nanofillers. *Applied Thermal Engineering*, *75*, 532–540.

54. Wen, R., Wu, M., Zhu, J., Zhu, S., & Chen, W. (2022). Preparation and characteristic of Ag nanoparticle modified expanded graphite for enhancing paraffin phase change material properties. *Fullerenes, Nanotubes and Carbon Nanostructures*, *30*(10), 1046–1053.

55. Şahan, N., Fois, M., & Paksoy, H. (2015). Improving thermal conductivity phase change materials—A study of paraffin nanomagnetite composites. *Solar Energy Materials and Solar Cells*, *137*, 61–67.

56. Zhichao, L., Qiang, Z., & Gaohui, W. (2015). Preparation and enhanced heat capacity of nano-titania doped erythritol as phase change material. *International Journal of Heat and Mass Transfer*, *80*, 653–659.

57. Zheng, Q., Kaur, S., Dames, C., & Prasher, R. S. (2020). Analysis and improvement of the hot disk transient plane source method for low thermal conductivity materials. *International Journal of Heat and Mass Transfer*, *151*, 119331.

58. Lecompte, T., Le Bideau, P., Glouannec, P., Nortershauser, D., & Le Masson, S. (2015). Mechanical and thermo-physical behaviour of concretes and mortars containing phase change material. *Energy and Buildings*, *94*, 52–60.

59. Wang, Y., Ji, H., Shi, H., Zhang, T., & Xia, T. (2015). Fabrication and characterization of stearic acid/polyaniline composite with electrical conductivity as phase change materials for thermal energy storage. *Energy Conversion and Management*, *98*, 322–330.

60. Alkan, C., Sarı, A., Karaipekli, A., & Uzun, O. (2009). Preparation, characterization, and thermal properties of microencapsulated phase change material for thermal energy storage. *Solar Energy Materials and Solar Cells*, *93*(1), 143–147.

61. Peiró, G., Gasia, J., Miró, L., & Cabeza, L. F. (2015). Experimental evaluation at pilot plant scale of multiple PCMs (cascaded) vs. single PCM configuration for thermal energy storage. *Renewable Energy*, *83*, 729–736.

62. Hu, B. W., Wang, Q., & Liu, Z. H. (2015). Fundamental research on the gravity assisted heat pipe thermal storage unit (GAHP-TSU) with porous phase change materials (PCMs) for medium temperature applications. *Energy Conversion and Management*, *89*, 376–386.

63. Atkin, P., & Farid, M. M. (2015). Improving the efficiency of photovoltaic cells using PCM infused graphite and aluminium fins. *Solar Energy*, *114*, 217–228.

64. Sarier, N., Onder, E., & Ukuser, G. (2015). Silver incorporated microencapsulation of n-hexadecane and n-octadecane appropriate for dynamic thermal management in textiles. *Thermochimica Acta*, *613*, 17–27.

65. Chang, I., & Cai, Q. (2022). From simple binary to complex multicomponent eutectic alloys. *Progress in Materials Science, 123*, 100779.
66. Cheng, X. M., Dong, J., Wu, X. W., Gong, D. Q. (2010). Thermal storage properties of high temperature phase transformation on Al-Si-Cu-Mg-Zn alloys. *Heat Treatment of Metals, 3*, 13–16.
67. Sun, J. Q., Zhang, R. Y., Liu, Z. P., & Lu, G. H. (2007). Thermal reliability test of Al–34% Mg–6% Zn alloy as latent heat storage material and corrosion of metal with respect to thermal cycling. *Energy Conversion and Management, 48*(2), 619–624.
68. Wang, X., Guo, Q., Zhong, Y., Wei, X., & Liu, L. (2013). Heat transfer enhancement of neopentyl glycol using compressed expanded natural graphite for thermal energy storage. *Renewable Energy, 51*, 241–246.
69. Oh, J., Jung, H., & Yoh, J. J. (2022). Observation of gunpowder-like thermochemical responses of a thermal energy storage system based on $KNO_3/NaNO_3$/Graphite exposed to a heat transfer fluid. *Applied Thermal Engineering, 207*, 118215.
70. Fang, X., Zhang, Z., & Chen, Z. (2008). Study on preparation of montmorillonite-based composite phase change materials and their applications in thermal storage building materials. *Energy Conversion and Management, 49*(4), 718–723.
71. Wei, H., Xie, X., Li, X., & Lin, X. (2016). Preparation and characterization of capric-myristic-stearic acid eutectic mixture/modified expanded vermiculite composite as a form-stable phase change material. *Applied Energy, 178*, 616–623.
72. Michels, H., & Pitz-Paal, R. (2007). Cascaded latent heat storage for parabolic trough solar power plants. *Solar Energy, 81*(6), 829–837.
73. Mei, D., Zhang, B., Liu, R., Zhang, Y., & Liu, J. (2011). Preparation of capric acid/halloysite nanotube composite as form-stable phase change material for thermal energy storage. *Solar Energy Materials and Solar Cells, 95*(10), 2772–2777.
74. Li, M., Wu, Z., & Tan, J. (2013). Heat storage properties of the cement mortar incorporated with composite phase change material. *Applied Energy, 103*, 393–399.
75. Xiao, X., Zhang, P., & Li, M. (2013). Thermal characterization of nitrates and nitrates/ expanded graphite mixture phase change materials for solar energy storage. *Energy Conversion and Management, 73*, 86–94.
76. Zhang, D., Tian, S., & Xiao, D. (2007). Experimental study on the phase change behavior of phase change material confined in pores. *Solar Energy, 81*(5), 653–660.
77. Hu, P., & Chen, Z. S. (2009). *Calorimetric Technology and Thermal Physical Properties Determination*. University of Science and Technology of China Press, Hefei.
78. Zielenkiewicz, W., & Margas, E. (2006). *Theory of Calorimetry* (Vol. 2). Springer Science & Business Media, Berlin.
79. Dubois, S., & Lebeau, F. (2015). Design, construction and validation of a guarded hot plate apparatus for thermal conductivity measurement of high thickness crop-based specimens. *Materials and Structures, 48*, 407–421.
80. Bukkambudhi, A. K., & Madhusudana, C. V. (1998). Accuracy of thermal conductivity measurement of low conductivity materials using a guarded hot plate. In *International Heat Transfer Conference Digital Library*. Begel House Inc., Danbury, CT.
81. Gustafsson, S. E. (1991). Transient plane source techniques for thermal conductivity and thermal diffusivity measurements of solid materials. *Review of Scientific Instruments, 62*(3), 797–804.
82. Motahar, S., Nikkam, N., Alemrajabi, A. A., Khodabandeh, R., Toprak, M. S., & Muhammed, M. (2014). Experimental investigation on thermal and rheological properties of n-octadecane with dispersed $TiO2$ nanoparticles. *International Communications in Heat and Mass Transfer, 59*, 68–74.

83. Nagai, H., Rossignol, F., Nakata, Y., Tsurue, T., Suzuki, M., & Okutani, T. (2000). Thermal conductivity measurement of liquid materials by a hot-disk method in short-duration microgravity environments. *Materials Science and Engineering: A, 276*(1–2), 117–123.

84. Joshi, G. P., Saxena, N. S., & Mangal, R. (2003). Temperature dependence of effective thermal conductivity and effective thermal diffusivity of Ni-Zn ferrites. *Acta Materialia, 51*(9), 2569–2576.

85. Bouguerra, A., Aït-Mokhtar, A., Amiri, O., & Diop, M. B. (2001). Measurement of thermal conductivity, thermal diffusivity and heat capacity of highly porous building materials using transient plane source technique. *International Communications in Heat and Mass Transfer, 28*(8), 1065–1078.

86. Saxena, N. S., Pradeep, P., Mathew, G., Thomas, S., Gustafsson, M., & Gustafsson, S. E. (1999). Thermal conductivity of styrene butadiene rubber compounds with natural rubber prophylactics waste as filler. *European Polymer Journal, 35*(9), 1687–1693.

87. Yang, X., Yue, K., Han, D., & Zhang, X. (2022). Effective thermal conductivity measurement of PCM composites during phase transition by using the transient plane source method. *Modern Physics Letters B, 36*(14), 2250034.

88. Salim, S. G. R. (2022). Thermal conductivity measurements using the transient hot-wire method: A review. *Measurement Science and Technology, 33*(12), 125022.

89. Prado, J. I., Calviño, U., & Lugo, L. (2022). Experimental methodology to determine thermal conductivity of nanofluids by using a commercial transient hot-wire device. *Applied Sciences, 12*(1), 329.

90. Vélez, C., Reding, B., de Zárate, J. M. O., & Khayet, M. (2019). Thermal conductivity of water Ih-ice measured with transient hot-wires of different lengths. *Applied Thermal Engineering, 149*, 788–797.

91. Panchal, M., Saraswat, A., Verma, S., & Chaudhuri, P. (2020). Measurement of effective thermal conductivity of lithium metatitanate pebble bed by transient hot-wire technique. *Fusion Engineering and Design, 158*, 111718.

92. Nourani, M., Hamdami, N., Keramat, J., Moheb, A., & Shahedi, M. (2016). Thermal behavior of paraffin-nano-Al_2O_3 stabilized by sodium stearoyl lactylate as a stable phase change material with high thermal conductivity. *Renewable Energy, 88*, 474–482.

93. Watanabe, H. (2002). Further examination of the transient hot-wire method for the simultaneous measurement of thermal conductivity and thermal diffusivity. *Metrologia, 39*(1), 65.

94. Hong, S. W., Kang, Y. T., Kleinstreuer, C., & Koo, J. (2011). Impact analysis of natural convection on thermal conductivity measurements of nanofluids using the transient hot-wire method. *International Journal of Heat and Mass Transfer, 54*(15–16), 3448–3456.

95. Lee, J., Lee, H., Baik, Y. J., & Koo, J. (2015). Quantitative analyses of factors affecting thermal conductivity of nanofluids using an improved transient hot-wire method apparatus. *International Journal of Heat and Mass Transfer, 89*, 116–123.

96. Turgut, A., Tavman, I., & Tavman, S. (2009). Measurement of thermal conductivity of edible oils using transient hot wire method. *International Journal of Food Properties, 12*(4), 741–747.

97. Assael, M. J., Antoniadis, K. D., Metaxa, I. N., Mylona, S. K., Assael, J. A. M., Wu, J., & Hu, M. (2015). A novel portable absolute transient hot-wire instrument for the measurement of the thermal conductivity of solids. *International Journal of Thermophysics, 36*, 3083–3105.

98. Gustafsson, S. E., Karawacki, E., & Khan, M. N. (1979). Transient hot-strip method for simultaneously measuring thermal conductivity and thermal diffusivity of solids and fluids. *Journal of Physics D: Applied Physics, 12*(9), 1411.

99. Belkerk, B. E., Soussou, M. A., Carette, M., Djouadi, M. A., & Scudeller, Y. (2012). Measuring thermal conductivity of thin films and coatings with the ultra-fast transient hot-strip technique. *Journal of Physics D: Applied Physics, 45*(29), 295303.

100. Wei, G., Du, X., Zhang, X., & Yu, F. (2009, January). Theoretical study on transient hot-strip method by numerical analysis. In *Heat Transfer Summer Conference* (Vol. 43567, pp. 481–489), AMSE Publishing, San Francisco, CA.

101. Wei, G., Liu, Y., Zhang, X., Yu, F., & Du, X. (2011). Thermal conductivities study on silica aerogel and its composite insulation materials. *International Journal of Heat and Mass Transfer, 54*(11–12), 2355–2366.

102. Parker, W. J., Jenkins, R. J., Butler, C. P., & Abbott, G. L. (1961). Flash method of determining thermal diffusivity, heat capacity, and thermal conductivity. *Journal of Applied Physics, 32*(9), 1679–1684.

103. Abdul Jaleel, S. A., Kim, T., & Baik, S. (2023). Covalently functionalized leakage-free healable phase-change interface materials with extraordinary high-thermal conductivity and low-thermal resistance. *Advanced Materials, 35*(30), 2300956.

104. Zhu, S., Li, C., Su, C. H., Lin, B., Ban, H., Scripa, R. N., & Lehoczky, S. L. (2003). Thermal diffusivity, thermal conductivity, and specific heat capacity measurements of molten tellurium. *Journal of Crystal Growth, 250*(1–2), 269–273.

105. Ohta, H., Ogura, G., Waseda, Y., & Suzuki, M. (1990). Thermal diffusivity measurements of molten salts using a three-layered cell by the laser flash method. *Review of Scientific Instruments, 61*(10), 2645–2649.

106. Ruoho, M., Valset, K., Finstad, T., & Tittonen, I. (2015). Measurement of thin film thermal conductivity using the laser flash method. *Nanotechnology, 26*(19), 195706.

107. Potenza, M., Coppa, P., Corasaniti, S., & Bovesecchi, G. (2021). Numerical simulation of thermal diffusivity measurements with the laser-flash method to evaluate the effective property of composite materials. *Journal of Heat Transfer, 143*(7).

108. Yu, F., Wei, G., Zhang, X., & Chen, K. (2006). Two effective thermal conductivity models for porous media with hollow spherical agglomerates. *International Journal of Thermophysics, 27*, 293–303.

109. Ouyang, Z., Rao, Q., Wang, J., & Peng, X. (2022). Investigation on thermal conduction of graphite film enhanced carbon fiber polymer composite by laser flash analysis. *Journal of Composite Materials*, 00219983221149795.

110. Karaman, S., Karaipekli, A., Sarı, A., & Bicer, A. (2011). Polyethylene glycol (PEG)/diatomite composite as a novel form-stable phase change material for thermal energy storage. *Solar Energy Materials and Solar Cells, 95*(7), 1647–1653.

111. Tao, Z., Liang, Z., & Dong, Z. (2010). Improvement of thermal properties of hybrid inorganic salt phase change materials by expanded graphite and graphene. *Inorganic Chemicals Industry, 5*, 009.

112. Lin, X., Zhang, X., Ji, J., Liu, L., Yang, M., & Zou, L. (2022). Experimental investigation of form-stable phase change material with enhanced thermal conductivity and thermal-induced flexibility for thermal management. *Applied Thermal Engineering, 201*, 117762.

113. Feng, Z., & Xiao, X. (2022). Thermal conductivity measurement of flexible composite phase-change materials based on the steady-state method. *Micromachines, 13*(10), 1582.

114. Cao, L., & Zhang, D. (2019). Styrene-acrylic emulsion/graphene aerogel supported phase change composite with good thermal conductivity. *Thermochimica Acta, 680*, 178351.

115. Gariboldi, E., Colombo, L. P., Fagiani, D., & Li, Z. (2019). Methods to characterize effective thermal conductivity, diffusivity and thermal response in different classes of composite phase change materials. *Materials, 12*(16), 2552.

116. Rathod, M. K., & Banerjee, J. (2015). Thermal performance enhancement of shell and tube Latent Heat Storage Unit using longitudinal fins. *Applied Thermal Engineering*, *75*, 1084–1092.

117. Zhao, C. Y., & Wu, Z. G. (2011). Heat transfer enhancement of high temperature thermal energy storage using metal foams and expanded graphite. *Solar Energy Materials and Solar Cells*, *95*(2), 636–643.

118. Wu, Z. G., & Zhao, C. Y. (2011). Experimental investigations of porous materials in high temperature thermal energy storage systems. *Solar Energy*, *85*(7), 1371–1380.

119. Acem, Z., Lopez, J., & Del Barrio, E. P. (2010). KNO3/NaNO3—Graphite materials for thermal energy storage at high temperature: Part I.—Elaboration methods and thermal properties. *Applied Thermal Engineering*, *30*(13), 1580–1585.

120. Mancin, S., Diani, A., Doretti, L., Hooman, K., & Rossetto, L. (2015). Experimental analysis of phase change phenomenon of paraffin waxes embedded in copper foams. *International Journal of Thermal Sciences*, *90*, 79–89.

121. Mesalhy, O., Lafdi, K., & Elgafy, A. (2006). Carbon foam matrices saturated with PCM for thermal protection purposes. *Carbon*, *44*(10), 2080–2088.

122. Elgafy, A., & Lafdi, K. (2005). Effect of carbon nanofiber additives on thermal behavior of phase change materials. *Carbon*, *43*(15), 3067–3074.

123. Fethi, A., Mohamed, L., Mustapha, K., & Sassi, B. N. (2015). Investigation of a graphite/paraffin phase change composite. *International Journal of Thermal Sciences*, *88*, 128–135.

124. Yu, Z. T., Fang, X., Fan, L. W., Wang, X., Xiao, Y. Q., Zeng, Y., & Cen, K. F. (2013). Increased thermal conductivity of liquid paraffin-based suspensions in the presence of carbon nano-additives of various sizes and shapes. *Carbon*, *53*, 277–285.

125. Fan, L. W., Zhu, Z. Q., & Liu, M. J. (2015). A similarity solution to unidirectional solidification of nano-enhanced phase change materials (NePCM) considering the mushy region effect. *International Journal of Heat and Mass Transfer*, *86*, 478–481.

126. Shi, J. N., Ger, M. D., Liu, Y. M., Fan, Y. C., Wen, N. T., Lin, C. K., & Pu, N. W. (2013). Improving the thermal conductivity and shape-stabilization of phase change materials using nanographite additives. *Carbon*, *51*, 365–372.

127. Frusteri, F., Leonardi, V., & Maggio, G. (2006). Numerical approach to describe the phase change of an inorganic PCM containing carbon fibres. *Applied Thermal Engineering*, *26*(16), 1883–1892.

128. Frusteri, F., Leonardi, V., Vasta, S., & Restuccia, G. (2005). Thermal conductivity measurement of a PCM based storage system containing carbon fibers. *Applied Thermal Engineering*, *25*(11–12), 1623–1633.

129. Warzoha, R. J., Weigand, R. M., & Fleischer, A. S. (2015). Temperature-dependent thermal properties of a paraffin phase change material embedded with herringbone style graphite nanofibers. *Applied Energy*, *137*, 716–725.

130. Warzoha, R. J., & Fleischer, A. S. (2015). Effect of carbon nanotube interfacial geometry on thermal transport in solid–liquid phase change materials. *Applied Energy*, *154*, 271–276.

131. Muñoz-Sánchez, B., Iparraguirre-Torres, I., Madina-Arrese, V., Izagirre-Etxeberria, U., Unzurrunzaga-Iturbe, A., & García-Romero, A. (2015). Encapsulated high temperature PCM as active filler material in a thermocline-based thermal storage system. *Energy Procedia*, *69*, 937–946.

132. Soares, N., Gaspar, A. R., Santos, P., & Costa, J. J. (2015). Experimental study of the heat transfer through a vertical stack of rectangular cavities filled with phase change materials. *Applied Energy*, *142*, 192–205.

133. Lee, D. Y., Nam, T. H., Park, I. J., Kim, J. G., & Ahn, J. (2013). Corrosion behavior of aluminum alloy for heat exchanger in an exhaust gas recirculation system of diesel engine. *Corrosion*, *69*(8), 828–836.

134. Bopanna, K. D., & Ganesha Prasad, M. S. (2020). Thermal characterization of aluminium-based composite structures using laser flash analysis. *Journal of The Institution of Engineers (India): Series C, 101,* 159–166.
135. Wang, X., Liu, J., Zhang, Y., Di, H., & Jiang, Y. (2006). Experimental research on a kind of novel high temperature phase change storage heater. *Energy Conversion and Management, 47*(15–16), 2211–2222.
136. Kenisarin, M. M. (2010). High-temperature phase change materials for thermal energy storage. *Renewable and Sustainable Energy Reviews, 14*(3), 955–970.
137. Tian, Y., & Zhao, C. Y. (2013). Thermal and exergetic analysis of metal foam-enhanced cascaded thermal energy storage (MF-CTES). *International Journal of Heat and Mass Transfer, 58*(1–2), 86–96.

5 Thermal Characterization Techniques for Phase Change Materials

Saeed Esfandeh and Mohammad Hassan Kamyab

HIGHLIGHTS

- Optimization of PCM's thermal characteristics by applying nanotechnology in TES systems.
- Optimization of phase change behavior (melting/solidification) in PCMs using nanotechnology.
- Introducing experimental techniques for PCM's thermal characteristic measurements.
- Introducing most applied nanomaterials in thermal energy systems.
- Introducing uncertainty sources in the measurement of PCM's thermal characteristics.

NOMENCLATURE

Latin letters

ΔH_{fus}	Heat of fusion
T_m	Melting point
Greek letters	
α	Thermal diffusivity
ρ	Density

ABBREVIATIONS

Al_2O_3	Aluminum oxide
Ag	Silver
CuO	Copper oxide
Cu	Copper
C_p	Heat capacity
DSC	Differential scanning calorimeter
FTIR	Fourier-transform infrared spectroscopy
Fe_3O_4	Magnetic iron oxide nanoparticles

DOI: 10.1201/9781003331957-5

FESEM	Field emission scanning electron microscopy
GO	Graphene oxide
GNP	Graphene nano-platelets
k	Thermal conductivity
LA	Lauric acid
MWCNT	Multi-wall carbon nanotube
MP	Methyl palmitate
NG	Nano-graphene
PCM	Phase change material
PVT	Photovoltaic thermal
SiO_2	Silica oxide
Si_3N_4	Silicon nitride
SEM	Scanning electron microscope
THB	Transient hot-bridge technique
TES	Thermal energy storage
TC	Thermal conductivity
TGA	Thermo-gravimetric analysis
TiO_2	Titanium oxide
TEM	Transmission electron microscopy
Wt	Weight
XRD	X-ray diffraction

Sub/superscript

| Fus | Fusion |
| M | Melting |

5.1 INTRODUCTION

Thermal energy storage (TES) systems can be an applicable host for phase change materials (PCM). PCMs can optimize the size, cost and capacity of a TES system. Emission reduction by about 40% [1] and also energy saving by about 50% [2] in TES systems can be among the results of applying PCMs in TES systems.

It should be considered that not every PCM can be a suitable choice to be applied in a TES system. A suitable PCM for applying in a TES system needs to have high latent heat, high thermal conductivity (TC), high specific heat capacity (high sensible heat), low volume change during phase change, high density, chemical stability, non-flammability, non-toxicity, availability and low cost that all mentioned features can be divided in to chemicals, kinetic and thermophysicals features [3–5]. Many PCMs do not have all of the aforementioned characteristics, so experts have tried and are trying to propose a new solutions to solve this deficiency. The present chapter has focused on the possible positive effects of nanotechnology on improving the thermal characteristics of PCMs.

5.2 NANOTECHNOLOGY EFFECT ON THE THERMAL CHARACTERISTICS OF PCMS

Nanoparticles can be applied as additives to pure conventional PCMs to improve the thermal characteristics of PCMs. Although applying nano-sized materials as thermal characteristic improvers is not the only choice, the aim of present section is to focus on adding nanoparticles to PCMs and studying the effect of the mentioned additions on the thermal characteristics of PCMs. Figures 5.1 and 5.2 show a schematic view of applied nano-enriched PCMs in photovoltaic thermal (PVT) systems and flat-plate solar collectors as TES systems. The following sections discuss the effects of added nanomaterials on thermal properties of PCMs in more detail.

5.2.1 OPTIMIZATION OF PCM THERMAL CONDUCTIVITY IN TES SYSTEMS

The main TES mechanisms in materials are chemical and physical mechanisms. The materials that apply physical mechanisms may use sensible heat or latent heat in the process of energy release or absorption. Because of their high latent heat, suitable phase change temperature and minimum volume change, PCMs are always among main potential candidates for applying in TES systems [6, 7]. Figures 5.3, 5.4 and 5.5 show the position of PCMs in the classification of energy storage mechanisms, the classification of PCMs based on different start phases and end phases in phase change process and the classification of solid–liquid PCMs, respectively.

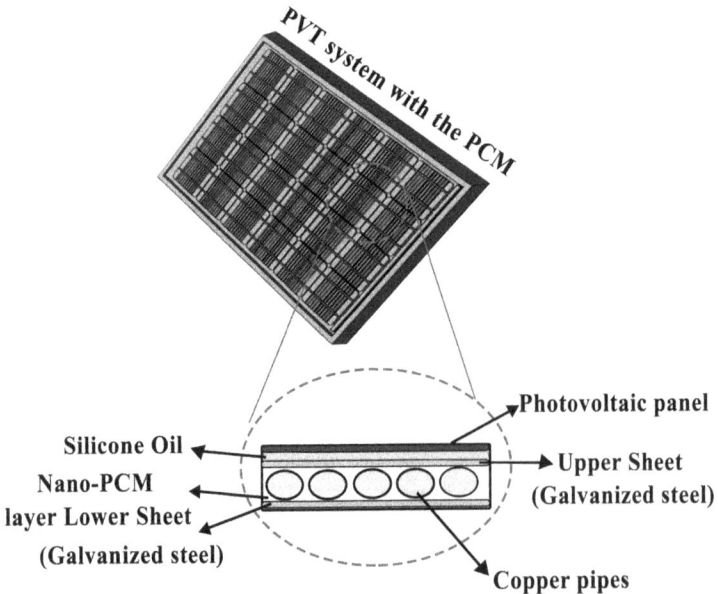

FIGURE 5.1   Details of inner layers of a PVT system enriched with nano-PCM.

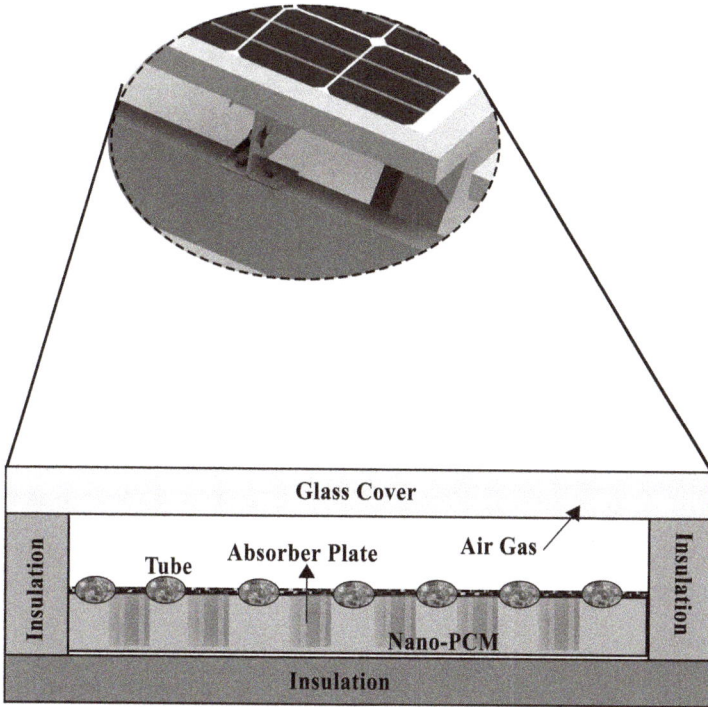

FIGURE 5.2 Details of inner layers of a flat-plate solar collector system enriched with nano-PCM.

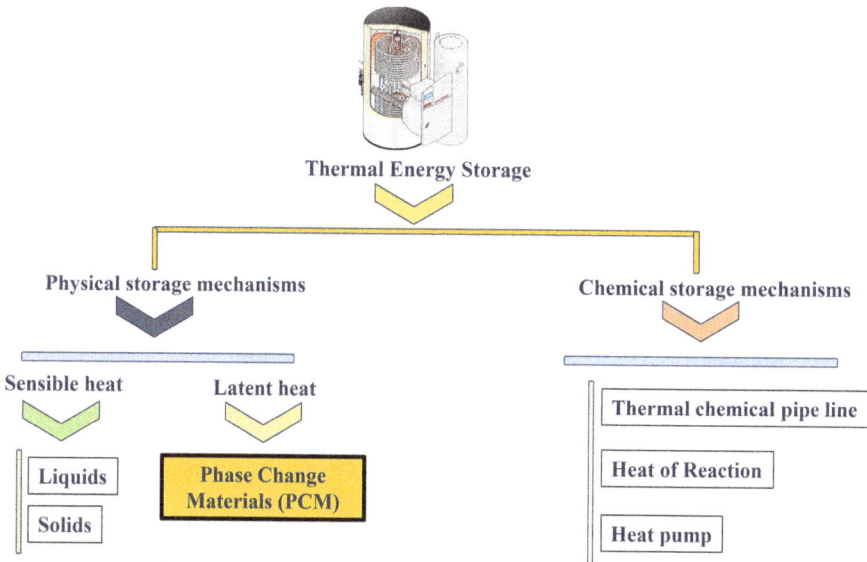

FIGURE 5.3 Classification of TES systems based on mechanisms.

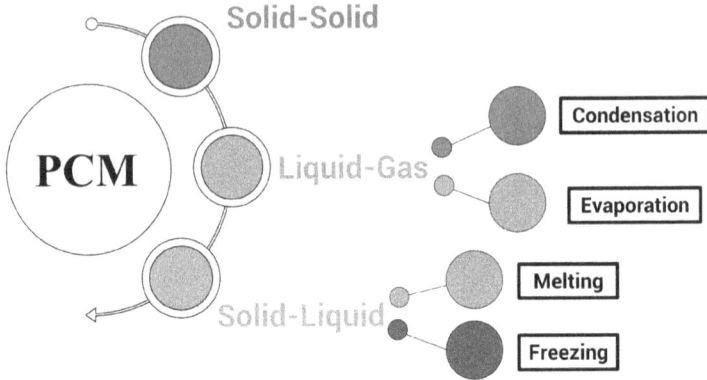

FIGURE 5.4   Classification of PCMs based on various start phase and end phase types.

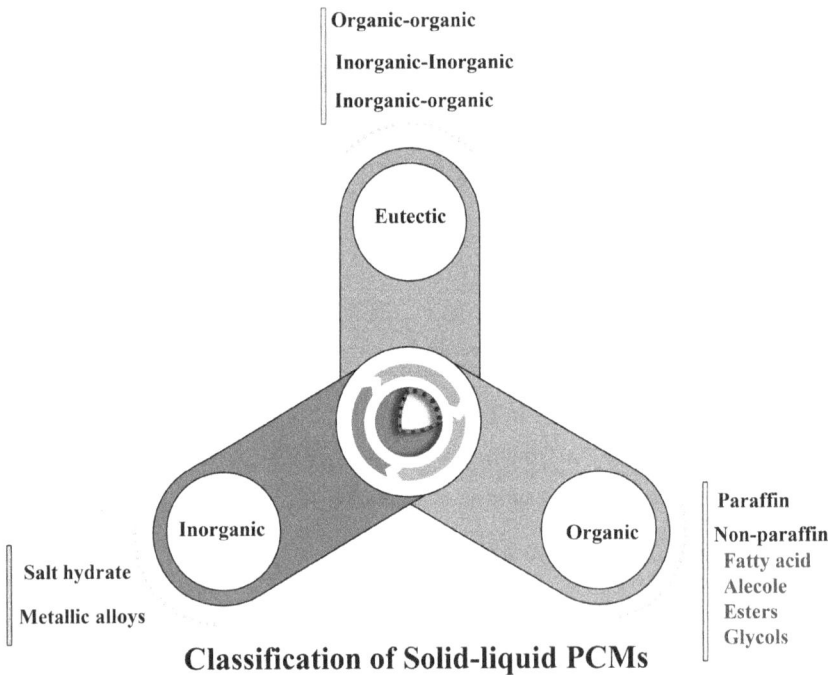

FIGURE 5.5   Classification of solid–liquid PCMs.

Next, to ensure the cost effectiveness and availability of PCMs, thermodynamic, physical, thermophysical and chemical properties of PCMs should be considered for the design of an applicable TES system. Also, system design temperature, operating temperature and system application are important issues in selecting PCMs because the compatibility between the temperature of the phase change of the material and the operating temperature of the system is of great importance in the cooling and heating processes. Two main thermal characteristics of a PCM are the latent heat and TC. A higher latent heat will provide a more compact TES system, and a higher TC will ease the charging and discharging of the system. Besides the advantages of PCMs, one of the limiting factors against using PCMs in TES systems is their low TC [8, 9], which needs to be solved. A low TC of PCMs will result in low rate of release and storage of the heat in a PCM-enriched TES system and will reduce the efficiency of the TES system. Therefore, this problem reduces the economic efficiency of using PCMs in TES systems.

To improve the TC of applied PCMs in a TES system, various techniques can be applied. The first technique is to extend the heat transfer surface, the second technique is to simultaneously use hybrid PCMs that consist of two or more PCM types and the third technique is to apply nanotechnology to the PCM structure [10–14]. By adding nanoparticles to the PCM structure, the TC of PCM will be increased [15–17]. Choosing a suitable nanoparticle in the process of PCM enrichment is crucial. For example, carbon-based and metallic nanoparticles have higher TC values compared to other types. Because of their lower density, higher stability and better dispersion in PCMs, carbon-based nanoparticles like carbon nanotubes and graphene are preferable over the metallic group. The effects of adding metal- and carbon-based nanoparticles on the TC of PCMs are presented and compared in Table 5.1. As can be seen, the TC improvement for enriched PCMs with carbon-based nanoparticles is absolutely higher than that of enriched PCMs with metallic nanoparticles. However, it should be noted that the focused PCM (paraffin) in Table 5.1 is not the best choice for all applications, and selecting the PCM type needs to be analyzed based on its specific application. In Table 5.1, paraffin is the base and host PCM for all nanosized additives to be able to make a better comparison about the result of adding each of the nanoparticle types to paraffin.

5.2.2 Most Common Nanoparticles in PCMs

Various nanoparticles can be applied as additives in PCMs. An overview shows that some nanoparticles are more common than others (Figure 5.6). Also, Table 5.2 shows the most commonly applied nanoparticles in PCMs with more details and related references. Based on the results, CuO, Cu, Al_2O_3 and graphene are among the most commonly applied nanoparticles for nano-enriched PCMs in various applications.

5.2.3 Optimization of PCM Heat Capacity in TES Systems

Next to TC, properties like melting point (T_m), heat of fusion (ΔH_{fus}), specific heat capacity of the solid and liquid phases (Cp) and density (ρ) of the solid and liquid phase are among other key thermophysical properties of PCMs. Heat capacity and specific heat capacity are the most important characteristics of PCMs. Saeed [64] in his master thesis proposed a novel mixed PCM that comprised methyl palmitate (MP) (60%), lauric acid (LA) (40%) and MHE-C and focused on improving thermal properties of the PCM by adding various concentrations of nano-graphene platelets (NGPs). They studied the effect of adding various concentrations of NGPs

TABLE 5.1

Effect of Various Nano-Sized Additives on the TC of Paraffin as a Conventional PCM

Researcher	PCM	Nanoparticle	Concentration (%)	Nanoparticle classification	TC improvement (%)
Nourani et al. [16]	Paraffin	Alumina (Al_2O_3)	10% wt	Metallic	• About 30% for solid state of PCM • About 13% for liquid state of PCM
Sahan [18]	Paraffin	Fe_3O_4	10% wt	Metallic	• About 48%
Jesumatty et al. [19]	Paraffin	CuO	10% wt	Metallic	• About 8%
Yang et al. [17]	Paraffin	Si_3N_4	10% wt	Carbon family	• About 35%
Li [20]	Paraffin	Nano graphite (NG)	10% wt	Carbon family	• About 6.50%
Goli et al. [21]	Paraffin	Graphene	20% wt	Carbon family	• Enhancement of means 14800–17900% depending on temperature and graphene type
Warzoha et al. [22]	Paraffin	Graphite nanofibers	11.4% wt	Carbon family	• About 180%
Shi et al. [23]	Paraffin	Graphite nano-platelets	10% wt	Carbon family	• About 980%
Fan et al. [24]	Paraffin	Graphene nano-platelets	5% wt	Carbon family	• About 164%

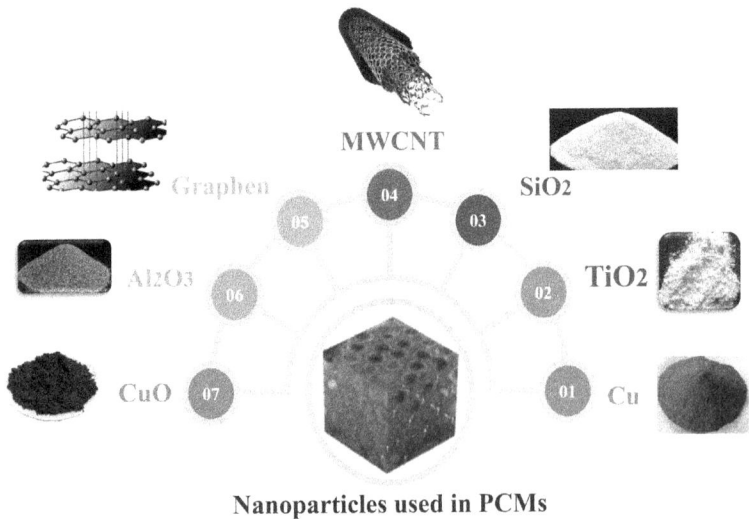

Nanoparticles used in PCMs

FIGURE 5.6 Most commonly applied nanoparticles in PCMs.

TABLE 5.2
Some of the Most Commonly Applied Nanoparticles in PCMs

Application	Applied nanoparticles	References
TES systems	CuO, Al_2O_3, Al_2O_3–Go, Al_2O_3–carbon black, Cu, Al, TiO_2, SiO_2, TiO_2–Cu, Ag–TiO_2, graphene, graphene-MWCNT, Cu	[25–41]
Heat exchangers	CuO, Al_2O_3	[42–49]
Electronic devices	CuO, graphene, MWCNT, graphite, Fe_3O_4	[50–55]
Photovoltaic/thermal systems	Silicon carbide, TiO_2–CuO	[56–61]
	For PVT applied in buildings: CuO, Cu, Al_2O_3	Building: [62–63]

on TC, specific heat capacity in solid and liquid phases and thermal diffusivity parameters in two liquid and solid phases. According to results, by adding 1–10% wt of NGPs to the PCM, the *Cp* in the solid phase was enhanced by 24–52%, and also the enhancement of *Cp* in liquid phase was reported to be between 47–64%. Also, based on the results of Saeed [64], TC enhancement was reported to be about 32–97%; that can be another achievement next to the aforementioned specific heat capacity improvement.

As can be seen, adding a suitable type of nanoparticles will result in a significant enhancement in specific heat capacity. By this significant increase in specific heat capacity, the amount of possible energy storage in the form of sensible heat will increase, which is very important, especially when the PCMs are working out of their phase change range. In fact, higher specific heat capacity will help the PCMs to be more efficient in storing energy in the form of sensible heat, next to efficient latent heat storage.

5.2.4 OPTIMIZATION OF PCM THERMAL DIFFUSIVITY IN TES SYSTEMS

The specific heat capacity and TC are not enough, and thermal diffusivity must be completely analyzed to understand the heat transfer behavior of a PCM in a TES system. Based on experimentally measured parameters like heat capacity, TC and density and by using Equation 5.1, the thermal diffusivity can be calculated. Based on Saeed [64], after adding nanoparticles, the diffusivity of PCM increased by 19–42% for the solid phase and 22–54% for the liquid phase. This means that adding appropriate nanoparticles to PCM can improve the thermal diffusivity of PCM. In fact, the thermal diffusivity enhancement is because of the TC enhancement after adding nanofluids. TC and thermal diffusivity are related based on Equation 5.1, that is named by thermal diffusivity. In fact, an increase in thermal diffusivity is more important than TC enhancement because it is a sign of TC improvement of a PCM next to its satisfactory specific heat capacity.

$$\alpha = k / \rho C_p \qquad (5.1)$$

5.2.5   OPTIMIZATION OF PCM MELTING AND
SOLIDIFICATION BEHAVIOR IN TES SYSTEMS

Two of the important and determinative characteristics of PCMs are the melting and solidification temperatures. To start the melting process, the ambient temperature needs to be above the melting temperature of the PCM. After that, the energy will start to absorb in the form of latent heat at an almost constant temperature. So, selecting a PCM that experiences a suitable melting process is very important and will result in best performance and highest efficiency of TES systems. An inappropriate choice of PCM may cause an incomplete phase change process, and the system will not be able to save the total possible heat storage. Also, discharging the PCM with the aim of preparing the PCM for the new energy storage cycle is important and shows the importance of complete solidification process management. Like other thermophysical properties, the melting and solidification temperatures will change by adding nanoparticles to PCMs. Figures 5.7 and 5.8 show a schematic for the solid–liquid transition of a PCM and the roles of latent and sensible heat in the energy storage process in PCMs, respectively.

According to the available literature, adding nanoparticles to PCMs can increase or decrease both solidification and melting temperatures simultaneously in such a way that either both increase or both temperatures decrease by adding nanoparticles to PCM. Based on Wang et al. [65], Mohamed et al. [66], Harikrishnan et al. [67], Singh et al. [68], Rufuss et al. [69] and Saeed et al. [70], adding various nanoparticles to different PCMs can change the phase change temperatures (melting and solidification) from 0.6°C to 6°C. Table 5.3 gives more details about solidification and melting

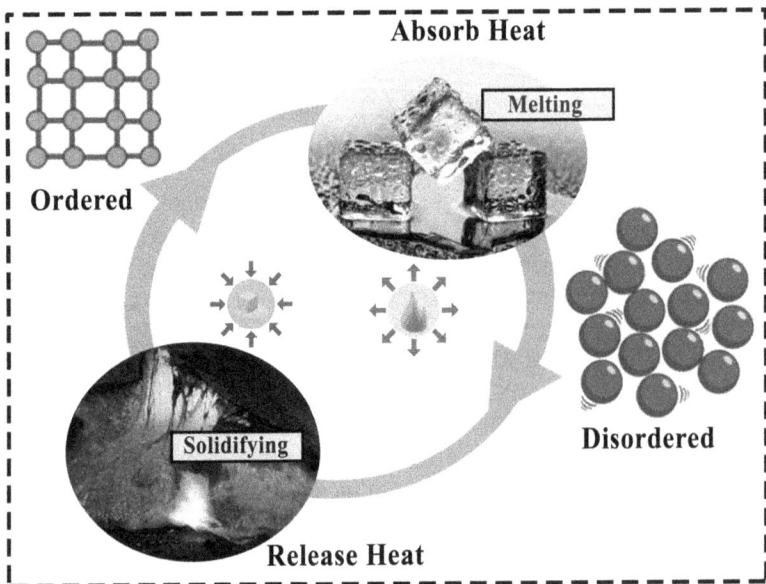

FIGURE 5.7 A schematic for solid–liquid transition in a PCM.

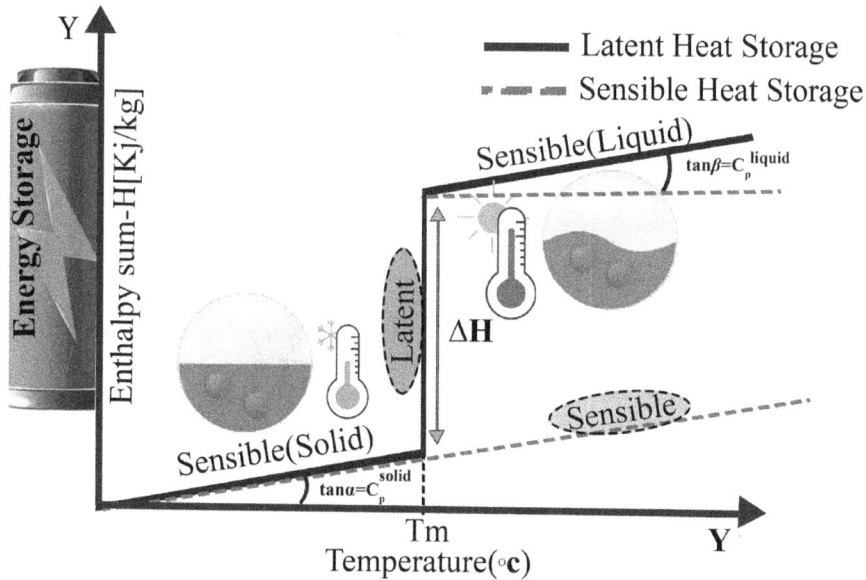

FIGURE 5.8 Roles of latent and sensible heat in energy storage in PCMs.

TABLE 5.3

Phase Change Temperature Variation after Adding Nanoparticles to PCMs

Reference	Nanoparticle+PCM	Melting temperature change	Solidification temperature change
[69]	GO+paraffin	About −6°C	About −3°C
[69]	TiO$_2$+paraffin	About −5°C	About −4°C
[69]	CuO+paraffin	About −4.5°C	About −4°C
[71]	Cu+paraffin	About −2.5°C	About +1°C
[67]	SiO$_2$+Myristic acid	About +0.5°C	About +0.5°C

temperature changes. Also, melting and solidification time is another thermophysical behavior of PCMs that may change after adding nanoparticles. The summarized results of some available studies are presented in Table 5.4.

5.2.6 Optimization of PCM Heat of Fusion and Density in TES Systems

In this section, the effects of adding nanoparticles to PCMs on the heat of fusion (ΔH_{fus}) and density (ρ) of solid and liquid phases are discussed.

Based on Saeed [64], after adding GNPs to the host PCMs, the density of the enriched PCM increased by about 0.6–16% in the solid phase and by about

TABLE 5.4
Melting and Solidification Times Affected by Added Nanoparticles

Studied parameter	Related references	Nanoparticle(s)/PCM
Melting time reduction	Ebrahimi and Dadvand [33], Singh et al. [38], Li et al. [72], Khan and Ahmad Khan [49]	—Al_2O_3/Paraffin wax —Graphene nanoparticles/sugar alcohol —CuO/coconut oil-based PCM —Expanded graphite/stearic acid —Aluminum nitride (AlN)/paraffin, and graphene nano-platelets (GnPs)/paraffin, Al_2O_3/Paraffin
Melting time increase	–	–
Solidification time reduction	Hosseinzadeh et al. [32], Sheikholeslami [28], Khan and Ahmad Khan [49], Kalaiselvam et al. [73]	—Al_2O_3–Go/water —Copper oxide nanoparticles/water —Aluminum oxide (Al_2O_3)/paraffin, aluminum nitride (AlN)/paraffin, graphene nano-platelets (GnPs)/paraffin —Aluminum and alumina nanoparticles/PCM (60% n-tetradecane plus 40% n-hexadecane)
Solidification time increase	–	–

0.7–17% for the liquid phase. Also, a comparison of density after melting/solidification is another important issue that is not to be ignored and was an issue that Saeed [64] analyzed. Based on Saeed [64], a 5% density decrease was reported in nano-enriched PCM after melting. A low density change is an important and positive characteristic for a PCM, because a high density change during phase change is associated with a high volume change. An uncontrolled volume increment during phase change from solid to liquid is a penalty that may cause any unforeseen problems, especially in the cases of PCMs that are encapsulated in containers or sheets for any applications.

Heat of fusion, also known as enthalpy of fusion or melting enthalpy, is another thermophysical property of PCMs. Heat of fusion means the enthalpy change of a PCM during phase change. In fact, the enthalpy of fusion is a type of latent heat that transfers in the melting/solidification process as latent heat of fusion. Based on Saeed [64], adding nanoparticles to a PCM may cause a ΔH_{fus} reduction in a nano-enriched PCM compared to the host PCM. Therefore, since a higher ΔH_{fus} is a positive point for a PCM material, achieving a minimum decrease in ΔH_{fus} after adding nanoparticles to PCM is very important and depends on the choice of a suitable nanoparticle.

5.2.7 Optimization of TES System Capacity by Nano-Enriched PCMs

As mentioned before, one of the main goals of adding nanoparticles to PCMs is improving the TC of conventional PCMs. By adding nanoparticles and other additives like gelling agents to PCMs, according to the study by Saeed [64], the latent heat or the heat of fusion may experience a reduction because of a reduction in volume or mass fraction of the PCM after adding nanoparticles or a gelling agent. This is not a desired reduction, but comparing the TES capacity of nano-enriched PCM with pure PCM is the key criterion to determine the success or failure of adding nanoparticles to pure PCM.

By adding nanoparticles, the first result that is expected is a TES reduction in the PCM because of PCM share reduction and a resulting reduction of latent heat of fusion. However, the result of Saeed [64] showed another result where a gelling agent and graphene nanoparticles were added to a PCM. The results showed a TES capacity reduction after adding the gelling agent to the pure PCM, but after adding the graphene nanoparticles, the TES capacity reduction was stopped and even enhanced, which was a good sign. They conducted their experiments in a temperature range of ±15°C and ±25°C around the melting point. Based on the wider temperature range around the melting point that was ±25°C, the TES capacity improvement was more satisfactory compared to the smaller temperature range (±15°C). In fact, a wider operating temperature range will enhance the share of sensible heat storage in the total TES after adding nanoparticles to the PCM. In other words the increase in the contribution of sensible heat compensates for the decrease in melting enthalpy or latent heat of fusion caused by a decrease in PCM mass or volume fraction after nanoparticle addition. Figure 5.9 shows the positive and negative effects of adding nanoparticles to PCMs.

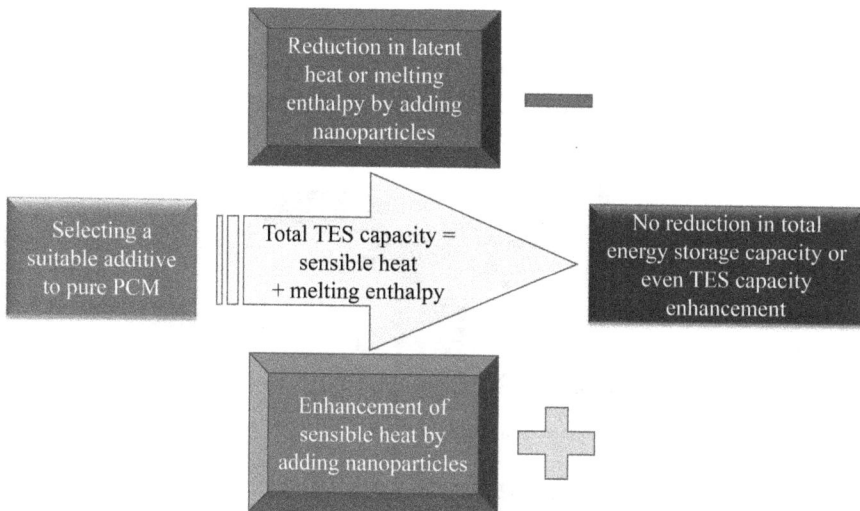

FIGURE 5.9   The positive and negative effects of adding nanoparticles to PCMs.

5.3 EXPERIMENTAL METHODS FOR DETERMINING THE THERMAL CHARACTERISTICS OF PCMS

The common analyses for PCMs can be divided into two categories that are: 1. structural determinations and constituent materials analysis and 2. heat storage properties analysis. The main common analyses in first category are XRD (X-ray diffraction), SEM (scanning electron microscopy), FESEM (field emission scanning electron microscopy), TEM (transmission electron microscopy) and FTIR (Fourier-transform infrared spectroscopy). This group of analyses is used to determine the exact structure of materials. On the other hand, the second category of analyses are the heat storage property analyses. The common analysis methods in this category are thermo-gravimetric analysis (TGA), DSC (differential scanning calorimetry) and T-history analysis. All three mentioned methods are based on heating the targeted PCM material and then comparing with a reference material. TGA is a method to check the material behavior against heat change. In this method, a certain amount of material needs to be heated, and weight changes in various temperatures will be recorded to determine the percent of material weight change. In this test, it is possible to detect thermal stability, moisture level, percentage of organic and mineral compounds, amount of weight enhancement caused by metal oxidation, etc.

Also, DSC uses an apparatus that can be applied for determining the heat ability and TC of a material. In this device, there are separate electric heaters for heating the unknown and control samples, and two thermocouples determine the temperature of the samples. After receiving the temperature signals related to the samples, the controller circuit determines and applies the necessary amount of energy to equalize the temperature. The DSC method is defined as a differential method, and therefore the behavior of the sample is compared with a reference material. Figure 5.10 shows the schematic chamber for DSC.

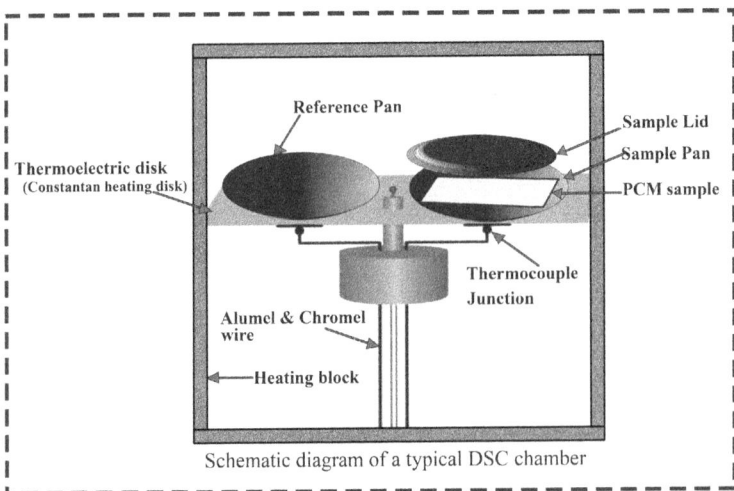

Schematic diagram of a typical DSC chamber

FIGURE 5.10 Differential scanning calorimetry chamber schematic.

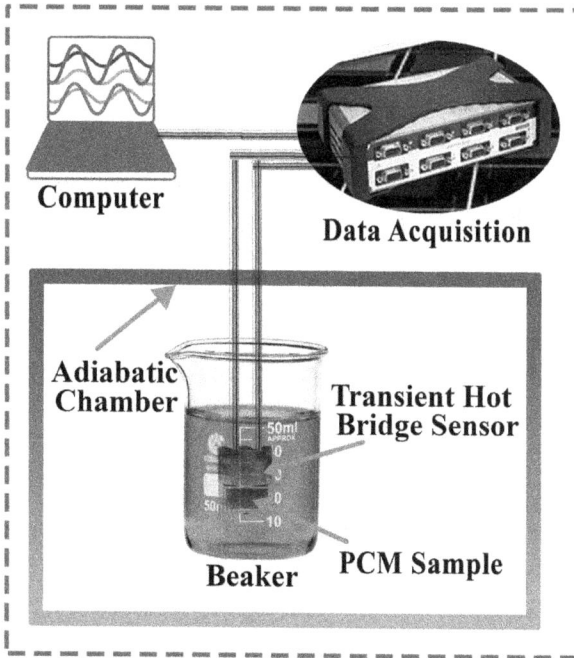

FIGURE 5.11 Schematic for a TC meter based on the THB method.

As third method, T-history analysis, provides the possibility of obtaining melting point temperature, supercooling temperature, TC and latent heat of phase change for several PCM samples simultaneously.

Next to the DSC method, there is another method for TC calorimetry that is the transient hot-bridge (THB) technique. In the THB method, the TC measuring instrument has a sensor that will be inserted in the sample and acts like a heater and a temperature probe. The probe needs about 1 minute to measure the TC of the PCM sample, although the time depends on the PCM type. An adiabatic chamber can be designed around the PCM sample and sensor to make the results independent from the ambient temperature. Other parts of the experimental setup include a data acquisition unit that needs to be connected to a computer to record data sets. Also, to ensure the repeatability of the results, experiments can be repeated five times or more based on the desired accuracy. Ten minutes or more is needed between two tests on the same sample with the aim of temperature reduction of the sample. Figure 5.11 shows a schematic of the TC meter instrument based on the THB method.

5.4 UNCERTAINTIES IN THE THERMAL CHARACTERISTIC DETERMINATION OF PCMS

Having an accurate understanding of a PCM's thermophysical properties is very important because it has a direct effect on the accuracy level of the design of

PCM-enriched TES systems [74]. Also, it is important for the application of PCMs in other industries. Parameters like heat capacity, TC, phase change temperature and other thermophysical behaviors of PCMs or nano-enriched PCMs need to be determined with the highest possible accuracy because they affect the TES system size and capacity. Dolado et al. [75] showed that improving the accuracy of determining melting or solidification temperatures of PCMs by about 0.75°C, may reduce the uncertainty in heat exchange rate estimation by about 25%. So, as a result, by reducing the uncertainties, we can hope for a more accurate design that matches the determined goals. The importance of melting and solidification temperature uncertainty will be clearer for applications like using PCMs in buildings with a low range of temperature change throughout a day.

As another error source is determining PCM properties without applying a specific standard. When measuring the PCM properties, codification of a clear standard for DSC measurements of PCMs is essential because the measurement process for PCMs is not the same as that for other materials based on literature [76, 77]. There are standards for calorimetry of other materials' properties, but using the same standards for PCMs may cause considerable errors in the process of calorimetry.

5.5   CONCLUSION

In this chapter, the heat storage mechanisms of PCMs were investigated, and the solutions to optimize the performance of PCMs were discussed, focusing on PCMs used in energy storage systems. The main focus of this chapter was on improving the performance of PCMs using nano-additives, and the effect of adding nano-sized particles on the characteristics and thermophysical and thermal properties of PCMs was studied. Among the most important results of this chapter, the following can be mentioned:

- The TC of an applied PCM in TES system that is enriched by nanoparticles can be improved by 6–18000%; the increment percent depends on base PCM types, selected nanoparticle type and weight fraction of added nanoparticle. A high TC of PCMs will result in high rate of release and storage of the heat in a nano-enriched PCM in a TES system and will enhance the efficiency of the TES system. In fact, a higher TC of the PCM will ease the charging and discharging of the system.
- The most applied nanoparticles in PCMs were listed based on available literature sources. According to gathered data, the most applied nanoparticles are metal, metal oxides and carbon nanotubes.
- Adding nanoparticles to a PCM can improve the specific heat capacity up to 50–60% or even more in solid and liquid phase of PCMs. Reaching a considerable increase in specific heat capacity and heat capacity is equivalent to enhancing the power of a PCM's thermal heat storage in the form of sensible heat, which is very important, especially for the conditions of PCMs working out of their phase change temperature range.
- Thermal diffusivity is another thermal characteristic that can be improved by about 50% or more depending on the applied nanoparticle type. Based

on the thermal diffusivity formula, it is the ratio of TC and specific heat capacity, and a higher thermal diffusivity is a sign of TC improvement of a PCM next to its satisfactory specific heat capacity.

- Melting/solidification temperature and duration are additional thermal characteristics that can be changed by adding nanoparticles to PCMs in a TES system. Based on the literature, by adding metal and oxide nanoparticles to PCM, a 0.5°C to 6°C melting/solidification temperature reduction can be expected, although this reduction is dependent on PCM and added nanoparticle type. Also, based on the literature, adding nanoparticles can reduce the melting/solidification duration of a PCM.

- The density change in a PCM material after adding nanoparticles and also the density change during melting/solidification was another PCM characteristic that was studied in the present chapter. The literature showed that adding nanoparticles increased the density of a PCM by about zero to higher percents as a function of nanoparticle type. Also, a nano-enriched PCM with minimum density changes during phase change would be an ideal choice because a low amount of density change is associated with a low volume change during phase change. An uncontrolled volume increment during phase change from solid to liquid is a penalty that may cause any unforeseen limitations, especially in the cases that PCMs are encapsulated in containers or sheets for any applications.

- Heat of fusion (ΔH_{fus}) may be faced with negative effects due to the addition of nanoparticles because of a decrease in PCM share after adding nanoparticles. So, selecting a suitable nanoparticle type with minimum storage reduction by the heat of fusion mechanism and an improved sensible heat storage mechanism by adding nanoparticles is a desired result.

REFERENCES

[1] Dincer I, Rosen MA. Energetic, environmental and economic aspects of thermal energy storage systems for cooling capacity. *Appl Therm Eng* 2001;21(11):1105–1117.

[2] Oró E, et al. Review on phase change materials (PCMs) for cold thermal energy storage applications. *Appl Energy* 2012;99:513–533.

[3] Heine D, Abhat A. Investigation of physical and chemical properties of phase change materials for space heating/cooling applications. In *SUN: Mankind's Future Source of Energy, Proceedings of the International Solar Energy Society Congress, New Delhi, India, January* 1978;1:500–506.

[4] Khudhair AM, Farid MM. A review on energy conservation in building applications with thermal storage by latent heat using phase change materials. *Energy Convers Manag* 2004;45(2):263–275.

[5] Tyagi VV, Buddhi D. PCM thermal storage in buildings: A state of art. *Renew Sust Energ Rev* 2007;11(6):1146–1166.

[6] Ma G, et al. Binary eutectic mixtures of stearic acid-n-butyramide/noctanamide as phase change materials for low temperature solar heatstorage. *Appl Therm Eng* 2017;111:1052–1059.

[7] Meng Z, Zhang P. Experimental and numerical investigation of a tube-in-tank latent thermal energy storage unit using composite PCM. *Appl Energy* 2017;190:524–539.

[8] Watanabe T, Kikuchi H, Kanzawa A. Enhancement of charging and discharging rates in a latent heat storage system by use of PCM with different melting temperatures. *Heat Recovery Syst CHP* 1993;13(1):57–66.

[9] Mehling H, Hiebler S, Ziegler F. Latent heat storage using a PCM-graphite composite material. In *Proceedings of TERRASTOCK 2000: 8th International Conference on Thermal Energy Storage, University of Stuttgart, Germany*; 2000.

[10] Elgafy A, Lafdi K. Effect of carbon nanofiber additives on thermal behavior of phase change materials. *Carbon* 2005;43(15):3067–3074.

[11] Alshaer W, et al. Numerical investigations of using carbon foam/PCM/Nanocarbon tubes composites in thermal management of electronic equipment. *Energy Convers Manag* 2015;89:873–884.

[12] Mesalhy O, et al. Numerical study for enhancing the thermal conductivity of phase change material (PCM) storage using high thermal conductivity porous matrix. *Energy Convers Manag* 2005;46(6):847–867.

[13] Py X, Olives R, Mauran S. Paraffin/porous-graphite-matrix composite as a high and constant power thermal storage material. *Int J Heat Mass Transf* 2001;44(14):2727–2737.

[14] Rao Z, Zhang G. Thermal properties of paraffin wax-based composites containing graphite. *Energy Sources A: Recovery Util Environ Eff* 2011;33(7):587–593.

[15] Khan Z, Khan ZA, Sewell P. Heat transfer evaluation of metal oxides based nano-PCMs for latent heat storage system application. *Int J Heat Mass Transf* 2019;144. http://doi.org/10.1016/j.ijheatmasstransfer.2019.118619.

[16] Nourani M, Hamdami N, Keramat J, Moheb A, Shahedi M. Thermal behavior of paraffin-nano-Al2O3 stabilized by sodium stearoyl lactylate as a stable phase change material with high thermal conductivity. *Renew Energy*. 2016;88:474–482. http://doi.org/10.1016/j.renene.2015.11.043.

[17] Yang Y, Luo J, Song G, Liu Y, Tang G. The experimental exploration of nano-Si3N4/paraffin on thermal behavior of phase change materials. *Thermochim Acta*. 2014;597:101–106. http://doi.org/10.1016/j.tca.2014.10.014.

[18] Sahan N, Fois M, Paksoy H. Improving thermal conductivity phase change materials—a study of paraffin nanomagnetite composites. *Sol Energy Mater Sol Cells* 2015;137:61–67.

[19] Jesumathy S, Udayakumar M, Suresh S. Experimental study of enhanced heat transfer by addition of CuO nanoparticle. *Heat Mass Transf* 2012;48:965e978.

[20] Li M. A nano-graphite/paraffin phase change material with high thermal conductivity. *Appl Energy* 2013;106:25–30.

[21] Goli P, et al. Graphene-enhanced hybrid phase change materials for thermal management of Li-ion batteries. *J Power Sources* 2014;248:37–43.

[22] Warzoha RJ, Weigand RM, Fleischer AS. Temperature-dependent thermal properties of a paraffin phase change material embedded with herringbone style graphite nanofibers. *Appl Energy* 2015;137:716e725.

[23] Shi JN, Ger MD, Liu YM, Fan YC, Wen NT, Lin CK, et al. Improving the thermal conductivity and shape-stabilization of phase change materials using nanographite additives. *Carbon* 2013;51:365e372.

[24] Fan LW, Fang X, Wang X, Zeng Y, Xiao YQ, Yu ZT, et al. Effects of various carbon nanofillers on the thermal conductivity and energy storage properties of paraffin-based nanocomposite phase change materials. *Appl Energy* 2013;110:163e172.

[25] Alomair M, Alomair Y, Tasnim S, Mahmud S, Abdullah H. Analyses of bio-based nano-PCM filled concentric cylindrical energy storage system in vertical orientation. *J Storage Mater* 2018;20:380–394.

[26] Ebadi S, Tasnim SH, Aliabadi AA, Mahmud S. Geometry and nanoparticle loading effects on the bio-based nano-PCM filled cylindrical thermal energy storage system. *Appl Therm Eng* 2018;141:724–740.

[27] Ebadi S, Tasnim SH, Aliabadi AA, Mahmud S. Melting of nano-PCM inside a cylindrical thermal energy storage system: Numerical study with experimental verification. *Energy Convers Manag* 2018;166:241–259. https://doi.org/10.1016/j.enconman.2018.04.016.

[28] Sheikholeslami M. Numerical simulation for solidification in a LHTESS by means of Nano-enhanced PCM. *J Taiwan Inst Chem Eng* 2018;86:25–41.

[29] Wu S, Wang H, Xiao S, Zhu D. Numerical simulation on thermal energy storage behavior of Cu/paraffin nanofluids PCMs. *Procedia Eng* 2012;31:240–244.

[30] Elbahjaoui R, El Qarnia H, El Ganaoui M. Solidification heat transfer characteristics of nanoparticle-enhanced phase change material inside rectangular slabs. *Energy Procedia* 2017;139:590–595.

[31] Colla L, Fedele L, Mancin S, Danza L, Manca O. Nano-PCMs for enhanced energy storage and passive cooling applications. *Appl Therm Eng* 2017;110:584–589.

[32] Hosseinzadeh K, Alizadeh M, Tavakoli M, Ganji D. Investigation of phase change material solidification process in a LHTESS in the presence of fins with variable thickness and hybrid nanoparticles. *Appl Therm Eng* 2019;152:706–717.

[33] Ebrahimi A, Dadvand A. Simulation of melting of a nano-enhanced phase change material (NePCM) in a square cavity with two heat source—sink pairs. *Alexandria Eng J* 2015;54(4):1003–1017.

[34] Venkitaraj K, Suresh S, Praveen B, Venugopal A, Nair SC. Pentaerythritol with alumina nano additives for thermal energy storage applications. *J Storage Mater* 2017; 13:359–377.

[35] Soni V, Kumar A, Jain V. Performance evaluation of nano-enhanced phase change materials during discharge stage in waste heat recovery. *Renew Energy* 2018;127:587–601.

[36] Alizadeh M, Hosseinzadeh K, Ganji D. Investigating the effects of hybrid nanoparticles on solid-liquid phase change process in a Y-shaped fin-assisted LHTESS by means of FEM. *J Mol Liq* 2019;287:110931.

[37] Parameshwaran R, Deepak K, Saravanan R, Kalaiselvam S. Preparation, thermal and rheological properties of hybrid nanocomposite phase change material for thermal energy storage. *Appl Energy* 2014;115:320–330.

[38] Singh RP, Kaushik S, Rakshit D. Melting phenomenon in a finned thermal storage system with graphene nano-plates for medium temperature applications. *Energy Convers Manag* 2018;163:86–99.

[39] Kant K, Shukla A, Sharma A, Biwole PH. Heat transfer study of phase change materials with graphene nano particle for thermal energy storage. *Sol Energy* 2017;146:453–463.

[40] Ramakrishnan S, Wang X, Sanjayan J, Wilson J. Heat transfer performance enhancement of paraffin/expanded perlite phase change composites with graphene nano-platelets. *Energy Procedia* 2017;105:4866–4871.

[41] Qu Y, Wang S, Zhou D, Tian Y. Experimental study on thermal conductivity of paraffin-based shape-stabilized phase change material with hybrid carbon nano additives. *Renew Energy* 2020;146:2637–2645.

[42] Gorzin M, Hosseini MJ, Rahimi M, Bahrampoury R. Nano-enhancement of phase change material in a shell and multi-PCM-tube heat exchanger. *J Storage Mater* 2019;22:88–97.

[43] Sheikholeslami M, Haq R-U, Shafee A, Li Z, Elaraki YG, Tlili I. Heat transfer simulation of heat storage unit with nanoparticles and fins through a heat exchanger. *Int J Heat Mass Transf* 2019;135:470–478.

[44] Elbahjaoui R, El Qarnia H. Thermal analysis of nanoparticle-enhanced phase change material solidification in a rectangular latent heat storage unit including natural convection. *Energy Build* 2017;153:1–17.

[45] Pahamli Y, Hosseini M, Ranjbar A, Bahrampoury R. Effect of nanoparticle dispersion and inclination angle on melting of PCM in a shell and tube heat exchanger. *J Taiwan Inst Chem Eng* 2017;81:316–334.

[46] Mahdi JM, Lohrasbi S, Ganji DD, Nsofor EC. Accelerated melting of PCM in energy storage systems via novel configuration of fins in the triplex-tube heat exchanger. *Int J Heat Mass Transf* 2018;124:663–676.

[47] Mahdi JM, Lohrasbi S, Ganji DD, Nsofor EC. Simultaneous energy storage and recovery in the triplex-tube heat exchanger with PCM, copper fins and Al2O3 nanoparticles. *Energy Convers Manag* 2019;180:949–961.

[48] Abdulateef AM, Abdulateef J, Al-Abidi AA, Sopian K, Mat S, Mahdi MS. A combination of fins-nanoparticle for enhancing the discharging of phase-change material used for liquid desiccant air conditioning unite. *J Storage Mater* 2019;24:100784.

[49] Khan Z, Khan ZA. Experimental and numerical investigations of nano-additives enhanced paraffin in a shell-and-tube heat exchanger: A comparative study. *Appl Therm Eng* 2018;143:777–790.

[50] Praveen B, Suresh S. Experimental study on heat transfer performance of neopentyl glycol/CuO composite solid-solid PCM in TES based heat sink. *Eng Sci Technol Int J* 2018;21(5):1086–1094.

[51] Krishna J, Kishore P, Solomon AB. Heat pipe with nano enhanced-PCM for electronic cooling application. *Exp Therm Fluid Sci* 2017;81:84–92.

[52] Praveen B, Suresh S, Pethurajan V. Heat transfer performance of graphene nanoplatelets laden micro-encapsulated PCM with polymer shell for thermal energy storage based heat sink. *Appl Therm Eng* 2019;156:237–249.

[53] Farzanehnia A, Khatibi M, Sardarabadi M, Passandideh-Fard M. Experimental investigation of multiwall carbon nanotube/paraffin based heat sink for electronic device thermal management. *Energy Convers Manag* 2019;179:314–325.

[54] Huang Z, et al. Experimental and numerical study on thermal performance of Wood's alloy/expanded graphite composite phase change material for temperature control of electronic devices. *Int J Therm Sci* 2019;135:375–385.

[55] Alimohammadi M, Aghli Y, Alavi ES, Sardarabadi M, Passandideh-Fard M. Experimental investigation of the effects of using nano/phase change materials (NPCM) as coolant of electronic chipsets, under free and forced convection. *Appl Therm Eng* 2017;111:271–279.

[56] Rufuss DDW, Kumar VR, Suganthi L, Iniyan S, Davies P. Techno-economic analysis of solar stills using integrated fuzzy analytical hierarchy process and data envelopment analysis. *Sol Energy* 2018;159:820–833.

[57] Al-Waeli AH, Kazem HA, Yousif JH, Chaichan MT, Sopian K. Mathematical and neural network modeling for predicting and analyzing of nanofluid-nano PCM photovoltaic thermal systems performance. *Renew Energy* 2020;145:963–980.

[58] Al-Waeli AH, Chaichan MT, Sopian K, Kazem HA, Mahood HB, Khadom AA. Modeling and experimental validation of a PVT system using nanofluid coolant and nano-PCM. *Sol Energy* 2019;177:178–191.

[59] Al-Waeli AH, Sopian K, Yousif JH, Kazem HA, Boland J, Chaichan MT. Artificial neural network modeling and analysis of photovoltaic/thermal system based on the experimental study. *Energy Convers Manag* 2019;186:368–379.

[60] Al-Waeli AH, Kazem HA, Chaichan MT, Sopian K. Experimental investigation of using nano-PCM/nanofluid on a photovoltaic thermal system (PVT): Technical and economic study. *Therm Sci Eng Prog* 2019;11:213–230.

[61] Al-Waeli AH, et al. Evaluation of the nanofluid and nano-PCM based photovoltaic thermal (PVT) system: An experimental study. *Energy Convers Manag* 2017;151:693–708.

[62] Ma Z, Lin W, Sohel MI. Nano-enhanced phase change materials for improved building performance. *Renew Sustain Energy Rev* 2016;58:1256–1268.

[63] Nada S, El-Nagar D, Hussein H. Improving the thermal regulation and efficiency enhancement of PCM-Integrated PV modules using nano particles. *Energy Convers Manag* 2018;166:735–743.

[64] Saeed RMR. Thermal characterization of phase change materials for thermal energy storage. Missouri University of Science and Technology. Masters Theses, 7521; 2016. https://scholarsmine.mst.edu/masters_theses/7521.

[65] Wang J, Xie H, Xin Z. Thermal properties of paraffin based composites containing multi-walled carbon nanotubes. *Thermochim Acta* 2009;488(1):39–42. https://doi.org/10.1016/j.tca.2009.01.022.

[66] Mohamed NH, Soliman FS, El Maghraby H, Moustfa YM. Thermal conductivity enhancement of treated petroleum waxes, as phase change material, by α nano alumina: Energy storage. *Renew Sustain Energy Rev* 2017;70:1052–1058. https://doi.org/10.1016/j.rser.2016.12.009.

[67] Harikrishnan S, Imran Hussain S, Devaraju A, Sivasamy P, Kalaiselvam S. Improved performance of a newly prepared nano-enhanced phase change material for solar energy storage. *J Mech Sci Technol* 2017;31(10):4903–4910. https://doi.org/10.1007/s12206-017-0938-y.

[68] Singh D, Suresh S, Singh H, Rose B, Tassou S, Anantharaman N. Myo-inositol based nano-PCM for solar thermal energy storage. *Appl Therm Eng* 2017;110:564–572.

[69] Rufuss DDW, Suganthi L, Iniyan S, Davies P. Effects of nanoparticle-enhanced phase change material (NPCM) on solar still productivity. *J Cleaner Prod* 2018;192:9–29.

[70] Saeed RM, Schlegel JP, Castano C, Sawafta R. Preparation and enhanced thermal performance of novel (solid to gel) form-stable eutectic PCM modified by nano-graphene platelets. *J Storage Mater* 2018;15:91–102. https://doi.org/10.1016/j.est.2017.11.003.

[71] Lin SC, Al-Kayiem HH. Evaluation of copper nanoparticles—Paraffin wax compositions for solar thermal energy storage. *Sol Energy* 2016;132:267–278. https://doi.org/10.1016/j.solener.2016.03.004.

[72] Li C, et al. Stearic acid/expanded graphite as a composite phase change thermal energy storage material for tankless solar water heater. *Sustain Cities Soc* 2019;44:458–464.

[73] Kalaiselvam S, Parameshwaran R, Harikrishnan S. Analytical and experimental investigations of nanoparticles embedded phase change materials for cooling application in modern buildings. *Renew Energy* 2012;39(1):375–387.

[74] Saeed RM, et al., Uncertainty of thermal characterization of phase change material by differential scanning calorimetry analysis. *Int J Eng Res Technol* 2016;5(1):405–412.

[75] Dolado P., et al., Experimental validation of a theoretical model: Uncertainty propagation analysis to a PCM-air thermal energy storage unit. *Energy Build* 2012;45:124–131.

[76] Mehling H, Cabeza LF. *Heat and Cold Storage with PCM*. Springer; 2008.

[77] Sharma A, et al. Review on thermal energy storage with phase change materials and applications. *Renew Sust Energ Rev* 2007;13(2):318–345.

6 Thermal Energy Storage in Phase Change Materials—A Bibliometric Approach

S. Senthilraja, Umit Gunes, and Mohamed M. Awad

6.1 INTRODUCTION

Energy is a crucial aspect that determines a nation's prosperity, standard of living, and level of technical advancement. The continuous increase in vehicles and human population is directly linked with energy demand and affects countries' gross domestic product. It has been predicted that by 2030, global energy consumption will have increased by 71% [1,2]. Most nations have relied on petroleum and natural gas to satisfy their energy needs in recent decades, despite the significant health problems caused by fossil fuel byproducts. With the increasing fossil fuel depletion rate and environmental degradation, it is necessary to create awareness of global warming among humans. This has led to a steady rise in the use of renewable energy sources over the last several decades. But solar energy is only available some of the time. To overcome this issue, the use of different energy storage systems has recently been increasing.

The ability to capture and store thermal energy for later use is a key feature of thermal energy storage devices. Thermal energy storage technologies are more cost-effective and efficient than their counterparts. Generally, thermal energy storage systems can be classified as 1. sensible heat energy storage (SHES), 2. latent heat energy storage (LHES), and 3. thermochemical energy storage (TCES). Heat is stored in a sensible heat storage system when the temperature of a solid or liquid medium is raised without a corresponding change in phase. The specific heat capacity and density of the material play crucial roles in this process. The amount of energy stored in such a system may be determined using the formula

$$Q = m C_P \Delta T \qquad (6.1)$$

where 'Q' is the amount of heat stored, 'm' is the mass of storage medium (kg), 'C_P' is the specific heat of the material (kJ/kg K), and 'ΔT' is the change in temperature in this process (K). Generally, many materials such as water, thermal oil, molten salts,

                     DOI: 10.1201/9781003331957-6

liquid metals, and earth metals are used as the media in this type of storage system. The energy is stored in an LHES system as the potential energy of the particle. This procedure involves three phase transitions: solid to liquid, liquid to solid, and liquid to gas. Generally, solid-to-liquid type phase change energy storage systems are used in numerous applications. The latent heat of the solid-to-solid and liquid-to-gas phase change is low and high, respectively. But the change in the volume of the material is also high in liquid-to-gas phase change systems. These benefits make energy storage systems based on the solid-to-liquid phase shift a popular choice. The amount of energy storage in such system can be calculated as

$$Q = m\,L \tag{6.2}$$

where 'm' is the mass of storage medium (kg) and 'L' is the specific latent heat (kJ/kg). Numerous energy storage systems employ a wide variety of PCMs as store media, including organic (paraffins, fatty acids, alcohols, glycols, etc.) and inorganic (salts, metals, alloys) materials. Using a high-energy chemical process, energy is stored in a TCES device. This system outperforms LHES and SHES in terms of energy storage capacity and efficiency, both of which are improved by using this method. Though it contains many advantages, it is challenging to maintain the stability of the chemical compound during the charging and storing period. These difficulties in SHES and TCES have led many researchers and businesses to focus on PCM-based thermal energy storage solutions. Researchers have utilized a wide range of phase change materials for energy storage throughout the years. Therefore, it is crucial to conduct a bibliometric study of thermal energy storage systems using PCMs so that upcoming researchers may become familiar with the breadth and depth of previous studies. In-depth findings from a bibliometric investigation of a thermal energy storage systems based on PCMs are presented in this paper.

This chapter is organized as follows: The first part examines the topic of thermal energy storage systems and their types. The details about PCMs and their classification are discussed in Section 6.2. The bibliometric analysis approach is detailed in detail in Section 6.3. Section 6.4 presents the findings of the bibliometric analysis. Finally, this chapter is ended with Section 6.5. In this section, the significant findings are presented.

6.2 PHASE CHANGE MATERIAL (PCM)

A phase change material is a substance that can change phases in order to store heat. It is also called a latent storage material. In contrast, storing heat energy increases the molecules' vibration and kinetic energy. This leads to the loosening of the bonds between the molecules, and solid-to-liquid transformation takes place. While releasing the heat, it comes back to its original state by strengthening the molecular bond. Practically, many PCMs are used in different applications. According to their material composition, PCMs are classified as organic PCMs, inorganic PCMs, and eutectic compound PCMs [3]. Generally, organic PCMs (paraffin and nonparaffin) are noncorrosive and used in low-temperature and medium-temperature applications. Organic PCMs (salt hydrates, nitrates, and metallics) are easily available at lower prices. Due to their high latent heat values, these PCMs are often employed in high-temperature applications. Despite many advantages, these PCMs have limitations, such as improper solidification and corrosiveness. Finally, eutectic PCMs are a type of PCM consisting of two or more organic and inorganic PCMs. Combining two different PCMs (organic–organic, inorganic–inorganic, and organic–inorganic) acts as a single crystal component during crystallization and simultaneously melts. Among the numerous eutectic PCMs used for thermal storage, solar salts are among the most popular. Solar salt consists of sodium nitrate ($NaNO_3$) and potassium nitrate (KNO_3) at 60% and 40%, respectively. This PCM works well in both high- and low-temperature settings.

From this context, PCMs are one of the essential elements for storing and controlling thermal energy in modern thermal storage systems such as solar power plants [4], solar hot water systems [5], and building thermal regulation systems [6]. The previously published research results [7–10] indicate the need for PCMs to enhance the performance of thermal energy storage systems. Therefore, a bibliometric examination of thermal energy storage systems utilising PCM has to be studied. An in-depth study of PCM-based thermal energy storage devices is conducted in this chapter, with data collected on the number of publications, authors, articles cited, and nations involved. The results are presented in section 6.4.

6.3   BIBLIOMETRIC ANALYSIS AND METHODOLOGY

Bibliometric analysis is a computer-assisted review method used to investigate scientific output quantitatively in terms of research authors, countries, article citations, and funding agencies. Though many tools are available to analyze academic outcomes, bibliometric analysis is gaining more attention among researchers to explore the enormous volume of data. This analysis consists of four critical stages. The stages of bibliometric analysis are given in Figure 6.1. In the first stage, the required data should be acquired from the Scopus and Web of Science (WoS)

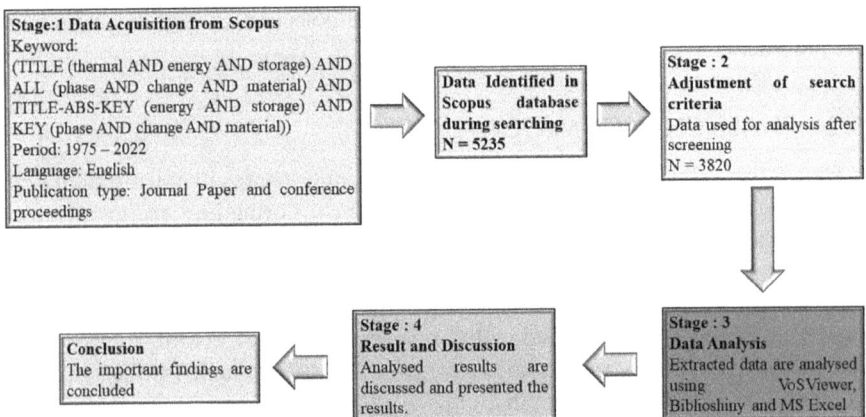

Stage:1 Data Acquisition from Scopus
Keyword:
(TITLE (thermal AND energy AND storage) AND ALL (phase AND change AND material) AND TITLE-ABS-KEY (energy AND storage) AND KEY (phase AND change AND material))
Period: 1975 – 2022
Language: English
Publication type: Journal Paper and conference proceedings

Data Identified in Scopus database during searching
N = 5235

Stage : 2 Adjustment of search criteria
Data used for analysis after screening
N = 3820

Conclusion
The important findings are concluded

Stage : 4 Result and Discussion
Analysed results are discussed and presented the results.

Stage : 3 Data Analysis
Extracted data are analysed using VoS Viewer, Bibhoshiny and MS Excel

FIGURE 6.1   Stages of bibliometric analysis.

databases. To collect the most related documents, a suitable keyword should be used, and the collected data should be further filtered to remove the unwanted articles. This study collected related articles from 1974 to 2022 using the following search query string:

(TITLE (thermal AND energy AND storage) AND ALL (phase AND change AND material) AND TITLE-ABS-KEY (energy AND storage) AND KEY (phase AND change AND material))

Using this query string, initially, 8255 articles were acquired from the Scopus database. Then, these were further filtered with article type and keywords, which resulted in 8212 documents. As per the search results in Scopus, the first article related to phase change material was published in 1975. The acquired data were further screened in the second stage using the PRISMA method. From the developed database, it was found that a few articles have incomplete data such as author details, volume number, page number, etc. Articles with insufficient data were removed at this stage. Then the remaining articles were further screened to remove the duplicate content in the database. After completing the data processing, finally, around 3820 articles were considered for further analysis.

During the third stage, the extracted data are analyzed in terms of publication, research, and development. To obtain the holistic and timely development details, the extracted data were exported to VOSviewer and Biblioshiny software. The obtained results provided details on the annual publication rate and the most relevant authors, countries, and institutions involved in this topic. The results are presented in the form of a knowledge map. The knowledge map indicates the collaboration between the authors, countries, and their funding agencies. Generally, such a study is carried out using two different methods. The first method suggests the co-authorship details. This provides the details of the link between the two countries and the quantity of publications. The second method (co-occurrence) provides the details of the link strength between the keyword and the related publications. Finally, the significant findings are discussed and presented in the final section.

6.4   RESULTS AND DISCUSSION

6.4.1   DOCUMENT TYPE AND YEARLY PUBLICATIONS

Generally, the development of any particular field is identified using publications. While using thermal energy storage as a keyword in Scopus, around 8212 documents were obtained. These included 6301 journal articles, 1246 conference proceedings, and 528 reviews. The details of the number of articles concerning document type are presented in Figure 6.2. This figure inferred that 11 document types were available in the extracted dataset. The journal articles (76.9%) and conference proceedings (15.2%) mainly contributed to sharing ideas and outputs in this field. The remaining documents, such as errata, lecture notes, books, and reports, contributed less than 1%. This result indicates that journal articles are the primary platform for communicating the research output among the researchers.

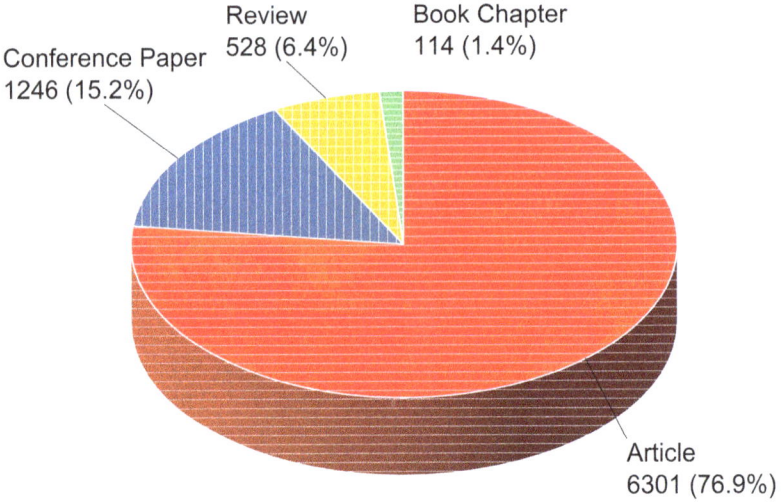

FIGURE 6.2 Distribution of publications by different document types.

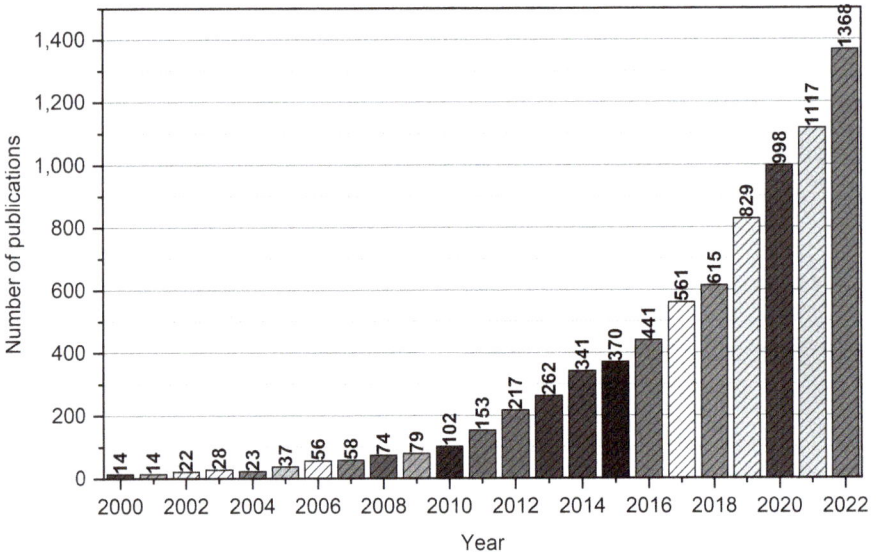

FIGURE 6.3 Yearly publication and citation details.

The details of the yearly publications are described in Figure 6.3. As per the extracted data, the first article related to thermal energy storage was published in 1975. From 1975 to 2000, only fewer than 10 articles per year were published. After 2000, the energy demand gradually increased, forcing researchers to develop new energy storage systems. This led to an increase in the number of publications gradually until 2016. During this period, fewer than 500 articles were published every year. In the last decade, energy needs per capita have increased rapidly. Hence, more researchers have focused on developing a system that stores available thermal energy using different PCMs. This is the reason for the rapid increase in the number of publications after 2016. The highest number of publications, around 1368, was obtained in 2022. This suggests that both the demand for and interest in energy storage systems that use PCMs are on the rise.

6.4.2 ANALYSIS OF COUNTRIES/REGIONS, INSTITUTIONS, AND FUNDING AGENCIES

The extracted articles from the Scopus database clearly show the involvement of different countries in this thermal energy storage research. Details of the number of studies conducted in various countries around the world are depicted in Figure 6.4. The density map of the publications in different countries/regions is illustrated in Figure 6.5. Based on the obtained details, around 100 countries have been involved in this research from January 1975 to December 2022. Among the articles, around 3045 (28.1%) were published in China. Next to China, India and the United States were the most productive and influential countries, with 994 (9.2%) and 775 (7.2%) articles, respectively. Next to these countries, 465 articles were from the United Kingdom, 399 articles were from Spain, and 381 publications were from Iran. At the

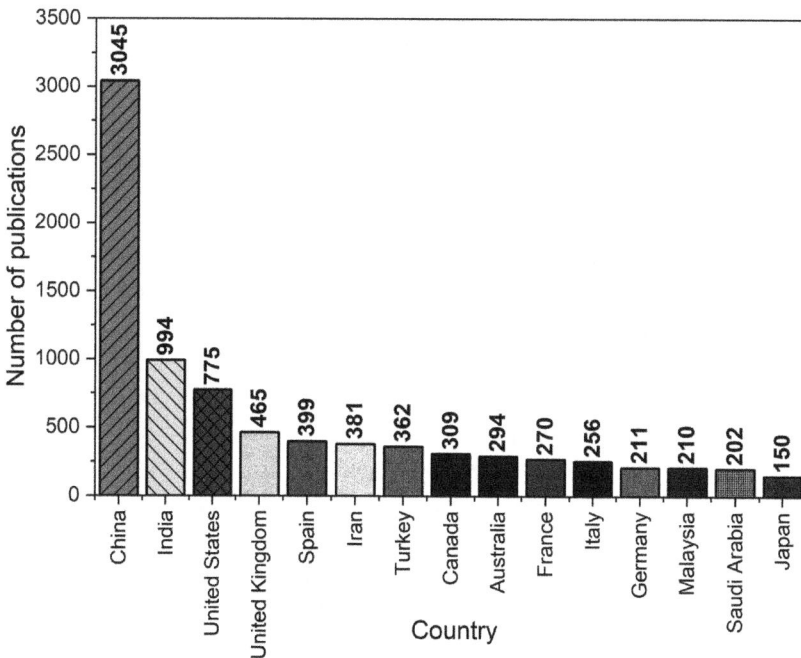

FIGURE 6.4 The publication distribution in the top 15 countries.

Number of publications

< 50
50–100
100–200
200–500
500–1,000
≥ 1,000

FIGURE 6.5 Density map of the publications in different countries/regions.

same time, 200 to 300 publications were from Turkey, Canada, Australia, France, Malaysia, and Saudi Arabia. The remaining countries contributed less than 1% (i.e. less than 200) of the articles during this period.

Due to human populations, vehicles, and the modernization of industries, the requirement for energy has dramatically increased in the last decade. This pushed the researchers to implement new energy storage systems with different PCMs. Especially in Asian countries, the environmental pollution caused by fossil fuels has rapidly increased recently. The government has motivated researchers to research this field to overcome this issue. This is the reason for obtaining more publications from many Asian countries, such as China, India, and Malaysia, in this study.

The density visualization of co-author details based on different organizations is presented in Figure 6.6. The details of the number of publications related to the organizations are given in Figure 6.7. While considering institutions, around 160 countries were from the top 85 countries. Based on the data, the Ministry of Education of China is the most productive institute in this field. This was followed by the Chinese Academy of Sciences (184), Universitat de Lleida (170), South China University of Technology (167), and Xi'an Jiaotong University (160). Of the top 15 institutions, nine are from China, two are from Turkey, one is from India, one is from Spain, one is from Saudi Arabia, and one is from the United Kingdom. The remaining institutions are from Portugal, Vietnam, Pakistan, Austria, Thailand, etc.

The details of article publications related to funding agencies are given in Figure 6.8. Concerning the funding agencies, the China-based National Natural Science Foundation has sponsored a considerable amount of research related to thermal energy storage. Hence, around 1563 publications have acknowledged this sponsoring agency. After this, the Central Universities received funding from the Fundamental Research Funds, which allowed for a greater investment in research

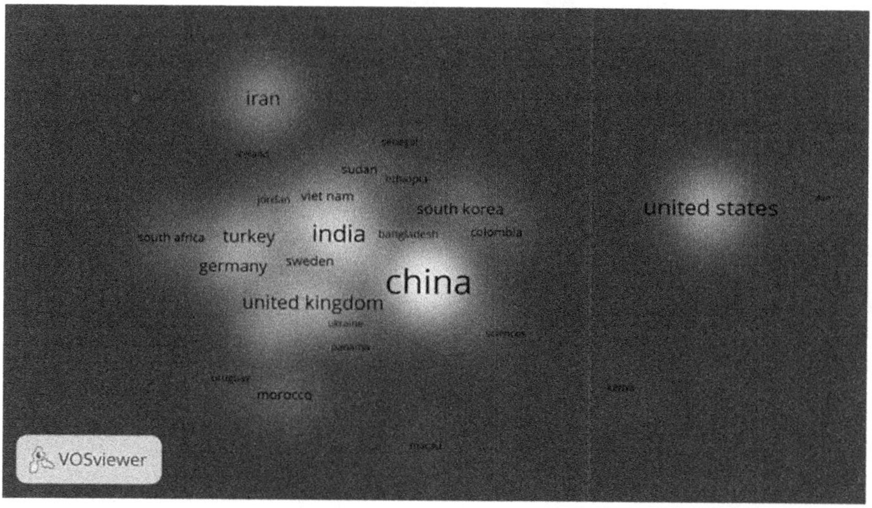

FIGURE 6.6   Density visualization of co-author details based on different organizations.

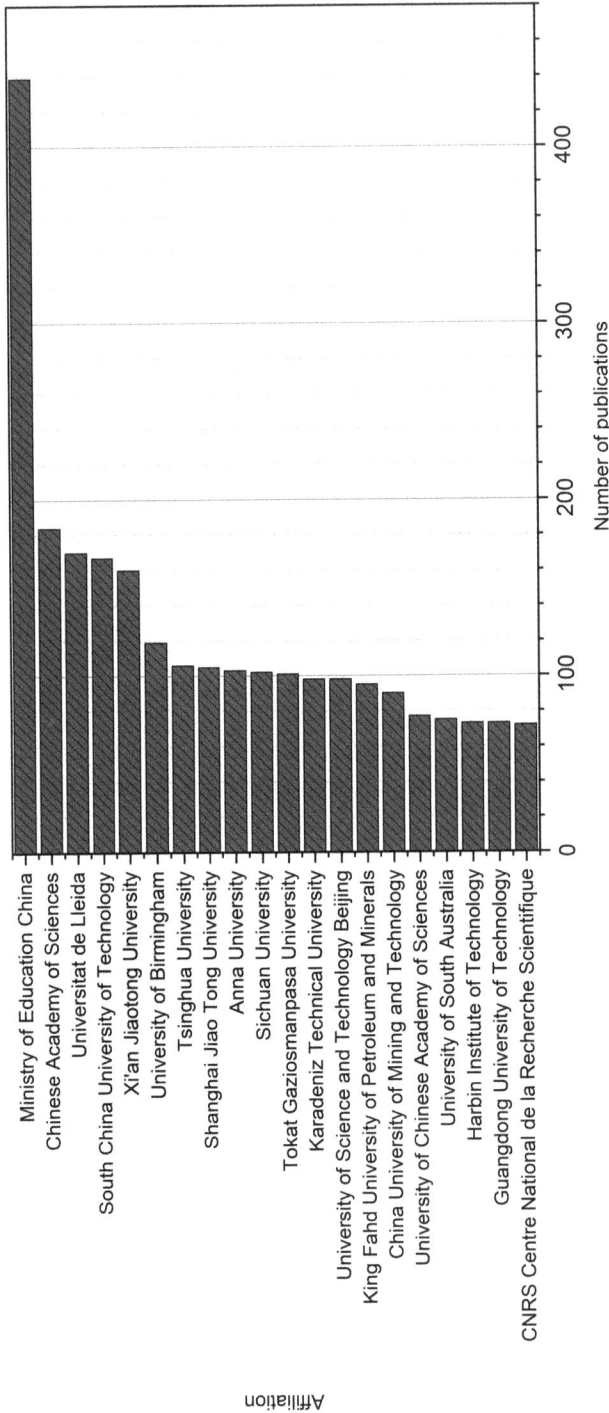

FIGURE 6.7 Publication distribution by organization.

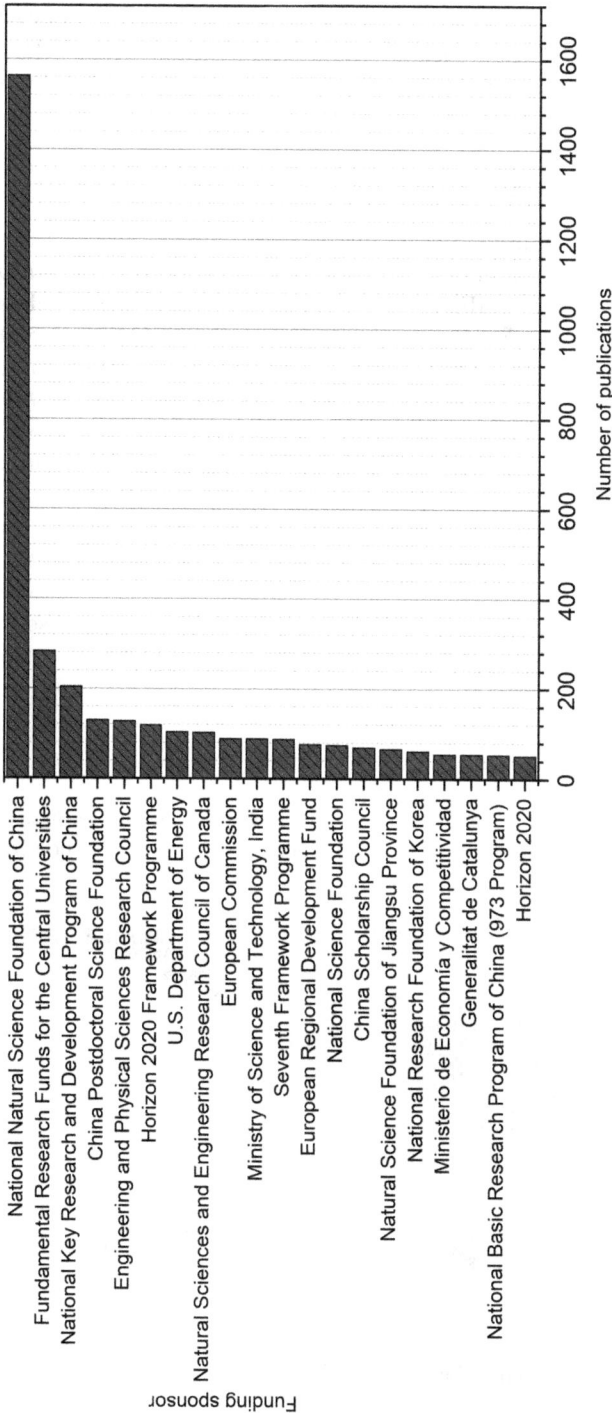

FIGURE 6.8 Publication distribution by funding agencies.

and the publication of 283 corresponding publications. Out of the top 5 funding agencies, four agencies were from China and one was from the United Kingdom. This implies that China encourages its researchers to conduct more PCM-based thermal energy storage systems. Therefore, China placed in the top position with more publications in this period.

6.4.3 ANALYSIS OF JOURNALS

Figure 6.9 displays information on the 25 most influential journals in the field of thermal energy storage. Based on the figure output, the *Journal of Energy Storage* (Q1, H-Index: 60, IF: 8.9) was the top journal with 579 articles and 7920 citations. *Applied Thermal Engineering* (Q1, H-Index: 173, IF: 6.4) placed in the second position with 453 articles and 24,110 total citations. Next to these two journals, the most articles were published in the *Applied Energy Journal* (Q1, H-Index: 235, IF: 9.7). During this time span, the journal has published around 360 publications on energy storage with PCMs. Other than these journals, *Solar Energy Materials and Solar Sells* (Q1, H-Index: 195, IF: 7.2), *Renewable Energy* (Q1, H-Index: 210, IF: 8.6), *Solar Energy* (Q1, H-Index: 194, IF: 7.1) and *Energy* (Q1, H-Index: 212, IF: 8.8) were the most influential journals with 360, 310, 251, and 217 articles, respectively. Not only these journals but numerous others have also published articles related to thermal energy storage. Also, this result indicates that this field is one of the most promising research fields among young researchers.

6.4.4 ANALYSIS OF AUTHORS

Author information for those who have actively contributed to the area of thermal energy storage is shown in Figure 6.10. The outcome of this study showed that more

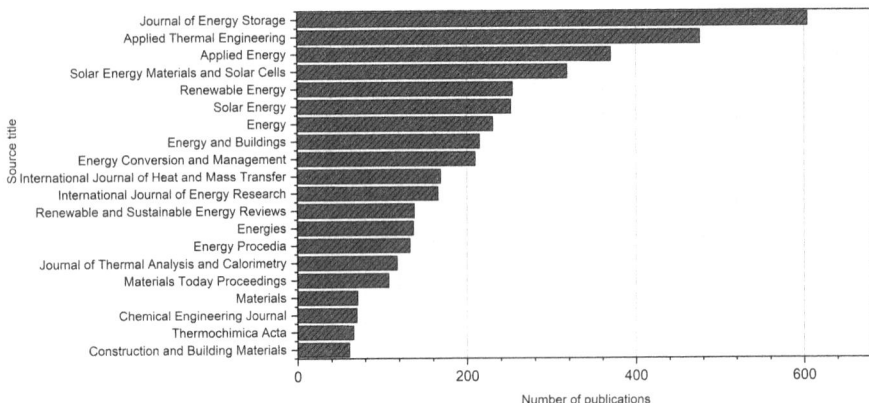

FIGURE 6.9 Publication distribution by source type.

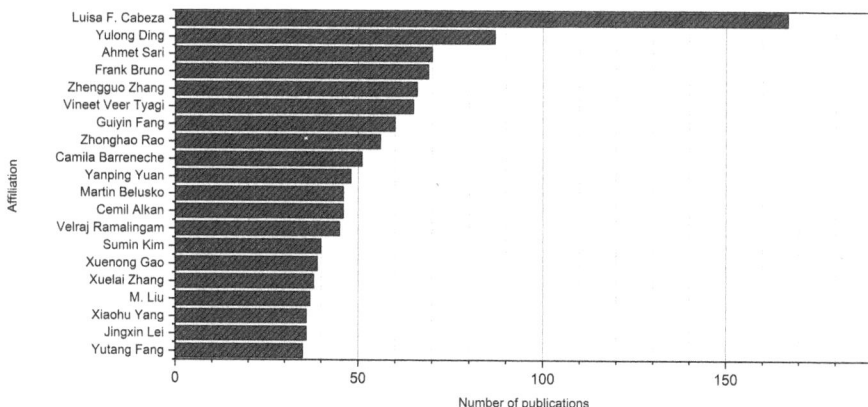

FIGURE 6.10 Article distribution by different authors.

than 20,000 articles were published in different reputed journals. The Price law was used in this study to find the criteria for core authors in this field. Based on the Price law, the requirements were calculated as

$$m \approx 0.749 \sqrt{N_{max}} \tag{6.3}$$

where 'm' denotes the minimum number of papers for a core author, and 'N_{max}' denotes the maximum number of articles published by a single author. As per Figure 6.10, Cabeza L. F. published a maximum of 167 articles. As per Equation 6.3, around 10 articles are required for a core author. As per the Scopus database, more than 150 authors published more than 10 articles from January 1975 to December 2022. Looking at the top 50 most influential authors, it can be found that the Spanish researcher Cabeza L. F. contributed the most publications (167) in this research field. However, researchers from other countries were involved in this research, and the scientists from China contributed more compared with other countries. Because more funding opportunities exist in China, more young researchers are highly involved in finding a suitable PCM for different energy storage applications.

The overlay visualization of co-author details is presented in Figure 6.11. This diagram clearly shows that three significant clusters were identified and labeled in blue, green, and yellow, respectively. In this diagram, the collaboration between two authors is indicated by the connection between two nodes. As per the figure, Cabeza L. F. and Zhan X. carried out more collaborative research works with other country researchers. Also, this indicates that more research needs to be conducted to identify suitable PCMs for different thermal storage applications.

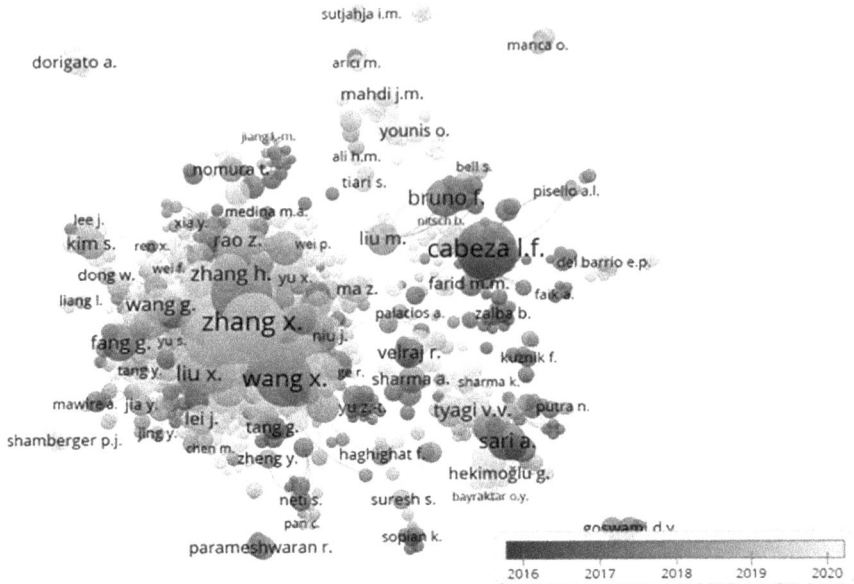

FIGURE 6.11   Overlay visualization of co-author analysis.

6.4.5   ANALYSIS OF CITATIONS AND SUBJECT AREAS

The details of the top 20 articles with the most citations are given in Table 6.1. Also, this table provides the authors' details with their country, journal name, title of the article, type of article, yearly total citation, etc. While analyzing this, it can be found that the article titled 'Review on thermal energy storage with phase change materials and applications,' published in *Renewable and Sustainable Energy Reviews* by Sharma et al. [11] in 2009, is the most influential publication in this field. Among the top 20 publications, around eight were published in *Renewable and Sustainable Energy Reviews*, many of which are reviews. Except for six articles, all of them were reviews.

The details of the subject areas of publications are presented in Figure 6.12. This figure revealed that many authors were focused on engineering and energy areas. The remaining subject areas focused on by different researchers are material science, physics and astronomy, chemical engineering and environmental science, etc. In detail, around 50% of the publications were found in engineering and energy areas. This also indicates that most researchers contributed effectively to storing thermal energy in different engineering and energy applications.

6.4.6   ANALYSIS OF KEYWORDS

Generally, the frequently used words in this research are mainly represented in a word cloud. These words can be used as keywords to search for the most relevant

TABLE 6.1

Top 20 Most Cited Publications with Article Details

S. No	Author details	Title of the paper	Journal Name	Article type	Year of publication	Total number of authors	Total Citation	Total citation per year	Number of countries who collaborated	Ref.
1	Sharma, A., Tyagi, V. V., Chen, C. R., Buddha, D.	Review of thermal energy storage with phase change materials and applications	Renewable and Sustainable Energy Reviews	Review	2009	4	3923	261.53	2	[11]
2	Zalba, B., Marin, J. M., Cabeza, L. F., Mehling, H.	Review of thermal energy storage with phase change: materials, heat transfer analysis and applications	Applied Thermal Engineering	Review	2003	4	3646	173.62	2	[12]
3	Farid, M. M., Khudhair, A. M., Razack, S. A. K., Al-Hallaj, S.	A review on phase change energy storage: materials and applications	Energy Conversion and Management	Review	2004	4	2416	120.8	2	[13]
4	Agyenim, F., Hewitt, N., Eames, P., Smyth, M.	A review of materials, heat transfer and phase change problem formulation for latent heat thermal energy storage systems (LHTESS)	Renewable and Sustainable Energy Reviews	Review	2010	4	1435	102.5	1	[14]
5	Pielichowska, K., Pielichowski, K.	Phase change materials for thermal energy storage	Progress in Materials Science	Review	2014	2	1255	125.5	1	[15]
6	Zhou, D., Zhao, C. Y., Tian, Y.	Review of thermal energy storage with phase change materials (PCMs) in building applications	Applied Energy	Review	2012	3	1229	102.42	2	[16]

(Continued)

TABLE 6.1

Top 20 Most Cited Publications with Article Details (Continued)

S. No	Author details	Title of the paper	Journal Name	Article type	Year of publication	Total number of authors	Total Citation	Total citation per year	Number of countries who collaborated	Ref.
7	Cabeza, L. F., Castell, A., Barreneche, C., De Gracia, A., Fernández, A. I.	Materials used as PCM in thermal energy storage in buildings: A review	Renewable and Sustainable Energy Reviews	Review	2011	5	1213	93.31	1	[17]
8	Kenisarin, M., Mahkamov, K.	Solar energy storage using phase change materials	Renewable and Sustainable Energy Reviews	Article	2007	2	1075	63.24	1	[18]
9	Hasnain, S. M.	Review of sustainable thermal energy storage technologies, Part I: heat storage materials and techniques	Energy Conversion and Management	Review	1998	1	974	37.46	1	[19]
10	Kenisarin, M. M.	High-temperature phase change materials for thermal energy storage	Renewable and Sustainable Energy Reviews	Article	2010	1	810	57.86	1	[20]
11	Sari, A., Karaipekli, A.	Thermal conductivity and latent heat thermal energy storage characteristics of paraffin/expanded graphite composite as phase change material	Applied Thermal Engineering	Article	2007	2	767	45.12	1	[21]
12	Kuznik, F., David, D., Johannes, K., Roux, J.-J.	A review on phase change materials integrated into building walls	Renewable and Sustainable Energy Reviews	Review	2011	4	764	58.77	1	[22]
13	Oró, E., de Gracia, A., Castell, A., Farid, M. M., Cabeza, L. F.	Review on phase change materials (PCMs) for cold thermal energy storage applications	Applied Energy	Review	2012	5	760	63.33	2	[23]

14	Fan, L., Khodadadi, J. M.	Thermal conductivity enhancement of phase change materials for thermal energy storage: A review	*Renewable and Sustainable Energy Reviews*	Review	2011	2	733	56.38	1	[24]
15	Soares, N., Costa, J. J., Gaspar, A. R., Santos, P.	Review of passive PCM latent heat thermal energy storage systems for buildings' energy efficiency	*Energy and Buildings*	Review	2013	4	693	63	1	[25]
16	Regin, A. F., Solanki, S. C., Saini, J. S.	Heat transfer characteristics of thermal energy storage systems using PCM capsules: A review	*Renewable and Sustainable Energy Reviews*	Review	2008	3	680	42.5	1	[26]
17	Pereira da Cunha, J., Eames, P.	Thermal energy storage for low and medium temperature applications using phase change materials—A review	*Applied Energy*	Review	2016	2	672	84	1	[27]
18	Nazir, H., Batool, M., Bolivar Osorio, F. J., Isaza-Ruiz, M., Xu, X., Vignarooban, K., Phelan, P., Inamuddin, Kannan, A. M.	Recent developments in phase change materials for energy storage applications: A review	*International Journal of Heat and Mass Transfer*	Article	2019	9	661	132.2	8	[28]
19	Cabeza, L. F., Castellón, C., Nogués, M., Medrano, M., Leppers, R. Zubillaga, O.	Use of microencapsulated PCMs in concrete walls for energy saving	*Energy and Buildings*	Article	2007	6	646	38	2	[29]
20	Hawlader, M. N. A., Uddin, M. S., Khin, M. M.	Microencapsulated PCM thermal-energy storage system	*Applied Energy*	Article	2003	3	634	30.19	1	[30]

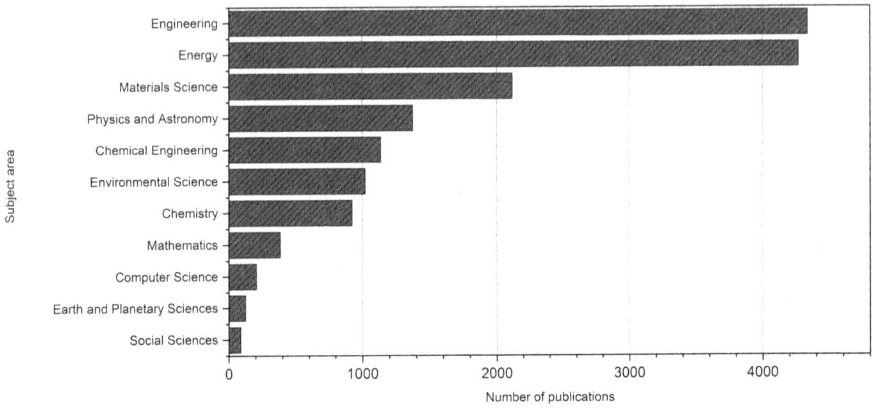

FIGURE 6.12   Article distribution in different subject areas.

FIGURE 6.13   Word cloud of the keywords included in the study.

articles in this particular field. The word cloud generated in this research is presented in Figure 6.13. From this figure, it can be understood that a word with a larger font size is the most frequent keyword in this research. This chart displays the relative prominence of many significant research concepts, including thermal energy storage, heat storage, and storage (materials). The details of the top 50 keywords identified in this study are given in Table 6.2.

TABLE 6.2

Top 50 Keywords in Thermal Energy Storage

S. No	Keyword	Frequency	S. No	Keyword	Frequency
1	Phase change materials	7,328	26	Temperature	531
2	Heat storage	6,216	27	Solidification	526
3	Thermal energy	4,175	28	Digital storage	518
4	Storage (materials)	3,981	29	Thermodynamic properties	506
5	Thermal energy storage	3,324	30	Phase change material (PCM)	474
6	Energy storage	2,828	31	Heat exchangers	460
7	Phase change material	2,373	32	Fins (heat exchange)	459
8	Thermal conductivity	2,279	33	Phase transition	453
9	Latent heat	1,893	34	Solar heating	453
10	Heat transfer	1,795	35	Enthalpy	442
11	Solar energy	1,342	36	Energy conservation	436
12	Melting	1,195	37	Thermogravimetric analysis	424
13	Differential scanning calorimetry	1,102	38	Graphite	417
14	Paraffins	874	39	Performance assessment	408
15	Energy efficiency	790	40	Phase change temperature	408
16	Latent heat thermal energy storage	757	41	Low thermal conductivity	406
17	Thermal energy storage systems	740	42	Fourier-transform infrared spectroscopy	403
18	Phase change	738	43	Nanoparticles	401
19	Thermal performance	605	44	Cooling	387
20	Thermal power	602	45	Heating	379
21	Composite phase change materials	601	46	Thermodynamic stability	363
22	Scanning electron microscopy	566	47	Thermal storage	352
23	Energy usage	565	48	Air conditioning	342
24	Specific heat	536	49	Microencapsulation	335
25	PCM	532	50	Latent heat storage	329

6.5   CONCLUSION

This chapter presented a comprehensive analysis of the literature on phase change materials (PCMs) for thermal energy storage. This work was mainly focused on the yearly publication, most influential authors, institutions, funding agencies, keywords, etc. As a whole, the total number of publications related to thermal energy storage increased from 1975 to 2022. China, India, the United States, the United Kingdom, and Spain are the top 5 most influential countries with the most publications. Consequently, researchers from China and the United States actively collaborate with Indian researchers.

Similarly, China-based institutions such as the Ministry of Education of China and the Chinese Academy of Sciences are the most commonly affiliated institutions in this research field. Cabeza L. F., Ding Y., Sari A., Bruno F., and Zhang Z. are the most influential authors with the most articles in this research area. These authors are mainly focused on studying the performance of energy storage with PCMs. Cabeza L. F. is the most powerful author with 167 publications. Similarly, the *Journal of Energy Storage*, *Applied Thermal Engineering*, and *Applied Energy* were the top three most productive journals during this period. Based on this study, 'phase change material' and 'thermal energy storage' are the best keywords to identify the most relevant articles in this field.

As a whole, the results of this bibliometric analysis can provide more details in terms of publications, authors, citations, etc., to encourage young researchers to continue their research in this field. Meanwhile, more details and new current techniques, such as deep learning and machine learning, must present the results more accurately.

REFERENCES

[1]. Sarbu, I. and Adam, M., 2011. Applications of solar energy for domestic hot-water and buildings heating/cooling. *International Journal of Energy*, 2(5), pp. 34–42.

[2]. Sarbu, I. and Sebarchievici, C., 2013. Review of solar refrigeration and cooling systems. *Energy and Buildings*, 67, pp. 286–297.

[3]. Kuravi, S., Trahan, J., Goswami, D.Y., Rahman, M.M. and Stefanakos, E.K., 2013. Thermal energy storage technologies and systems for concentrating solar power plants. *Progress in Energy and Combustion Science*, 39(4), pp. 285–319.

[4]. Aydin, D., Casey, S.P. and Riffat, S., 2015. The latest advancements on thermochemical heat storage systems. *Renewable and Sustainable Energy Reviews*, 41, pp. 356–367.

[5]. Seddegh, S., Wang, X., Henderson, A.D. and Xing, Z., 2015. Solar domestic hot water systems using latent heat energy storage medium: A review. *Renewable and Sustainable Energy Reviews*, 49, pp. 517–533.

[6]. Al-Saadi, S.N. and Zhai, Z.J., 2013. Modeling phase change materials embedded in building enclosure: A review. *Renewable and Sustainable Energy Reviews*, 21, pp. 659–673.

[7]. Yang, X., Wang, X., Liu, Z., Guo, Z., Jin, L. and Yang, C., 2021. Influence of aspect ratios for a tilted cavity on the melting heat transfer of phase change materials embedded in metal foam. *International Communications in Heat and Mass Transfer*, 122, p. 105127.

[8]. Sierra, V. and Chejne, F., 2022. Energy saving evaluation of microencapsulated phase change materials embedded in building systems. *Journal of Energy Storage*, 49, p. 104102.

[9]. Lakhdari, Y.A., Chikh, S. and Campo, A., 2020. Analysis of the thermal response of a dual phase change material embedded in a multi-layered building envelope. *Applied Thermal Engineering*, 179, p. 115502.

[10]. Tian, W., Dang, S., Liu, G., Guo, Z. and Yang, X., 2021. Thermal transport in phase change materials embedded in metal foam: Evaluation on inclination configuration. *Journal of Energy Storage*, 33, p. 102166.

[11]. Sharma, A., Tyagi, V.V., Chen, C.R. and Buddhi, D., 2009. Review on thermal energy storage with phase change materials and applications. *Renewable and Sustainable Energy Reviews*, 13(2), pp. 318–345.

[12]. Zalba, B., Marın, J.M., Cabeza, L.F. and Mehling, H., 2003. Review on thermal energy storage with phase change: Materials, heat transfer analysis and applications. *Applied Thermal Engineering*, 23(3), pp. 251–283.

[13]. Farid, M.M., Khudhair, A.M., Razack, S.A.K. and Al-Hallaj, S., 2004. A review on phase change energy storage: Materials and applications. *Energy Conversion and Management*, *45*(9–10), pp. 1597–1615.

[14]. Agyenim, F., Hewitt, N., Eames, P. and Smyth, M., 2010. A review of materials, heat transfer and phase change problem formulation for latent heat thermal energy storage systems (LHTESS). *Renewable and Sustainable Energy Reviews*, *14*(2), pp. 615–628.

[15]. Pielichowska, K. and Pielichowski, K., 2014. Phase change materials for thermal energy storage. *Progress in Materials Science*, *65*, pp. 67–123.

[16]. Zhou, D., Zhao, C.Y. and Tian, Y., 2012. Review on thermal energy storage with phase change materials (PCMs) in building applications. *Applied Energy*, *92*, pp. 593–605.

[17]. Cabeza, L.F., Castell, A., Barreneche, C.D., De Gracia, A. and Fernández, A.I., 2011. Materials used as PCM in thermal energy storage in buildings: A review. *Renewable and Sustainable Energy Reviews*, *15*(3), pp. 1675–1695.

[18]. Kenisarin, M. and Mahkamov, K., 2007. Solar energy storage using phase change materials. *Renewable and Sustainable Energy Reviews*, *11*(9), pp. 1913–1965.

[19]. Hasnain, S.M., 1998. Review on sustainable thermal energy storage technologies, Part I: Heat storage materials and techniques. *Energy Conversion and Management*, *39*(11), pp. 1127–1138.

[20]. Kenisarin, M.M., 2010. High-temperature phase change materials for thermal energy storage. *Renewable and Sustainable Energy Reviews*, *14*(3), pp. 955–970.

[21]. Sarı, A. and Karaipekli, A., 2007. Thermal conductivity and latent heat thermal energy storage characteristics of paraffin/expanded graphite composite as phase change material. *Applied Thermal Engineering*, *27*(8–9), pp. 1271–1277.

[22]. Kuznik, F., David, D., Johannes, K. and Roux, J.J., 2011. A review on phase change materials integrated in building walls. *Renewable and Sustainable Energy Reviews*, *15*(1), pp. 379–391.

[23]. Oró, E., De Gracia, A., Castell, A., Farid, M.M. and Cabeza, L.F., 2012. Review on phase change materials (PCMs) for cold thermal energy storage applications. *Applied Energy*, *99*, pp. 513–533.

[24]. Fan, L. and Khodadadi, J.M., 2011. Thermal conductivity enhancement of phase change materials for thermal energy storage: A review. *Renewable and Sustainable Energy Reviews*, *15*(1), pp. 24–46.

[25]. Soares, N., Costa, J.J., Gaspar, A.R. and Santos, P., 2013. Review of passive PCM latent heat thermal energy storage systems towards buildings' energy efficiency. *Energy and Buildings*, *59*, pp. 82–103.

[26]. Regin, A.F., Solanki, S.C. and Saini, J.S., 2008. Heat transfer characteristics of thermal energy storage system using PCM capsules: A review. *Renewable and Sustainable Energy Reviews*, *12*(9), pp. 2438–2458.

[27]. Da Cunha, J.P. and Eames, P., 2016. Thermal energy storage for low and medium temperature applications using phase change materials—a review. *Applied Energy*, *177*, pp. 227–238.

[28]. Nazir, H., Batool, M., Osorio, F.J.B., Isaza-Ruiz, M., Xu, X., Vignarooban, K., Phelan, P. and Kannan, A.M., 2019. Recent developments in phase change materials for energy storage applications: A review. *International Journal of Heat and Mass Transfer*, *129*, pp. 491–523.

[29]. Cabeza, L.F., Castellon, C., Nogues, M., Medrano, M., Leppers, R. and Zubillaga, O., 2007. Use of microencapsulated PCM in concrete walls for energy savings. *Energy and Buildings*, *39*(2), pp. 113–119.

[30]. Hawlader, M.N.A., Uddin, M.S. and Khin, M.M., 2003. Microencapsulated PCM thermal-energy storage system. *Applied Energy*, *74*(1–2), pp. 195–202.

7 Heat Transfer Augmentation Techniques for Phase Change Materials

Sanaz Akbarzadeh, Maziar Dehghan, and Hafiz Muhammad Ali

HIGHLIGHTS

- This study explores two main categories for improving the thermo-physical properties of phase change materials (PCMs): heat transfer enhancement methods and thermal conductivity enhancement methods.
- One of the techniques for enhancing the thermal conductivity of PCMs is adding nanoparticles, such as graphene, which has high thermal conductivity (2000–4000 W/m·K) and high specific area (> 2000 m^2/g).
- Other techniques include using fins, metal foams, mixing PCMs, and modifying the heat transfer surface.
- The performance of PCMs in heat transfer systems is influenced by several parameters, such as the latent heat of fusion, the mass fraction, and the Stefan number. The optimal impact of PCMs is achieved when the Stefan number is below 1.

NOMENCLATURE

Symbols

a	Fraction of PCM melted
c_m	Mass fraction of PCM in suspension
C_p	Specific heat capacity, (J/kg°C)
h	Heat fusion of PCM per unit mass, (W/kg)
k	Thermal conductivity, (W/m°C)
m	Mass, (kg)
q	Heat, (J)
q_w	Heat flux across the pipe wall, (W/m^2)
R	Pipe radius, (mm)

DOI: 10.1201/9781003331957-7

S(t)	Distance between heat pipes, (m)
Ste	Stefan number
t	Time, (s)
T	Temperature, (°C)

GREEK SYMBOLS

λ	Latent heat of fusion of PCMs, (J/kg)
ρ	Density, (kg/m³)

SUBSCRIPTS

f	Fluid
i	Inlet
l	Liquid
m	Middle, Melting
s	Solid

ABBREVIATIONS

CNTs	Carbon nanotubes
COP	Coefficient of performance
CPCM	Composite PCM
CPV/T	Concentrated photovoltaic/thermal
EG	Expanded graphite
HDH	Humidification-dehumidification
HSB	Hollow steel ball
LHTES	Latent heat thermal energy storage
MPCM	Micro-encapsulated PCM
MWCNT	Multi-walled carbon nanotubes
PCB	Printed circuit board
PCL	Polycaprolactone
PCM	Phase change materials
PLA	Polylactic acid
PPI	Pore per inch
PS	Polystyrene
PTC	Parabolic trough collector
PV/T	Photovoltaic/thermal system
rGO	Reduced graphene oxide
RT	Rubitherm
SAT	Sodium acetate trihydrate
TES	Thermal energy storage (system)

7.1 INTRODUCTION

Phase change materials (PCMs) have a high potential as thermal storage and control media because they can absorb and release a large amount of heat. PCMs can enhance the performance of cooling and heating systems by reducing their dimensions and expenses [1]. Thermal energy storage can bridge the mismatch between energy availability and energy consumption. In latent heat thermal energy storage (LHTES or LTES), the storage material undergoes a phase change. PCMs are the common material for LHTES systems. PCMs can be categorized into three main types: organic, inorganic, and eutectic, as shown in Figure 7.1 (a) [2]. PCMs, which are materials that can store and release heat, have a wide range of working temperatures. They can be used for cryogenic applications with negative melting points or for solar applications with temperatures up to 1000 °C. Organic PCMs have a stable latent heat of fusion, but they conduct heat poorly, do not separate into phases, and melt at low temperatures. Their melting temperature varies from −10 °C to 100 °C. Inorganic PCMs have good thermal conductivity, high latent heat density per volume, and cycle stability. However, salt hydrates are prone to corrosion, which makes the containers more expensive. The melting temperature of inorganic PCMs can be as high as 1000 °C. Eutectic PCMs do not show any phase segregation [3].

The choice of PCMs relies on the aim of their applications. The material which is applied in an LHTES should have high latent heat in the phase change process, high

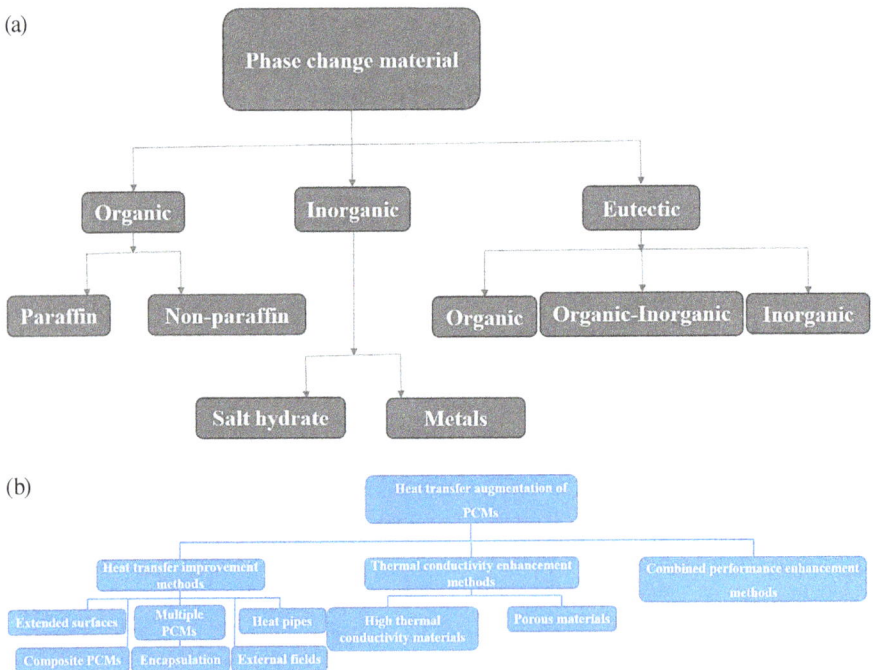

FIGURE 7.1 (a) The classifications of PCMs; (b) techniques for the thermo-physical behavior improvement of PCMs.

thermal conductivity, high density, low vapor pressure, and small volume changes, and is preferred to be non-toxic, non-flammable, non-corrosive, and cost effective. PCMs including all of these specifications are not available; as a result, many analyses have been carried out to enhance their thermo-physical properties [4]. The performance of the latent thermal energy systems is affected by the low thermal conductivity of PCMs, which lowers the heat transfer rate between the PCMs and the heat transfer fluids. The techniques for the thermo-physical behavior improvement of PCMs can be categorized into two main groups: (i) heat transfer improvement methods and (ii) thermal conductivity enhancement methods. Heat transfer enhancement methods include the utilization of extended surfaces, multiple PCMs, heat pipes, composite PCMs, encapsulated PCMs, and external fields (See Figure 7.1 (b)). Highly thermally conductive, low-density, and porous materials are used to enhance the thermal conductivity of a PCM.

In this chapter, heat transfer methods and thermal conductivity improvement methods to increase the performance of systems applying PCMs are explained in more detail. In addition, the use of PCMs, their benefits and drawbacks, and the types of PCMs that are commercially available are discussed.

7.2 PCM PROPERTIES

Latent heat of fusion, thermal conductivity, specific heat capacity, chemical stability, melting temperature, and volume change in the phase change process are important properties which must be considered in the selection of PCMs. As a characterizing parameter, the Stefan number (*Ste*) is important for PCMs, especially when the Stefan number < 1 [5]:

$$Ste = \frac{c_p (q_w \frac{R}{k})}{c_m \lambda} \tag{7.1}$$

where q_w is heat flux across the pipe wall, R is the pipe radius, c_m is the mass fraction of PCM in suspension, and λ is the latent heat of fusion of PCMs. It can be found that PCMs with high mass fractions are preferred. However, it should be controlled to prevent a significant pumping power because of increasing the viscosity of the slurry [5]. PCMs should have a high latent heat of fusion, high thermal conductivity, high specific heat capacity, chemical stability, convenient melting temperature, and low volume change in the phase change process [6]. More importantly, the transition temperature is an important parameter in the selection of PCMs. If it is not in the temperature range of the application, PCMs cannot release the absorbed heat and would add thermal inertia to the system without any advantage. The high thermal conductivity of PCMs results in easy charging and discharging processes in PCMs. A high heat capacity of PCMs enhances the ability of PCMs to store sensible and latent heat [7]. Multicomponent PCMs including components with different densities may be separated into different phases. The changes in the volume of PCMs during the melting process should be considered in the design of their container. Their flammability and toxicity should be checked in domestic applications. To increase the form stability of PCMs, a porous matrix is utilized. Polymers, minerals, ceramics, and metal foams are common matrix materials.

7.3 HEAT TRANSFER IMPROVEMENT METHODS

In the melting and solidification processes of PCMs, the interphase boundary change is introduced as the Stefan problem. The heat transfer process in the melting and solidification steps includes conduction and/or convection. The primary heat transfer mechanism is conduction. In convection phase change, it is assumed that the PCM is at the melting point in the transmission stages. In the conduction/convection phase change, the impacts of conduction and convection are considered during the transformation stages. The transition process is described using the energy equation at the solid–liquid interface [2]:

$$\lambda \rho \left(\frac{dS(t)}{dt} \right) = k_s \left(\frac{\partial T_s}{\partial t} \right) - k_l \left(\frac{\partial T_l}{\partial t} \right) \tag{7.2}$$

where T_s and T_l are the liquid and the solid phase temperatures. The heat storage capacity of an LHTES system is calculated from [2]:

$$Q = \int_{T_i}^{T_m} mc_p dT + ma_m \Delta h_m + \int_{T_m}^{T_f} mc_p dT \tag{7.3}$$

where a_m and Δh_m are the fraction of PCM melted and the heat of fusion of PCM per unit mass, respectively.

The low thermal conductivity of PCMs causes a low heat transfer rate. To improve heat transfer, many techniques like extended surfaces, multiple PCMs, heat pipes, composite PCMs, encapsulated PCMs, and external fields are utilized, which are explained in more detail in the following subsections.

7.3.1 EXTENDED SURFACES

To enhance the heat transfer of PCMs, extended surfaces are applied. Metallic particles, powders, wires, meshes, and fins are added to PCMs to increase their thermal conductivity. Among these, fins are widely used due to their efficiency and simplicity [8]. Moreover, they offer more enhancement than nanoparticles, as well as a lower price than other methods [9]. Their selection relies on their thermal conductivity, density, corrosion resistance, and cost. Graphite, aluminum, stainless steel, carbon steel, and copper are widely applied as extended surfaces [10]. The properties of PCMs are affected by fins in two ways, including heat conduction and natural convection [11]. Using fins in latent heat storage results in increasing the rate of melting and solidification due to heat transfer area enhancement of the hot/cold surface and the PCM [9]. However, the natural convection is limited by utilizing fins. In addition, the volume of the PCM and the storage capacity decrease [12]. A numerical simulation based on a two-dimensional transient method was performed to examine the impact of a PCM, namely n-eicosane, in an aluminum-finned heat sink to improve the cooling process in an electronic device. The time needed for melting the PCM increased with the decrease in the power intensity [13]. COMSOL-MULTIPHYSICS software was employed to investigate the effects of a PCM, namely Rubitherm GmbH (RT-27), and fins with an inclination angle of 0°, 45°, and 90° on the thermal storage of a rectangular capsule. The results demonstrated that the maximum rate of thermal storage was 36.88% at the angle of 0° [14]. To decrease the time of releasing energy, the PCM and longitudinal fins were applied. The geometry of fins is shown in Figure 7.2. Using discontinuous fins improves the time of releasing energy by up to 89% and 84% in comparison to continuous and aluminum fins, respectively [15].

To enhance the melting time of PCMs (RT-42), internal–external fins were attached to the back of a PCM container. At the inclination angle of 30°, the Nusselt number was 6–9 times more than that without fins [16]. Different configurations of fins and PCM (paraffin wax–RT-55) in a thermal energy storage (TES) system, which are presented in Figure 7.3, were studied numerically. In the case of fin 4, the melting time increased by 65% [17].

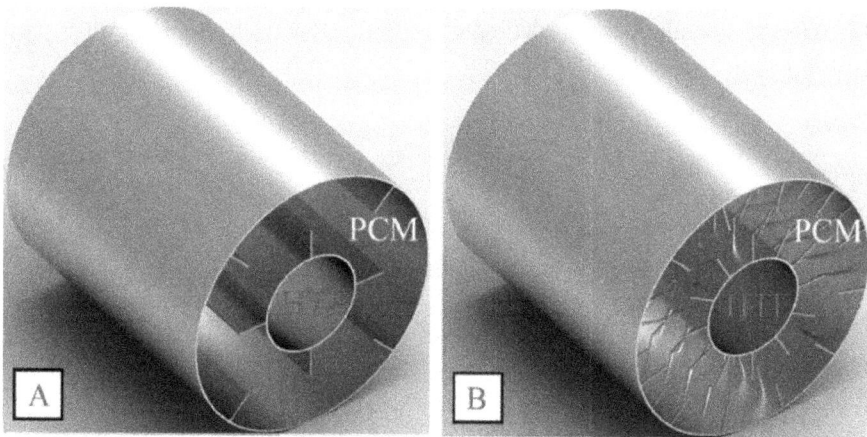

FIGURE 7.2 (a) Continuous fins and (b) discontinuous fins [15].

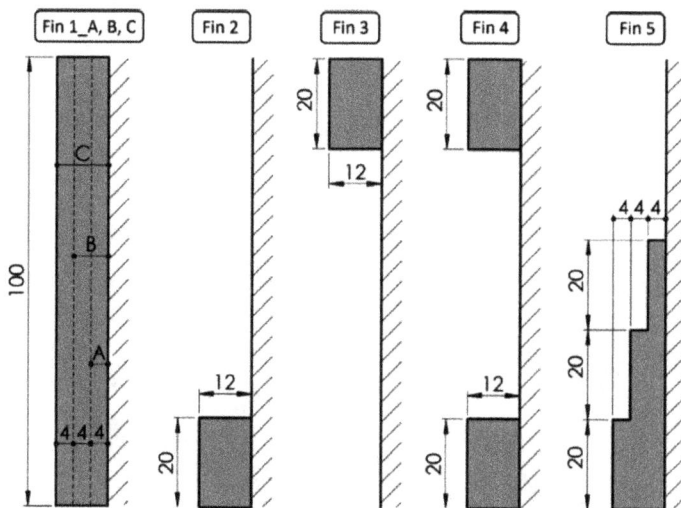

FIGURE 7.3 Different configuration of fins [17].

In a heat exchanger, rectangular copper fins were installed radially to surrounded the greywater and cold water pipes to increase the thermal conductivity of the PCM. The effective thermal conductivity of the PCM increased by 1.38 times and 4.75 times for melting and freezing, respectively [18]. Three layers of PCM (RT-42, RT-50, and RT-60) in a latent thermal energy storage system were utilized. Different arrays of fins, which are shown in Figure 7.4, were investigated. In cases 2, 3, and 4 in comparison to case 1 (with no fins), the melting time was saved by 55.3%, 66.1%, and 71%, respectively [19].

The study of impacts of different length-to-height ratios of rectangular fins and tree-like fins in an enclosure containing a PCM showed that the tree-like fins were not effective compared to the rectangular ones. At an aspect ratio of 1, the natural convection increased by 20% [20]. The investigation of the melting process of a lauric acid PCM in a rectangular cavity including fins revealed that the melting time decreased with a length increase in the fins and a decrease in the aspect ratio of fins [21].

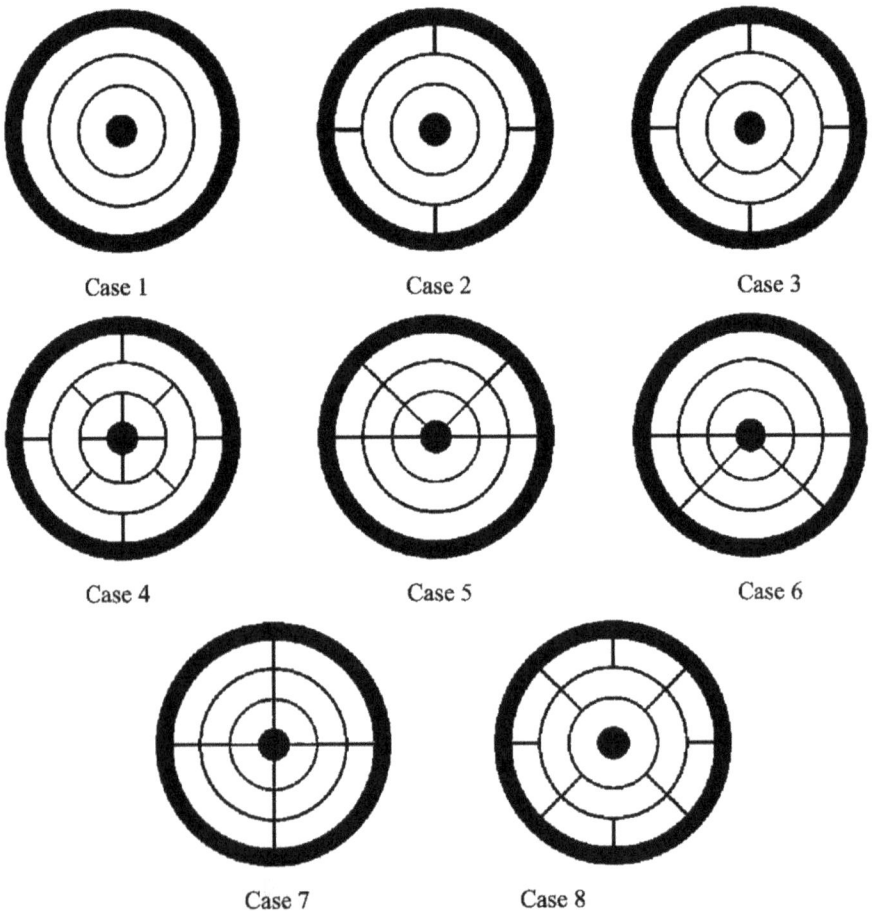

Case 1   Case 2   Case 3

Case 4   Case 5   Case 6

Case 7   Case 8

FIGURE 7.4   Different arrangements of fins [19].

L-shaped fins in a vertical closure containing lauric acid as a PCM were utilized with different coincidence angles which ranged from −90° to 90°. The solidification time increased by 45% at 0°, and the melting time decreased by 14% at −30° [22]. In an electronic device, a PCM was utilized to check the peak temperature, and the peak temperature decreased by 9% using fins, as shown in Figure 7.5 [23].

In a shell-and-tube heat exchanger, the combinations of fins (See Figure 7.6) and the PCM (RT-82) with a melting temperature of 82 °C decreased the melting time up to 33% using tee fins in comparison to the longitudinal fins because of increasing the area of heat transfer [24].

Different stepped fins with step ratio ranges of 0.66–4 were applied in an LHTES system (see Figure 7.7). The enhancement in the melting process using downward-stepped fins increased by 65.5% compared to the horizontal ones at t = 3600 s. The

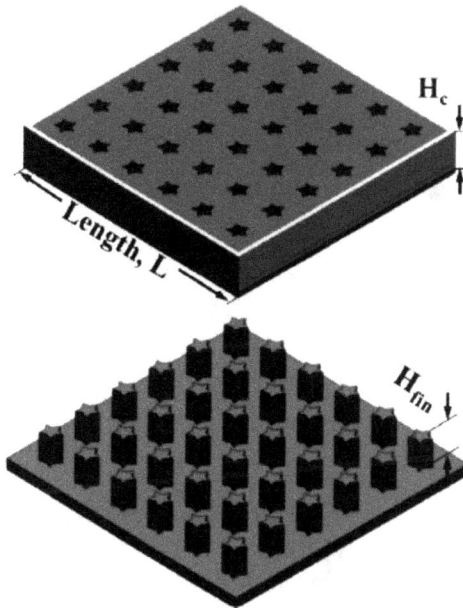

FIGURE 7.5   Container including PCM and fins on the plate [23].

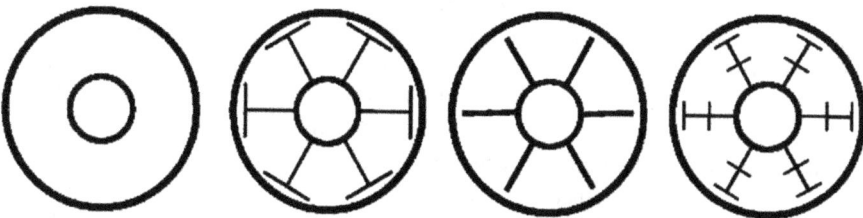

FIGURE 7.6   Different fins on the shell side of a heat exchanger [24].

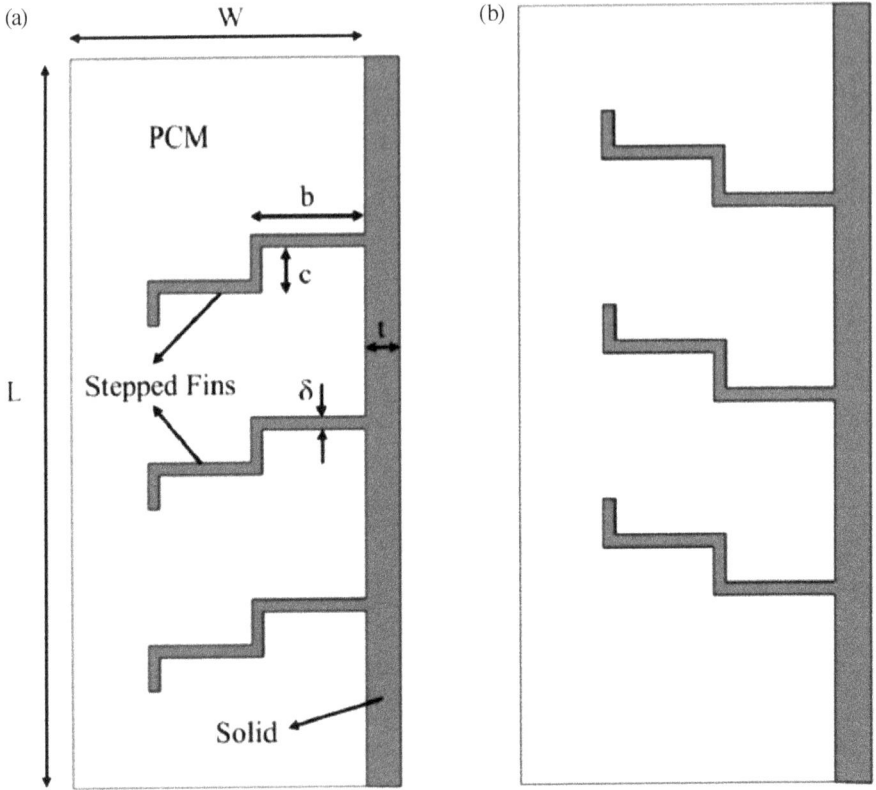

FIGURE 7.7   A schematic view of fins: (a) downward-stepped fins and (b) upward-stepped fins [25].

rise in step ratio resulted in a uniform melting process; as a consequence, the performance of melting was enhanced [25].

An investigation of a heat sink including aluminum plate fins filled with n-eicosane, paraffin, and salt hydrate PCMs showed that although salt hydrate has high thermal conductivity and low cost, it is chemically and thermally unstable. Paraffin has a low capacity for thermal energy storage. In addition, a large space is needed to contain paraffin due to its low density and thermal conductivity. Using n-eicosane increased the thermal performance of the unit [26]. Different shapes of fins presented in Figure 7.8 were employed to enhance the thermal conductivity of PCMs and to improve the melting process. The area enhancement ratio, which is the ratio of the surface area of fins in connection with the PCM to the base cross-sectional area, is a key parameter in the performance of the thermal control module. Triangular prism fins exhibited the maximum thermal performance enhancement due to their highest value of area enhancement ratio [27]. In addition, utilizing triangular fins (Figure 7.9) in a PCM as a TES system caused a 30.98% reduction in the solidification time [28].

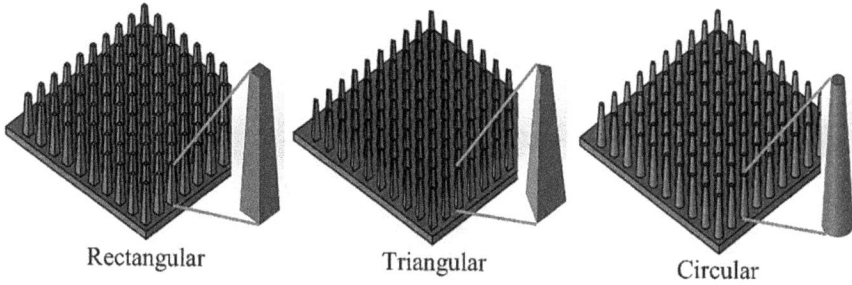

Rectangular   Triangular   Circular

FIGURE 7.8 The geometry of different fins [27].

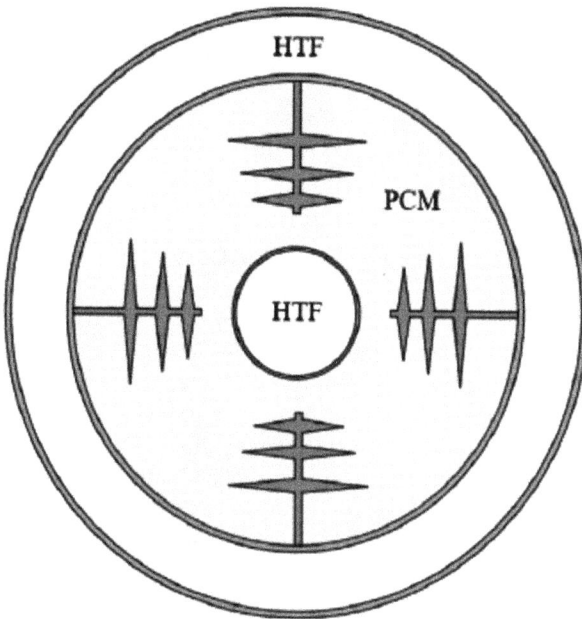

FIGURE 7.9 Triangular fins in a PCM [28].

The impacts of different fin configurations in a PCM (Figure 7.10) were numerically studied. The rate of solidification in the top side was more than that on the bottom side due to its higher temperature. The rise in the number of fins and their length has a more significant impact on the solidification process than on the melting process. At the same total length of fins, using long fins led to more heat transfer enhancement compared to the short ones [9]. The number of fins and the inclination angle are crucial parameters in designing fins for the thermal performance improvement of a heat sink using PCMs. To investigate the impact of these parameters, five plate fins with inclination angles between 0° and 90° were installed in a heat sink. The optimum number of fins and the inclination angle were three and 60°,

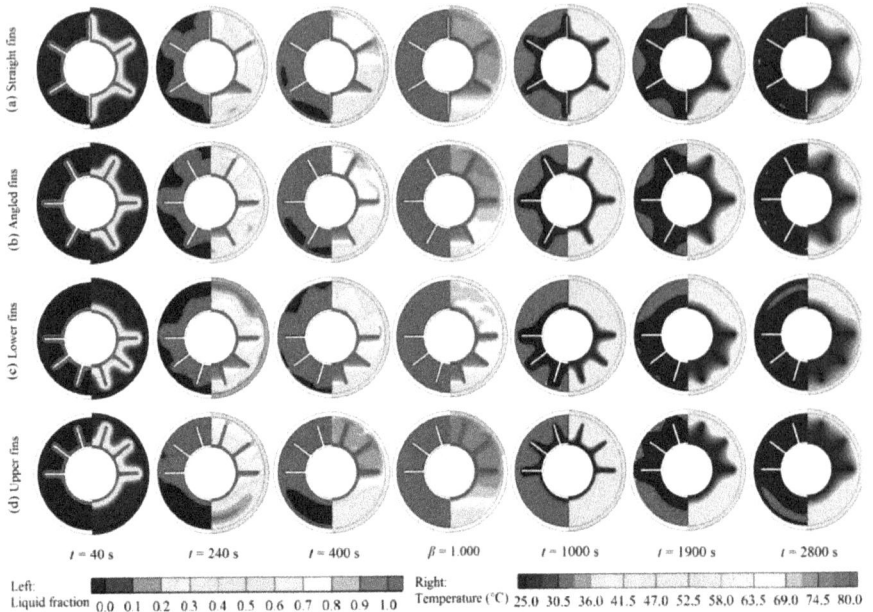

FIGURE 7.10   The contours of the liquid–solid interface [9].

respectively [29]. The impacts of wall temperature, fin area, and fin diameter on the thermal performance of a heat exchanger using finned tubes and a PCM were studied. The fin diameter enhancement increased the interface velocity and decreased the solidification time. The decrease in the temperature of the tube wall resulted in a greater temperature gradient; as a result, the process of the phase changes improved [69]. Applying H-shaped, tree-like, and tee fins to PCM RT-82 increased the melting rate by 69.14%, 35.58%, and 22.33%, respectively [30].

7.3.2   Multiple (or Cascaded) PCMs

To improve the charging and discharging processes, multiple PCMs were proposed. The fractions of PCMs and their melting temperatures are crucial parameters in multiple PCMs [31]. In a heat exchanger, multiple PCMs (m-PCMs) were applied, in which the natural convection is a critical parameter in heat transfer enhancement. The melting time decreased by 15.5% in comparison to the single PCM [32]. The charging and discharging time decreased by 30% and 9% using a non-uniform distribution of fins and multiple PCMs including three different layers, namely potassium hydroxide, potassium nitrate, and sodium nitrate, with melting temperatures of 360 °C, 335.8 °C, and 305.4 °C, respectively, compared to the single PCM [33]. The m-PCMs, RT-25, RT-20, and RT-18 with melting temperatures of 25 °C, 20 °C, and 18°C, respectively, were used in an air conditioning system. The melting points were adopted according to the thermal comfort conditions in a building. The combination of RT-20 and RT-25 was the best option for the air conditioning unit [34]. The m-PCMs (calcium chloride hex hydrate, $CaCl_2.6H_2O$, paraffin C_{18}, and RT-25) were used as a thermal energy storage system. A coefficient of performance (COP) of 7 was obtained for the system using m-PCMs $CaCl_2.6H_2O$ and RT2-5 [35]. In cooling applications using LHTES with m-PCMs, the exergy increased by 6.2% with an 11% reduction in the inlet temperature of the air. In contrast, the exergy decreased by 5.6% with the increase in flow rate from 800 to 1600 m^3/h [36]. The m-PCMs (hydroquinone and d-mannitol) with a melting temperature of 150–200 °C were applied in a TES system. The increase in the effectiveness parameter, the ratio of the actual heat discharged to the theoretical maximum heat discharged, was 19.36% using m-PCMs [37]. The PV module's maximum temperature was reduced by 4 °C and 7.2 °C when using the multilayered PCM system, including a system with two layers of OM37 and one layer of OM42 and a system with one layer of OM37 and two layers of OM42, compared to the single-layer PCM and the PV reference, respectively. The multilayered PCM system also helped to keep the PV module's operating temperature within a suitable range throughout the year, resulting in a 3.3% increase in the annual electricity output. Moreover, the PV module's life was increased by about 10 years and its lifetime earnings almost doubled [38].

7.3.3 HEAT PIPES

By using heat pipes, the heat transfer surface in TES systems can be enlarged. The heat transfer in heat pipes includes the phase change of working fluids, which results in reaching high thermal conductivity. Heat pipes have an evaporator, an adiabatic section, and a condenser. The schematic of a heat pipe is shown in Figure 7.11. [39]. They can transfer the battery heat to the environment, and the PCM can absorb the heat. Fins were added to the condenser part of a heat pipe presented in Figure 7.12. Beeswax and RT-44HC with melting temperatures of more than and equal to the desired working temperature of the battery, respectively, were selected. At heat loads

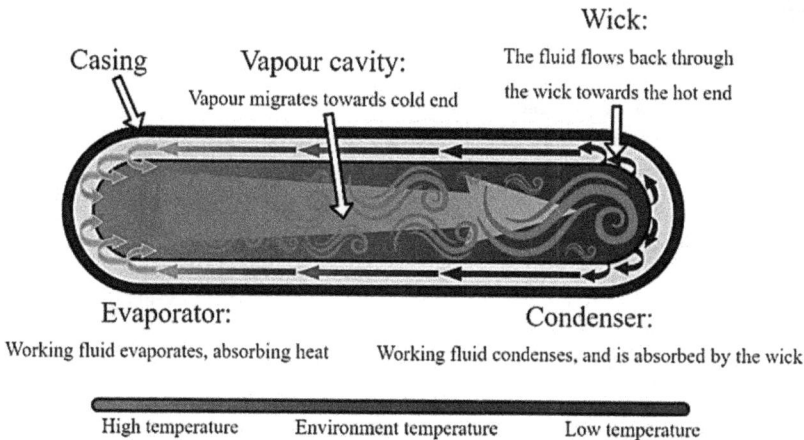

Wick:
The fluid flows back through the wick towards the hot end

Casing

Vapour cavity:
Vapour migrates towards cold end

Evaporator:
Working fluid evaporates, absorbing heat

Condenser:
Working fluid condenses, and is absorbed by the wick

High temperature   Environment temperature   Low temperature

FIGURE 7.11   The schematic of a heat pipe [39].

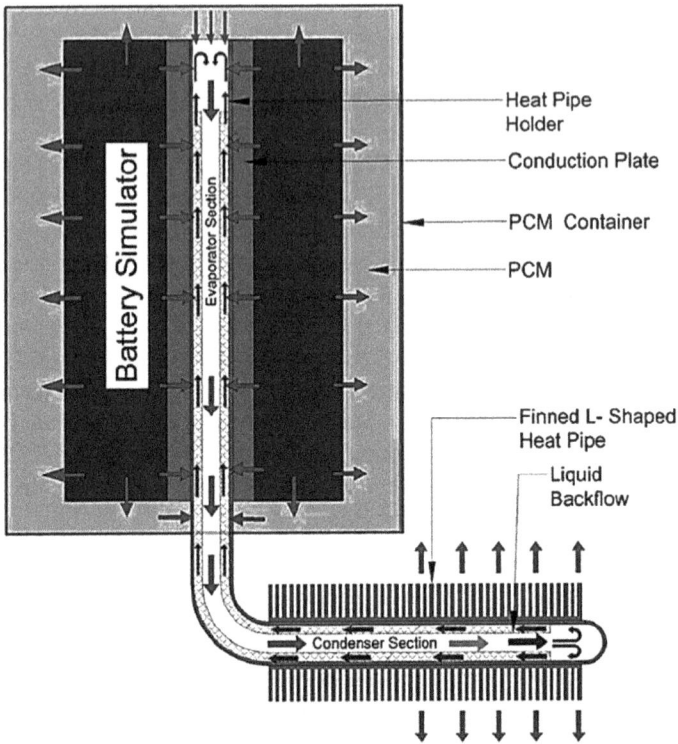

FIGURE 7.12 Using a heat pipe and PCM in the thermal management unit of a battery [40].

of 60 W, using beeswax and the heat pipe, the battery temperature decreased by 31.9 °C compared to the RT-44HC, which was 33.2 °C. The RT-44HC showed more heat absorption due to involving latent and sensible heat [40]. In a latent heat storage system, composite PCMs (10% and 20% expanded graphite and octadecanol) and oscillating heat pipes were used. The charging efficiency had a 32% increase using oscillating heat pipes. The charge time difference between conventional LHTES and composite PCM was negligible [41].

An experimental and numerical study of a thermal management system of a lithium-titanate battery cell consisting of a heat pipe and PCM resulted in the maximum temperature of the cell using the heat pipe and PCM decreasing by 17.3% and 40.7%, respectively [42]. The impacts of PCMs and a vapor chamber on the thermal performance of a heat sink were studied numerically. The temperature of the heat source was reduced by 33.1% and 9.5% using PCMs and a vapor chamber, respectively [43].

7.3.4 COMPOSITE PCMs

Among PCMs, inorganic PCMs draw considerable attention due to having high thermal conductivity, high volumetric latent heat density, and stability in cycles [3]. But they face a leakage risk at temperatures above their melting temperature, as well as low heat transfer rates. One of the methods to increase the stability of their form is using them in the form of composite PCMs [44].

To investigate the impact of wood and boron nitride (BN) in a PCM, a wood–PCM composite including polyethylene glycol 6000 was studied. The melting and freezing enthalpy had an increase of 8% compared to that of the pure PCM [25]. A biochar–PCM composite was applied in an LTES. The high thermal conductivity and porosity of biochar due to including high carbon content make it an appropriate option to use in PCM. The best mixing ratio was obtained at the ratio of PCM to the biochar of 6:4 (wt/wt %). Adding aluminum as a metal powder and water hyacinth biochar increased the thermal conductivity 17.27 times and 13.82 times more than the pure PCM [45]. The floor, ceiling, and west wall of the room used SP29, a PCM with a melting temperature range of 28–30 °C. The north and east walls of the room used RT-18, another PCM with the same melting temperature range. In summer, the temperature of the room decreased by about 4.28–7.7 °C. In winter, the temperature increased by 6.93–9.48 °C [46]. MG22 with a melting temperature of 22.4 °C and MG30 with a melting temperature of 29.8 °C were placed on the east and north walls and the south and west walls, respectively. The room temperature showed a decrease of 2.9–7.3 °C and an increase of 0.5–4 °C during the day and night, respectively [47].

The combination of high-density polyethylene, expanded graphite, and paraffin wax with weight fractions of 15% wt, 15% wt, and 70% wt, respectively, were arranged in building walls. The cooling load on summer days increased by 43.23% [48]. Natural waxes, including palm wax, beeswax, golden soy wax, and nature soy wax, containing coffee wastes were selected as a composite PCM. The high heat storage and stability of the new composite made it suitable for use in building materials [49]. Bio-based PCMs containing cellulose fibers, clay powder, and graphite were used in the envelope of a building because they are eco-friendly and low in cost. The best thermal properties were obtained when the ratio of cellulose fibers, clay powder, and graphite were 5:14, 2:7, and 3:7, respectively [50]. Applying blast furnace slag modified by $Ca(OH)_2$ increased the adsorption capacity of paraffin in the composite PCM to be used in building and solar TES. Using 10% of $Ca(OH)_2$ increased paraffin absorption by modified blast furnace slag by 13.2% [51]. Lauryl alcohol and stearic acid were prepared as composite PCM to be applied in a building. To increase the thermal conductivity of the composite PCM, Al_2O_3, Fe_2O_3, and TiO_2 nanoparticles were added. Adding 0.5% wt Al_2O_3 increased the thermal conductivity of the composite PCM up to 43.3%. The composite PCM was encapsulated by a styrene–acrylic emulsion and dry powder of cement to have a stable shape. The concrete consisting of encapsulated composite PCM including Al_2O_3 nanoparticles not only had the best mechanical performance but also had a great ability to increase indoor temperature stability [52]. Paraffin wax, silicon carbide, and slag aggregate were combined to prepare a composite PCM. The temperature of the room decreased by 3 °C [53].

The combination of paraffin and carbon nanotubes (CNTs) was utilized as a composite PCM in the heat sink of a microchannel. The impacts of the pulsating flow of graphene oxide particles, square pin-fins in the PCM cavity, and pyramid pin-fins in a heat exchange cavity, as shown in Figure 7.13, were considered. Using nanofluids and composite PCM increased the Nusselt number by 34.9% at a 20% proportion of CNTs in comparison to a decrease of 15.2% in the Nusselt number using pulsating

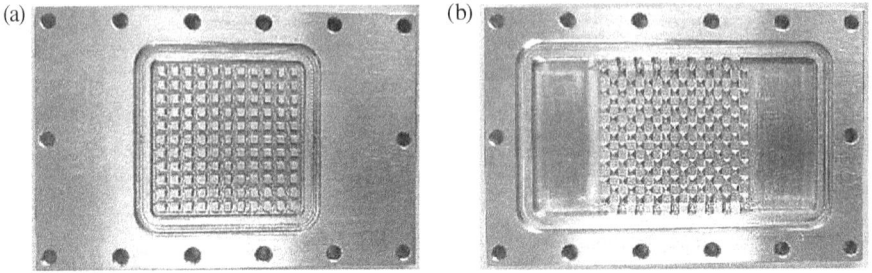

FIGURE 7.13 The schematic of (a) a PCM cavity and (b) a heat exchange cavity [54].

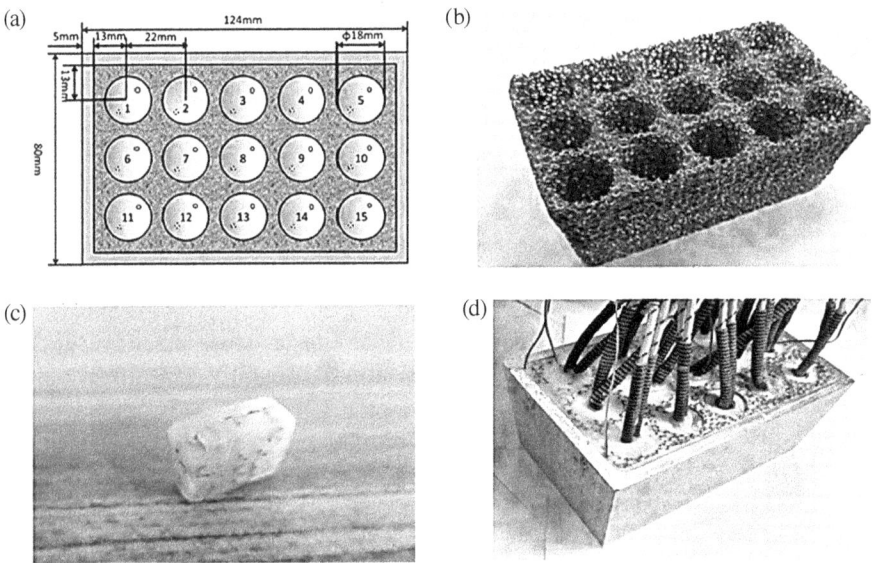

FIGURE 7.14 The schematic of the (a) battery module, (b) copper foam utilized in the battery module, (c) PCM and copper foam composite, and (d) structure of the module [55].

flow with the frequency of 6 Hz [54]. To manage the thermal process in a battery module, copper foam and organic PCM were used (Figure 7.14). Using metal foams and PCM increased the working life of the battery and decreased the temperature of the battery. The results demonstrated that to reach the highest cooling power, the current was 6–6.5 A [55].

7.3.5 ENCAPSULATION

Encapsulated forms of PCMs prevent the leakage of PCMs during the phase change stage. PCMs can be applied to concrete using different methods including macroencapsulation, micro/nanoencapsulation, and shape stabilization. The limitations of using the micro/nanoencapsulation method are costs, thermal conductivity, and mechanical strength in building materials [56,57]. In contrast, in the macroencapsulation method, PCMs are stored in tubes or spheres without any impact on the structure of buildings. Their low cost and direct usage in concrete make them applicable for practical usage [58,59]. A hollow steel ball (HSB) can be used as a carrier for an organic PCM, namely octadecane. The strength of concrete using the PCM showed an increase of 14% and 22% using 50% and 75% PCM–HSB attached with a metal clamp, respectively [60].

The capsules in an LHTES system used an organic PCM, A164, which melts at 168.7 °C. As the inlet temperature increased from 120 °C to 140 °C, the thermal efficiency dropped by 25.81%. The increase in the inlet temperature led to a decrease in the temperature difference; as a result, the extraction of energy decreased [61]. The impacts of a plain sphere encapsulated by polyethylene, a sphere including 32 pin-fins, and a sphere including 32 pin-fins covered by a copper plate layer on the heat transfer of TES were investigated. Using pins and simultaneous use of pins and copper cover decreased the phase change time by 27% and 37%, respectively [62]. A metallic capsule including 80% organic paraffin wax was applied to the air channel of a solar dryer. When the weight of PCM was 6 kg, and hot air was provided for 10 hours of drying. The cost of heating 1 kg of hot air was \$0.0074 [63]. Microencapsulation of a PCM including 54–56% wt paraffin and graphene nanoplatelets surrounded by polyurethane shell in a TES system was examined. The thermal conductivity of the PCM had an increase of 97.39% using graphene nanoplatelets. The rate of response increased by 29.6% [64]. Methyl hexadecanoate encapsulated in a polymeric shell with a melting temperature of 23–27 °C was utilized in cement mortars. The decrease in the hydration degree and the increase in macrovoids because of leaked wax led to the weakening of the mechanical performance of the PCM mortar [65]. Polyethylene balls filled with calcium chloride hexahydrate PCM were used as a TES system in buildings for cooling applications. The payback period of coal, diesel, and solar energy for heating loads of 3 kW was 3.43, 1.9, and 1.26 years. Keeping thermal comfort for a 2 kW heating load took 3.5 hours and 10 min with the thermal management system and without it, respectively [66].

7.3.6 External Fields

The heat transfer enhancement methods for PCMs can be categorized into two main groups: (1) passive methods, and (2) active methods. In passive techniques, foreign materials such as nanoparticles or fins are added to PCMs to enhance the heat transfer rate. The increase in the volume of these materials causes a decrease in the density of the thermal storage, which cannot satisfy some requirements in thermal management systems such as spacecraft [67]. Ultrasonic waves, magnetic fields, mechanical vibrations, and electrohydrodynamics are categorized as active methods.

7.3.6.1 Ultrasonic Waves

The frequency range of ultrasound is 2×10^4 to 10^7 Hz [68]. Ultrasonic waves have a wide range of effects ranging from chemical and thermal to cavitation and mechanical.

The potential of ultrasound to increase the convection heat transfer makes it an effective method to apply in PCMs. A mixture of palmitic acid and stearic acid placed in a steel container showed a reduction in the charging time of the PCMs with an increase in heating temperature [69]. Because ΔT works as a driving force in the phase change, the increase in ultrasonic power can provide an increase in the temperature difference. The heat transfer rate increased from 10.64% to 31.91% when the power increased from 60 W to 150 W [69]. Using paraffin for latent heat storage under ultrasound led to a 60.69% decrease in the charging time and a 1250.97% increase in the average heat transfer coefficient [70]. A polystyrene/n-dotriacontane composite nano-encapsulated PCM was used in a TES system. The impacts of time, power, and temperature of ultrasound treatment were investigated. The increase in ultrasound time resulted in a decrease in the z-average particle size. When the temperature increased from 30 to 55 °C, the z-average particle size decreased from 222.8 mm to 168.2 mm and the encapsulation efficiency was 61.23% [71]. The effect of ultrasonic vibration on the stability of melamine–urea–formaldehyde/paraffin nanocapsules was examined. The maximum encapsulation efficiency was obtained at ultrasonic power of 600 W and was 35.8% [72].

7.3.6.2 Magnetic Fields

A magnetic field, as a factor affecting the convective heat transfer, can be applied to PCMs, as seen in batteries and electronic devices. The convective heat transfer and properties of PCMs are influenced by the Lorentz force. PCMs with high electrical conduction, such as gallium, have the potential to improve their convective energy transport properties under the magnetic field. Nano-PCM (paraffin with Al_2O_3 nanoparticles) was used in a storage tank under the magnetic field. At a Hartmann number of 500, the dimensionless melting time increased by up to 266% [73]. The magnetic field was applied to a trapezoidal cavity including nano-PCMs. The magnetic field angle, λ, changed from −90° to 90°. The flow patterns were symmetrical at the angles of −90°, 0°, and 90°, as shown in Figure 7.15. At the magnetic field angle of 0°, the circulation of the PCM resulting from convection decreased. The increase in the magnetic field angle from 0° to 45° caused a 3% increase in heat transfer. Moreover, the melting process slowed down with the increase in the Hartmann number [74]. RT-82 was utilized in a thermal energy storage unit under a magnetic field. Strip fins and Fe_3O_4 nanoparticles were used to increase the thermal conductivity of the PCM. Melting time improved by 22% at 5% vol. nanoparticles. Moreover, the melting process showed a 39% improvement using the magnetic field. In addition, the melting time showed a 51% decrease using strip fins [75]. Composite PCM (CPCM) was employed in a battery thermal management system under a magnetic field. The amplitude and frequency of vibration were in the range of 2 mm to 4 mm and 10 Hz to 30 Hz, respectively. Graphene with a mass fraction of 0–20% was added to a PCM to improve the thermo-physical properties. The best performance was reached at the frequency of 20 Hz and the mass fraction of 20% [76]. The effect of PCM including inorganic nanoparticles in a thermal energy storage system under the magnetic field was investigated. CuO/water was added to the PCM to improve its thermal conductivity. The increase in Hartmann number from 0 to 10 decreased the solidification time up to 23.5%. Furthermore, a 14% decrease in the solidification time was observed using 4% vol. nanoparticles [77].

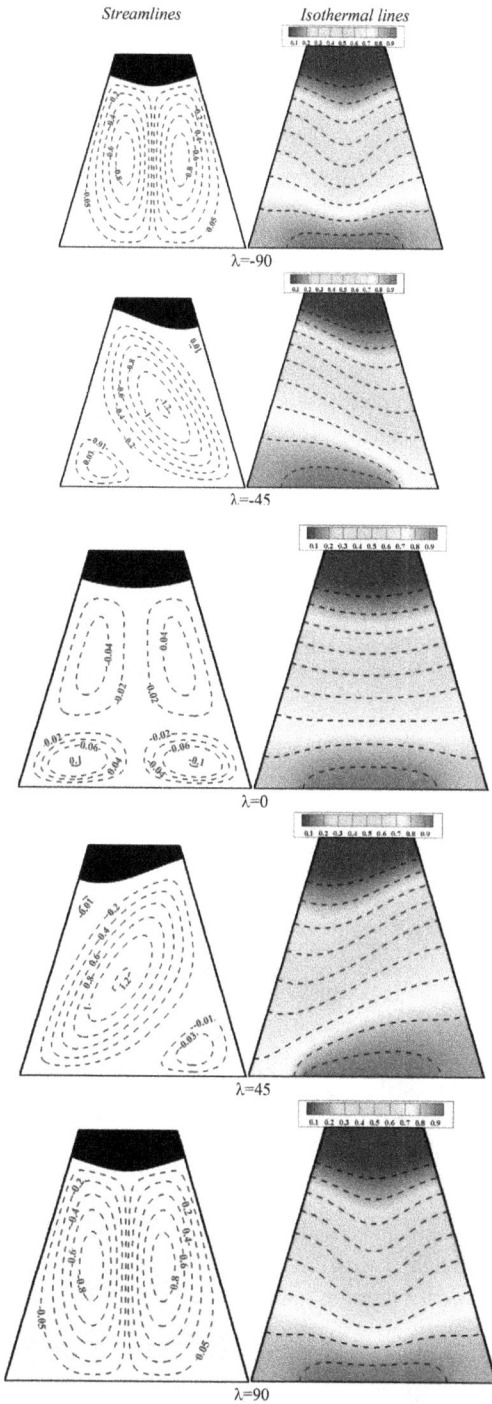

FIGURE 7.15 The streamlines at the left side and isothermal lines at the right side at the Harttman number of 100 and Rayleigh number of 3×10^5 [74].

In a conic cavity, a PCM including Al_2O_3–Cu was installed under a non-uniform magnetic field. The Reynolds number, the Hartmann number, the volume fraction of particles, the amplitude of the magnetic field, and the aspect ratio of the cavity were in the range of 100–500, 0–80, 0.001–0.02, 0.2–1, and 0.4–2, respectively. At the volume fraction of 0.02 of nanoparticles, a 42.9% decrease in the transition time was reached. At an amplitude of 1 and a Hartmann number of 80, the decrease in transition time was 15% [78]. Magnetic microcapsules made of n-hexadecane/Fe_3O_4 as the core and poly methyl methacrylate as the shell were introduced as a new PCM. The results demonstrated that this new PCM had great potential to be used in thermal energy storage systems due to its high thermal conductivity [79]. Cyclohexane–copper was applied as a nano-PCM in a porous cavity under the magnetic field. The speed of melting increased with the decrease in porosity due to easier penetration of heat via conductivity in the solid region. The conduction had a dominant impact compared to the convection heat transfer by increasing the Hartmann number. More importantly, porous media had more impact on the melting process compared to the using nanoparticles [80].

7.3.6.3 Electric Fields

To improve the melting temperature of PCMs, the electrohydrodynamic technique is a promising active heat transfer enhancement method utilized due to its fast response, easy control, and no moving parts [81]. Dielectric materials such as organic PCMs have a desirable potential to bear the electric field [82]. The melting time in an elliptical cylinder surrounded by a circular cylinder and the annulus filled with a dielectric substance was decreased by 85% [83]. The impact of electrohydrodynamic forces on the melting process of PCM, namely octadecane, was investigated. The Rayleigh number and Stefan number were 9×10^7 and 0.0445, respectively. In addition, the voltage was +3 kV and +6 kV. The increase in voltage resulted in a 1.7-fold increase in the heat transfer enhancement factor [81]. An RT-5 PCM was encapsulated in polystyrene (PS), polylactic acid (PLA), and polycaprolactone (PCL) using an electrohydrodynamic process. PCL with 43% wt RT-5 was the most efficient matrix. The room temperature was reached in 80 s and 3.5 min for PCL fibers and the PCM within the polymer fibers, respectively [84]. The melting process of n-octadecane under an electric field in a cavity was studied. The melting time decreased by 33.3% and 3.7% when the electric field with +20 KV was applied from the right side wall in the cavity without fins and with fins, respectively. It increased by 40% and 48.1% when the electric field was applied from the left side wall in the cavity without fins and with fins, respectively [82].

7.3.6.4 Mechanical Vibration

Vibration can improve the heat transfer rate as an external source. In thermal energy systems, vibration plays a critical role in heat transfer enhancement. To investigate the impact of mechanical vibration, a PCM in a tubular heat exchanger was applied under mechanical vibration. In addition, the effect of the inclination angle of the vibration axis was considered. The results demonstrated that at low frequencies of vibration in the range of 10–20 Hz, melted PCM had better mixing; as a consequence, heat transfer improved. The heat transfer reached its highest value when the inclination angle was 30° [85]. PCMs can be applied to cool the lithium-ion battery pack of electric vehicles. In electric devices, the thermal management section is under mechanical vibration. In a numerical study, the impacts of the PCM thickness and the mechanical vibration were considered simultaneously. At the PCM thickness of 7 mm, the minimum and maximum temperature of the battery module under mechanical vibration greatly decreased. At the PCM thickness of 7 mm, there was the smallest temperature difference between the minimum and maximum temperatures, but the weight of the battery increased considerably. The considered vibration frequency was in the range of 0–200 Hz. The maximum temperature had the smallest value at the vibration frequency of 50 Hz [86]. The impacts of fins and mechanical vibration on the melting process of a PCM in a cubic tank were studied. The frequency was in the range of 0–100. The ratio of fin length to the height of the tank (δ) was 1/5–4/5. The impacts of fin on the melting process appeared at $\delta \geq 1/3$. Fins with $\delta > 3/4$ were not beneficial under the vibration [87]. Paraffin was applied to a cubic heat sink under mechanical vibration. The vibration frequency was in the range of 0–1000 Hz. To make a comparison between the effects of natural convection and vibration, the Rayleigh number and vibration Grashof number were selected. The Rayleigh number of 10^4 and Grashof number of 10^6 can effectively demonstrate the impacts of mechanical vibration on transient convection heat transfer. To investigate the variation in time, the Fourier number was selected. Mechanical vibration shows its impact when enough space in the cavity is filled with PCM. The mechanical vibration had the maximum impact on the melting process improvement at low and moderate frequencies [88]. PCM (Rubitherm 35HC) in a battery thermal management system under mechanical vibration was experimentally investigated. The PCM was applied to keep the temperature of the battery in the range of 25–40 °C. PCMs absorb heat released from batteries; as a result, they can decrease the temperature of batteries. When the PCM passed its melting point, the maximum increase in the temperature of batteries was obtained. The range of frequency and amplitude were 20–30 Hz and 30–50 mm/s, respectively. The surface temperature of the battery increased with the increase in frequency and amplitude of vibration. The increase in the temperature of the batteries can lead to devastating impacts on them, including shortening their lifetime [89]. Applying mechanical vibration led to a decrease in the number of fins from 10 to 8 in a thermal management system including PCM and fins [90].

7.4 THERMAL CONDUCTIVITY ENHANCEMENT METHODS

The heat-releasing and storing rate of PCMs is low due to their low thermal conductivity. As a result, there is a basic need to study the thermal conductivity enhancement methods for PCMs. The use of fins, carbon nanomaterials, metal foams, and ceramics can help to overcome this challenge [91]. Some of these enhancement methods are presented here.

7.4.1 HIGH-THERMAL-CONDUCTIVITY MATERIALS

Nanoparticles, which are widely used to increase the thermal conductivity of PCMs, are practically metallic and carbon-based. The low density and excellent stability of carbon-based nanoparticles have made carbon-based nanoparticles superior to metallic nanoparticles [91]. The high thermal conductivity (2000–4000 W/m·K) and high specific area (~2630 m^2/g) of graphene make it an appropriate candidate for PCMs [92]. Using aluminum powder in paraffin wax decreased the charging time by about 60% [93]. The erythritol PCM is a candidate for waste heat recovery applications, and graphite and nickel particles are appropriate additives for the erythritol PCM to increase its thermal conductivity. Expanded graphite has extremely low density due to its high porosity (> 99%) [94]. Using expanded graphite (EG) as filler increased the thermal conductivity of the PCM up to 640% [95]. Graphene has a high intrinsic thermal conductivity which is more than that of CNTs [96]. Graphene nanoplatelets in water as a PCM in a spherical capsule as a TES system yielded a lower solidification time by 25% [97]. Adding functionalized graphene to paraffin wax at 1% volume led to a 20–30% reduction in the charging time [98]. Beeswax using graphene nanoplatelets was investigated as a candidate for thermal storage applications. The results demonstrated that the latent heat had a 22.5% increase [99].

The salt hydrate PCM, CaCl$_2$·6H$_2$O, had a supercooling degree of 25.5 °C. Applying hydrophilic graphene oxide sheets and SrCl$_2$·6H$_2$O resulted in reducing the supercooling degree to 0.2 °C. Additionally, the solidification enthalpy reached 207.9 J/g [100]. Reduced graphene oxide (rGO) was mixed with PCMs including paraffin, octadecanol, and stearic acid. In this method, the thermal conductivity reached as high as 3.21 W/m·K compared to 1–2 W/m·K obtained by other methods [101]. Adding graphene particles into the erythritol PCM at 1% wt. increased the thermal conductivity by 53.1%. In contrast, the latent heat enthalpy decreased by 6.1%. In addition, the melting and solidification temperatures had a 5.8% reduction and 18.76% enhancement, respectively [102]. Graphene nanoplatelets decreased the supercooling degree of water from −7 to −2.5 °C. Moreover, the solidification time decreased by 25% [97]. Graphene and silver nanoparticles were added into paraffin wax to utilize in a concentrated photovoltaic/thermal (CPV/T) system. The thermal conductivity of the PCM showed an 11% improvement. Also, the thermal efficiency of the CPV/T had a 4.16% enhancement in comparison to that of the pure PCM [103].

Cu and Al$_2$O$_3$ nanoparticles were combined with paraffin. The charging time decreased by 25.3% using 0.165% Cu compared to the 10.8% reduction using 0.165% Cu and 0.816% Al$_2$O$_3$ nanoparticles simultaneously [104]. Paraffin (RT-35) PCM was applied on the shell side of a vertical shell-and-tube heat exchanger. Cu nanoparticles were added to the PCM. When the ratio of PCM thickness to height (R) was below 0.05, the melting time in the top injection of heat transfer fluid was shorter than that in the bottom injection. The melting time decreased from 12.3% to 5.2% with an increase in R from 0.02 to 0.1 [105].

The effect of fins and nanoparticle shapes of CuO in water, which was considered a PCM, in a storage unit was studied. The maximum solidification time was for sphere-shaped nanoparticles. More importantly, a platelet shape decreased energy storage. For PCM, the discharging time was 2300 s and 1897 s using fins and adding nanoparticles, respectively [106]. In a window unit, paraffin wax PCM and CuO nanoparticles with 1% vol. and 15 nm size were applied. The overall time of melting increased from 26 min to 47 min with the increase in the size of nanoparticles from 5 nm to 25 nm compared to the overall time of solidification, which increased from 45 min to 63 min [107]. The effects of angle (38°, 45°, 52°), size, and length (2, 2.3,

2.6 cm) of V-shaped fins as well as CuO nanoparticle size (30–50 nm) in water on the thermal performance of a TES system were investigated. At the particle size of 30 nm, the length of 2.6, and the angle of 52°, the charging process was 1.29 times faster than without using nanoparticles. Solidification time increased with the increase in the angle of the fin [108]. The addition of multi-walled carbon nanotubes (MWCNTs) into paraffin PCM with 0.3 and 0.9% vol. decreased the melting time by 30% and 43%, respectively [109]. An RT-82/Al_2O_3 nano-PCM was used on the shell side of a heat exchanger. Long fins with low thickness decreased the solidification time due to more heat penetration. The increase in the volume fraction of fins showed better performance compared to the use of nanoparticles alone and the combination of nanoparticles and fins [110]. Cu/paraffin nano-PCM in an LHS system decreased the melting time up to 19.6% [111].

Metallic foams increase the performance of PCMs more than metallic nanoparticles due to lower density and larger specific area. To increase the heat transfer rate of a poor thermally conductive material, metal foams with low porosities are utilized. Meanwhile, for heat storage capacity improvement, high porosities are preferred to keep the heat storage capacity as high as possible. Using metallic foams decreased the time of charging and discharging due by increasing the thermal conductivity [91]. CuO nanoparticles and copper foams were applied to improve the performance of an RT-44HC PCM. In the melting process, the heat transfer rate increased by about 13%, 17%, and 24% for CuO/PCM, Cu metallic foam, and nanoparticle and metallic foam/PCM, respectively, in comparison to a 24%, 26%, and 65% enhancement in the solidification process [112]. Copper and nickel foams were utilized with a RT-54HC PCM. The temperature of the copper foam and nickel foam with PCM was 15% and 19% lower, respectively, than that without PCM. Copper foam with a volume fraction of 0.8 showed the best performance [113].

7.4.2 Porous Materials

The purpose of altering the surface of a porous medium is to enhance its wetting properties and compatibility with PCM, as well as to create suitable surface features for the condition in which other molecules are involved. The wetting properties of a porous medium are essential for developing shape-persistent PCMs. The material should have a balanced level of water attraction and repulsion; otherwise, it might leak or not absorb enough PCM. Therefore, the surface alteration method should produce appropriate attributes based on how well the PCM and the porous medium match. The effect of the energy of a surface and its connection with fluids are essential to be taken into account for a material selection. Any surface to reach balances reduces its contact area. The water-repelling property of the liquid makes it form spherical shapes that have the least surface area in contact. Increasing the hydrophilicity of a surface results in higher surface energy. When the surface energy is high, the liquid spreads more easily on the surface, resulting in a low angle of contact between the liquid and the surface. The molecules on the surface have a tendency to avoid contact with water, which is called hydrophobicity. The surface energy of hydrophobic surfaces is low. The water-repelling or water-attracting property of a material can be determined by measuring the angle of contact between water and the material, which is called the water contact angle test. A porous medium can be modified to be either water-repelling or water-attracting depending on the purpose of use. Building materials need to have a high degree of compatibility with PCM, which means they have to be highly wettable. For building applications, it is preferable to make the supporting material more water-attracting, which is called hydrophilic modification. To make textiles more comfortable for temperature regulation, the surface has to be modified to repel water more, which is known as increasing the hydrophobicity. The surface can resist leakage and maintain its shape by creating forces that pull or push the liquid, depending on whether the surface is more water-attracting or water-repelling. Applying a paraffin/hydrophobic expanded perlite mixture cannot only increase hydrophobicity but also enhance thermal stability. As a result, it can be an appropriate option for building applications to create leak-resistant cementitious composites [114]. Porous materials with high thermal conductivities are used to improve the thermal response of PCMs. In addition, their high porosity and large surface area per unit volume result in being an appropriate option to improve the thermal performance of different systems. A copper foam with a porosity of 0.7 and 0.9, and RT-35 as the PCM were utilized on the shell side of a heat exchanger. Melting time had a decrease of 14% and 55% for porosities of 0.9 and 0.7, respectively. The efficiency of the melting process was more than that of the solidification one [115]. Paraffin PCM and aluminum foam with porosities of 0.88 and 0.95 were used in a TES unit. A metallic foam with a higher pore per inch (PPI) value could store more energy. A lower porosity speeds up the melting process [116]. In a numerical study, porous materials made of copper, aluminum, nickel, and graphite were applied to a PCM composite. In addition, the impacts of nanomaterials and fins were considered. A helical heat exchanger including ethylene glycol as a heat transfer fluid was used in a cold LTES system (Figure 7.16). CuO

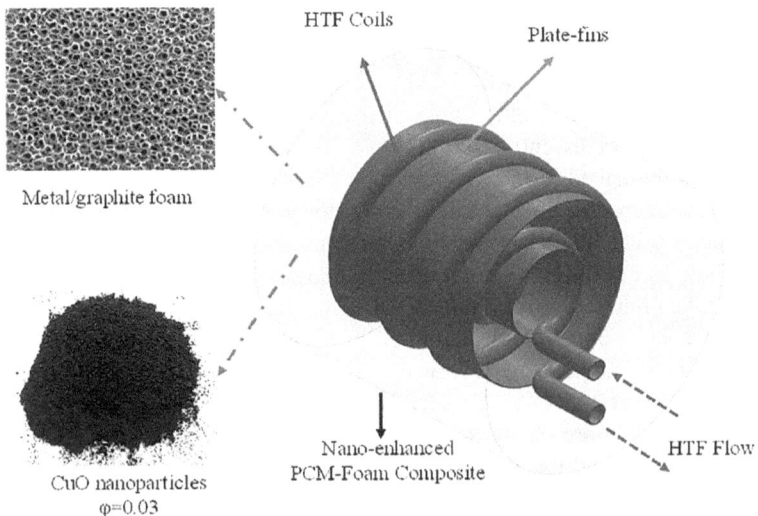

FIGURE 7.16   The schematic of a composite latent heat storage system [118].

nanomaterial was utilized to improve the thermal conductivity of the PCM. Ethylene glycol passed through the coils and exited from the enclosure after flowing through the heat exchanger. PCM changed to solid when ethylene glycol absorbed heat from it. The progress in the solidification process had an increase of 39.66%, 52.06%, 52.23%, and 66.77% for nickel, aluminum, graphite, and copper porous material in comparison to nano-PCM without porous materials [117].

A copper foam with a porosity of 95% in combination with RT-82 PCM was applied in an air heat exchanger. RT-82 had great potential to use in domestic heaters due to its appropriate melting point. The discharge time had a decrease of 56%; in contrast, heat recovery had an increase of 64% for the system with air channel dimensions of 15 cm×30 cm×2 cm [119].

7.5 COMBINED PERFORMANCE ENHANCEMENT OF PCMS

In combined enhancement methods of PCMs, a combination of the thermal conductivity enhancement and heat transfer improvement methods mentioned earlier are used. The combination of heat pipes (HPs), fins, and copper foams (CFs) in a PCM caused the melting and solidification time for HP–fin, HP–CF, and HP–Fin–CF to decrease by 82.7, 89.03, and 93.34%, respectively [120]. The combination of using 4, 8, and 16 radial fins with rotational speeds of 0.1, 0.5, and 1 rpm were applied in an LHTES system. N-eicosane PCM with a melting temperature of 309 K was used. Using 16 fins at rotational speeds of 0.1, 0.5, and 1 rpm, melting time decreased by 24.44%, 51.41%, and 63.43% compared to that with zero rotation, respectively. The heat transfer ratio using 16 fins was 4.13 times more than that without fins during the charging process in comparison to the discharging process, in which it was 4.9 times greater. It should be noted that the natural convection decreases by increasing the number of fins. In the discharge process, increasing the number of fins had a negative effect on the impact of rotational speed [121].

The impacts of using three methods including multi-tubes, Cu nanoparticles, and a nickel–steel porous matrix on the melting of ice were considered. On the upper side of the closure, the natural convection plays a critical role, while on the bottom side of that, the conduction is dominant. Using nanoparticles could have a negative impact on natural convection due to increased viscosity. The decrease in melting time changed from 29% to 34% with a reduction in porosity from 1 to 0.95 and 0.9, respectively. The melting time decreased by 50%, 17.3%, and 39.2% for cases (I), (II), and (III) when the volume concentration of nanoparticles increased from 0% to 2% [122].

7.6 APPLICATIONS OF PCMS

Following the aforementioned studies, PCMs have great potential and a wide range of applications in thermal systems. Today, PCMs are used in different applications, and some of them are described herein with a look at their advantages and disadvantages.

7.6.1 BUILDINGS

Buildings contribute to 30% of the energy demand in the world in 2022. Buildings have the most astounding capacity to slash energy intensity by a staggering 38% that would result in a phenomenal 12% reduction in the world's energy demand, if the full potential of these savings were realized [123]. To increase the energy efficiency of buildings, the thermal properties of materials used in the construction of buildings should be improved. Among them, concrete is widely used in the building industry. To improve its heat storage capacity, PCMs can be utilized [124]. PCMs can not only reduce temperature fluctuations but also decrease thermal loads [125]. To apply PCMs in concrete, they should have high thermal conductivity, high specific heat capacity, no toxicity, no flammability, small volume change, and a comparable price [126]. Moreover, the freezing and melting points of PCMs are selected based on their applications. Temperature ranges for cooling, human comfort, and hot water applications are 21 °C, 22 °C to 28 °C, and 29 °C to 60 °C, respectively [127].

PCMs in combination with concrete can be categorized into two main groups including organic (paraffin and non-paraffin) and inorganic PCMs, especially salt hydrates [128]. Stability, favorable melting point, low cost, and high heat capacity make organic paraffin wax an appropriate PCM to be used in concrete [129]. The disadvantages of paraffin are flammability, low thermal conductivity, and high volume changes. Non-paraffin PCMs such as bio-based fatty acids could provide non-flammable characteristics and also be recycled easily, but they are more expensive than paraffin PCMs [130].

PCMs can be used in combination with concrete in two ways: (i) direct methods, and (ii) indirect methods. In direct methods, PCMs are directly in contact with the concrete. Liquid PCMs can be added to the concrete mix. In addition, concrete is immersed in a tank including the liquid PCM. The PCM penetrates the concrete through capillary action [131]. The disadvantages of direct methods, including the risk of leakage and changes in the properties of PCMs in contact with the surrounding matrix, cause limitations [132]. In indirect methods, encapsulated PCMs or lightweight aggregates including PCMs (LWA-PCM) are used [133]. The weight of microencapsulated PCM in concrete should be less than 6% of the concrete weight [134].

In addition, PCMs can be used in floor heating systems to increase the heat storage capacity [135], like that of PCM Products Ltd. [136]. Cooling ceilings are utilized to cool spaces through cold water passes pipes in the ceilings. The companies Datum Phase Change Ltd. [137] and PCMTechnology sell these systems [138]. Moreover, ice banks as cold storage can be coupled with an air conditioning system for efficient cooling of space. Christopia Company, BAC Baltimore Aircoil Company, and Calmac Company provide such ice storage systems in which water is the main PCM [139]. Encapsulated PCMs can also be used for heating applications, as we see a growing industry with successful examples around the worldwide [140]. PCMs were employed in the ceiling and the walls of a house. In addition, the effect of using Al_2O_3/water nanofluid in a solar collector coupled with the house was examined. Using PCMs in the structure of the building saved energy of up to 620–6600 kWh/year in warm weather. Meanwhile, using nanofluid and the solar collector saved energy of up to 3400–4500 kWh/year [141].

7.6.2 Cooling of Electronic Devices

The development of microelectronic devices required to meet computing demands is accompanied by a thermal development to decrease the size of electronic chips

[142]. A decrease of 1 °C in the temperature of the electronic chips reduces the rate of failure of the electronic chips by about 4% [143]. Paraffin, fatty acids, and alcohols are widely used to cool the electronic chips. Paraffin wax has high latent heat in the phase change process, is chemically stable, and has low vapor pressure during melting. More importantly, paraffin waxes contain paraffin with different melting points. They have the potential to be used at different melting points in combination with other materials [144]. Fatty acids including palmitic acid [145], lauric acid [146], myristic acid [147], and stearic acid [148] can be utilized in the thermal management of electronic chips due to their high latent heat during the phase change and their appropriate melting temperature. The use of encapsulated PCMs draws considerable attention due to solving leakage, increasing the specific surface area, and overcoming corrosion and phase separation [149]. Nanomaterials such as carbon nanotubes [150], metal nanoparticles [151], and graphene [152] in PCMs increase the thermal conductivity, cyclic thermal stability, and efficiency of the storage system. Organic PCMs have relatively low to moderate phase change temperatures. In addition, they can be combined to have an engineered melting point.

Adding 1% wt SiO_2 to the paraffin improved the thermal management of the heat sink up to 220% compared to the PCM [153]. Combinations of four organic PCMs lead to a 0.3 °C supercooling degree [154]. Adding inorganic materials such as graphite [155] to PCMs can improve their thermal conductivity and be used in electronic chips. In addition, using metal foams can improve the thermal conductivity of organic PCMs [156].

Simultaneous use of air and PCM (Figure 7.17) in a heat sink was investigated to find the amount of decrease in the temperature of the electronic device. The melting time decreased from 649.7 s to 387.1 s when the PCM-based electronic device was replaced with the combination of air and PCM. Furthermore, the heat transfer coefficient enhancement caused an increase in heat loss between the device and the environment, especially in the air and PCM-based electronic device; as a consequence, the melting time increased [157]. The use of 10% and 20% volume fractions of PCMs in a plate-fin heat sink with fin heights of 10, 15, and 20 mm improved heat

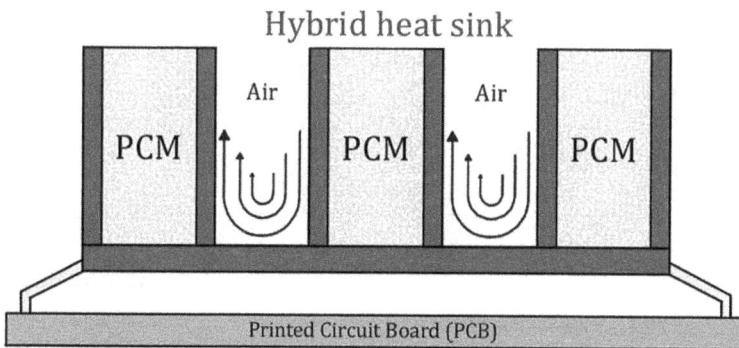

FIGURE 7.17 The combination of PCM and air in a heat sink [157].

transfer. The maximum decrease in the temperature of the heat sink was observed at a 20% volume fraction of PCM with a fin height of 20 mm [158].

Paraffin wax was applied to the printed circuit board (PCB) of a mobile phone and showed a significant drop of 48 °C in temperature, which is lower than 55 °C for the effective working of electronic elements suggested by the manufacturer [159]. Different heat fluxes, weight fractions of nanoparticles, and heat sinks were examined to reach the best thermal performance for an electronic device. A decrease of 21.3%, 25.03%, and 36.2% in the temperature of simple, circular pin-finned, and Cu foam heat sinks were obtained at a heat flux of 0.98 kW/m^2 and 0.25% wt Al$_2$O$_3$ nanoparticles [160]. In a heat sink including triangular aluminum fins (Figure 7.18), RT-58, RT-44, and n-eicosane were selected as PCMs. RT-44, with the highest latent heat, had the lowest peak temperature. The peak temperature showed a decrease of 6.7–12% using both RT-44 and n-eicosane compared to only n-eicosane and a decrease of 3.3–7.7% compared to only RT-44. Using the combinations of PCMs led to an increase in the slope of the latent heating and cooling stages. The shortest time for the melting and solidification processes belonged to RT-58 due to its lowest values of enthalpy [161].

7.6.3 SOLAR ENERGY SYSTEMS

Solar energy is not necessarily available when is required. So, LHTS systems are used in these systems to overcome this basic challenge. In addition, PCM can

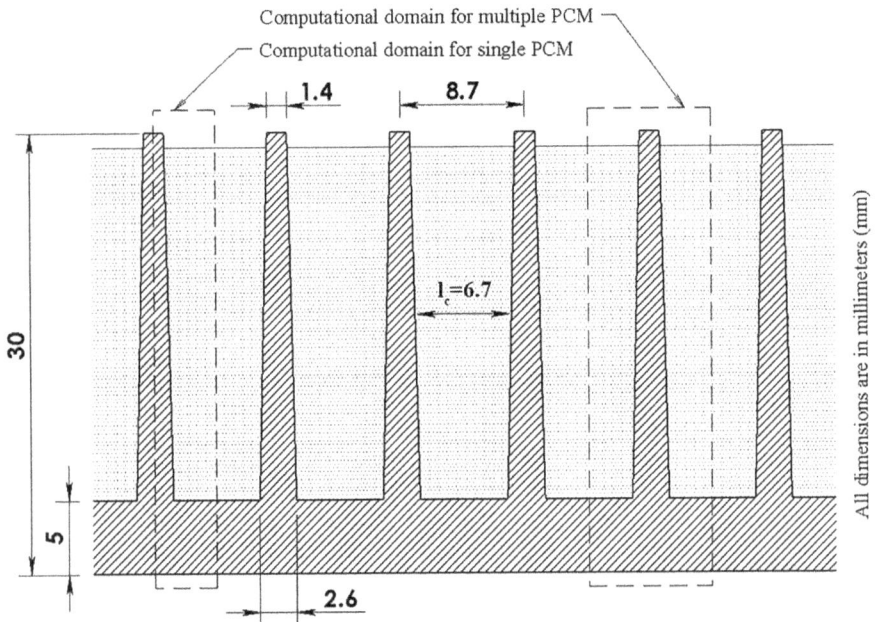

FIGURE 7.18   The schematic diagram of a triangular heat sink improved by a PCM [161].

improve the electrical performance of photovoltaic (PV) systems by controlling the peak temperature [162].

7.6.3.1 Desalination Units

To meet water demands, desalination units, in which salts are removed from the water, are extensively applied. Desalination techniques include solar stills, humidification–dehumidification, vapor compression distillation, and multi-stage flash distillation. In addition, PCMs can be applied in solar stills as latent heat storage, as they have a high capacity to store heat that is 5–14 times more than that in sensible heat storage [163]. The flexibility, low cost, and fewer needs for maintenance in the humidification–dehumidification (HDH) process make them an appropriate option to produce fresh water [164]. Using PCMs under the absorber plate of a solar air collector integrated with a HDH system can provide warm air consistently [165]. One of the important issues of using PCMs is their low thermal conductivity. Adding nanomaterials including SiO_2, Al_2O_3, and CuO increased the thermal conductivity by up to 71% [166]. Among nanoparticles added to PCMs to increase the productivity of solar stills, graphite, and graphene oxide nanoparticles showed the maximum enhancement in water production, which was 13.62 L/m^2/day and 1.55 L/m^2/h, respectively [167].

The results of some studies in the field of using PCMs in desalination systems and applying methods to improve their properties are as follows. To increase the freshwater yield of solar stills, LHES is utilized, providing higher temperature stability. Paraffin wax has great potential to use as a material for storing energy in LHES systems [168]. Furthermore, the combination of a PCM and pin-fins to increase its thermal conductivity was applied in a pyramid solar distiller in which the cumulative yield increased from 4085 and 4171 mLm^2/day to 9885–10015 mL/m^2/day [169]. A multi-stage desalination system comprised many evaporators, an ejector, and a condenser to convert water vapor into liquid. Flat plate collectors and a stearic acid PCM as energy storage systems were integrated into the multi-stage desalination system. The rate of exergy destruction had its maximum value of 344.05 kW for flat plate collectors. In contrast, it was 128.5 kW for PCM units. Using PCM resulted in increasing the storage capacity of the energy storage system [170]. When integrating a PCM (namely paraffin wax) storage tank to a solar slope still with a basin area of 1 m^2, PCMs weighed 17 kg and had a melting point of 56 °C. As a result, the annual energy increased by 10% compared to the solar still without PCMs. But using PCMs caused an enhancement in CO_2 by more than 400%; consequently, they did not have the potential to apply in this system [171].

The impact of utilizing paraffin wax weighing 17 kg in a water tank with a capacity of 100 L on the performance of a double solar still with a basin area of 0.5 m^2 was investigated. To improve the thermal conductivity of PCMs, copper chips and nanomaterials were applied. Using the combination of CuO and PCM led to water production and energy efficiency increasing by 113% and 112.5%, respectively. In contrast, the cost of water production had a decrease of 35.3% [172]. In a solar slope still with a basin area of 1 m^2, the effect of melting temperature of PCMs on the water production was considered. The higher the melting temperature is, the greater the water production is because increasing the melting temperature of PCMs allows water to be warmer. As a result, the vapor pressure and productivity increased. The rise in solar radiation from 400 to 1000 W/m² increased daily water production by 1.35 L [173]. The aqueous solution of NaCl in combination with microencapsulated PCM was applied as the working fluid in a spray flash evaporation (SFE) system and in a thermal energy storage. The SFE system included a condensation cycle, solar collectors, and a storage process. Using PCMs, water production had an increase of 23.1%, while energy consumption had a decrease of 18.3% [174]. Two parabolic trough collectors (PTCs) were linked to a solar slope still. Water, oil, and CuO/oil were used as working fluids in the PTCs. In addition, paraffin was used as a PCM in the evacuated absorber tube of the PTCs, as well as in solar still with a basin area of 1 m². The water productivity was 3.182, 6.21, and 11.14 L/m²/day for the conventional solar still, the solar still using PCM and oil, and the solar still using PCM and nano-oil, respectively [175]. In a pyramid solar still with a basin area of 1 m², the impacts of water depth (10–40 mm), the height of fins (20, 40, and 60 mm), and paraffin wax including fins were considered. The optimum height of fins and the depth of water were 40 mm and 10 mm, respectively. Using encapsulated PCM increased the overall efficiency from 34.6% to 49.9% as well as water production by 44.4%. in addition, the cost of water production was $0.043/L compared to $0.047/L for conventional stills [176].

7.6.3.2 Photovoltaic Systems

Photovoltaic (PV) systems that convert solar energy to electrical energy are growing increasingly. But PV systems change only 17–19% of solar energy to electricity, while the rest of the energy is lost as heat to the environment. As a result, applying cooling processes plays a crucial role in improving their efficiency. Among cooling methods, utilizing PCMs can be effective to cool and store thermal energy [177]. PCMs are usually applied under the PV panel to absorb heat [178]. To improve the thermal conductivity of PCMs, many studies on using nanoparticles, fins, metal foams, mixing PCMs, and surface modification of the channel in PV/T modules have been performed.

To increase the thermal conductivity of PCMs applied under a PV system, aluminum fins were applied. The temperature of PV cells decreased with the increase in the length and thickness of fins due to absorbing more heat from the PV system [179]. To find the appropriate range of melting temperature for applying PCMs in PV systems, paraffin wax, Vaseline petroleum jelly, and mixtures of them were analyzed. The melting temperature of paraffin wax, Vaseline petroleum jelly, and the mixture with 75%, 50%, and 25% paraffin were 45, 25, 41, 37, and 32 °C, respectively. PCMs were applied under the PV cells surrounding pipes including water to cool the PV module. The decrease in melting temperature of the PCMs (in the case of Vaseline) caused a reduction in the temperature difference between PCMs and the PV system; as a result, the performance of the PV system decreased. In the case of using paraffin, the melting point was high. But the solidification process was done quickly [180]. Using 3% vol. Ag nanoparticles in water as a working fluid in wavy tubes applied to cool a PV module, the energy absorption from the PV panel increased from 249.29 W to 251.27 W. Using nanofluids led to increasing the thickness of the boundary layer close to the tube wall; as a consequence, the temperature gradient increased. Furthermore, using microencapsulated PCM (MPCM) slurry including PCM, namely hydrocarbon n-eicosane, surrounded by a TiO_2 shell with a volume fraction of 5% increased the electrical efficiency from 11.19% (water) to 11.25% due to increasing heat transfer. Mixing Ag nanoparticles at 9% and MPCM at 15% in water increased the thermal conductivity and heat capacity by 156% and 323%, respectively. The plate temperature was about 313 °C, 311.7 °C, 311.3 °C, and 310.5 °C for water, Ag/water, MCPCM 15%, and nano MPCM, respectively [181]. Applying ZnO/water and PCM, namely paraffin wax, increased the thermal output of a PV/T system by about 9% in comparison to about a 5% increase for the PV/T system using only nanofluids [182]. Utilizing paraffin wax PCMs and fins on the back side of the PV/T system that surrounds copper risers containing water, the temperature of the PV module decreased by 53% [183]. Using SiC/water as a working fluid in risers that were in contact with SiC–PCM under the PV cells decreased the temperature of the PV/T system from 68.3 °C to 39 °C in comparison to the conventional PV/T system [184]. More importantly, the impact of adding graphene and CuO nanoparticles to the PCM, namely polyethylene glycol 1500, applied under the PV module was investigated. The results demonstrated that the combination of graphene and CuO 3% wt caused the maximum decrease in the PV temperature up to 6.6 °C. In addition, the thermal conductivity enhancement was 91.3%. The thermal conductivity enhancement enhances heat transfer, while the increase in viscosity decreases natural convection. The combination of CuO and graphene not only leads to a lower viscosity compared to graphene but can also increase thermal conductivity due to separating graphene layers and forming a wide network of conduction [185].

Using a combination of different PCMs can improve their thermo-physical properties. For instance, different arrangements of RT-26, RT-35, and RT-42 PCMs were applied under the Tedlar in a photovoltaic (PV) system. The arrangements of PCMs include a PCM with the melting temperature of 308 K, a combination of PCMs with melting temperatures of 299 K and 316 K, and a combination of PCMs with melting temperatures of 299 K, 308 K, and 316 K. Using three PCMs increased the melting time compared to using two PCMs. In addition, the melting time increased with the decrease in the inclination angle of the PV system. The melting time increased using m-PCMs [186].

Surface modification methods comprise corrugating, waving, and dimpling surfaces and are among the heat transfer enhancement methods. An aluminum dimpled surface in contact with RT-35 or RT-35–RT-26 with a thickness of 20 mm, 10–10 mm, and 15–5 mm was applied on the bottom of PV cells. Using dimples resulted in increasing the heat transfer surface. Moreover, a vortex flow was generated in the dimpled region, which increased the rate of PCM melting. The increase in the melting temperature of PCMs increased the heat absorption in the PV panel and decreased its temperature. The melting time of RT-35–RT-26 with a thickness of 10–10 mm and 15–5 mm increased by 5 min and 10 min compared to that of RT-35 [187].

Utilizing microencapsulated PCMs including liquid paraffin for encapsulation and a polymer as a shell in a photovoltaic thermal system (PV/T) increased the electrical and thermal efficiency of the PV/T by 0.8% and 13.5%, respectively [188].

OM35 PCM was utilized at the surface of the Tedlar in a PV system. The average electrical efficiency of the PV system using the PCM increased by 12.28%. Water circulation from top to bottom decreased the top surface temperature of the PV panel by 12%. The combination of PCM and copper foam was employed in a PV system. The porosity of the copper foam was 85%, 90%, and 95%. The melting process was affected by thermal conduction and natural convection at different inclination angles of the PV panel. Temperature changes had a negligible impact on the melting process at high porosity. At a lower value of porosity, the inclination angle had small impacts on the charging process due to the weak natural convection. The time of cooling at porosity values of 85% and 90% was two times more than that at 95% [159]. Paraffin PCMs (RT-27) with melting temperatures of 25 °C and 28 °C were chosen to apply to a PV system. The yearly power generation increased by 5.7% [189]. A paraffin PCM (OP35E) in a PV system was employed. OP35E was melted and added to the micelle solution. OP35E nanoemulsions with 10% wt and 20% wt were prepared. The PV panel temperature decreased by 5.3% for 20% wt for OP35E nanoemulsions, and the output power had an 18.4% increase [190]. Four organic PCMs (RT-44HC, RT-47, RT-50, and RT-55) with a melting temperature range of 42–50 °C were employed in a PV system. RT-44HC was the best option due to the high capacity of heat storage. MWCNT was added to the PCM to increase its thermal conductivity. The overall efficiency of nano-PCM PV/T, PCM PV/T, and PV/T was 85.3%, 82.6%, and 75.1%, respectively [191]. Al_2O_3 nanoparticles were added to the Rubitherm (RT) series of organic PCM. In addition, the working fluid in a PV system was Al_2O_3/water. The mass flow rate had the maximum impact on the thermal and electrical power. The minimum entropy generation was obtained for the volume fraction of 8.05% and 8% for nano-PCM and nanofluid [192]. Applying ribbed plates and MPCMs (RT-35 and RT-27) decreased the temperature of the PV system by 13 °C. Furthermore, the electrical efficiency showed an increase of 1.18% in comparison to that of a conventional PV system [193].

7.6.3.3 Solar Collectors

PCMs are used in integration with solar collectors to conserve excess energy and absorb heat from PV thermal (PV/T) collectors. Their temperature ranges for use in flat plate collectors, evacuated tube solar collectors, PV solar collectors, PV/T solar collectors, solar cookers, and parabolic trough collectors (PTCs) are 50–60 °C, 50–90 °C, 30–40 °C, 40–50 °C, 70–150 °C, and 120–220 °C, respectively [194]. PCMs can be utilized in flat plate collectors under the absorber plate, between the insulation layer and risers, as an anti-freeze in thermal energy storage units [195,196]. In addition, they can be used in absorber tubes of evacuated tube solar collectors and thermal energy storage tanks of evacuated tube solar collectors [197] and PTCs. The thermal conductivity of common PCMs is low, which causes poor performance of the latent thermal energy systems. This performance can be improved using several techniques, especially thermal conductivity improvement methods. For this aim, helical fins were applied around the heat pipes in an evacuated tube collector filled with paraffin wax. During the discharge process, efficiency increased by 15.13% at a flow rate of 0.5 L/min. The conventional fins had more impact on performance enhancement because helical fins prevented free convection in the liquid layers of the PCM [58].

An evacuated tube collector including a PCM storage tank is shown in Figure 7.19. Daily energy consumption decreased by 13.5% using PCM in the top part of the tank [198]. Sodium acetate trihydrate (SAT) PCM was applied in a storage system. Hydrogen phosphate dodecahydrate and expanded graphite were added to the SAT PCM. The PCM tank in series with the collector had an increase of 5–12% in the solar fraction compared to that in parallel with the collector due to decreasing the inlet temperature of the working fluid entered the collector [199]. The use of copper fins in an evacuated tube collector had a significant influence on heat transfer enhancement. To reach the best performance, the thickness and spaces between fins should be small. Adding 2% Cu to PCM (paraffin wax) had no impact on the storage process [200]. In a solar air collector using PCMs, the proper range of phase change in the winter season and transition season to reach better thermal performance was 22–24 °C and 19–21 °C [201]. Paraffin was applied to a storage tank integrated into the solar collectors. The energy consumption was reduced by 44.16% daily [202]. The performance of the Rankine cycle integrated into parabolic trough collectors and a PCM storage tank was analyzed. Six PCMs including isomalt, adipic acid, dimethylol propionic acid, KNO_3–$NaNO_2$–$NaNO_3$, Salicylic acid, and A164 were selected. KNO_3–$NaNO_2$–$NaNO_3$ composite PCM had the maximum energy storage due to low melting temperature. The maximum solar fraction was 89.61% for A164. In addition, the highest thermal and exergy efficiencies were obtained for R245fa/R500. From the aspects of energy, exergy, and economics, the combination of R245fa/R500 with dimethylol propionic acid was the best option [203]. PCMs were used in a storage tank integrated into a solar collector (see Figure 7.20). The thermal efficiency of the collector increased by 19% [204]. The heat pipe evacuated tube solar collector used tritriacontane

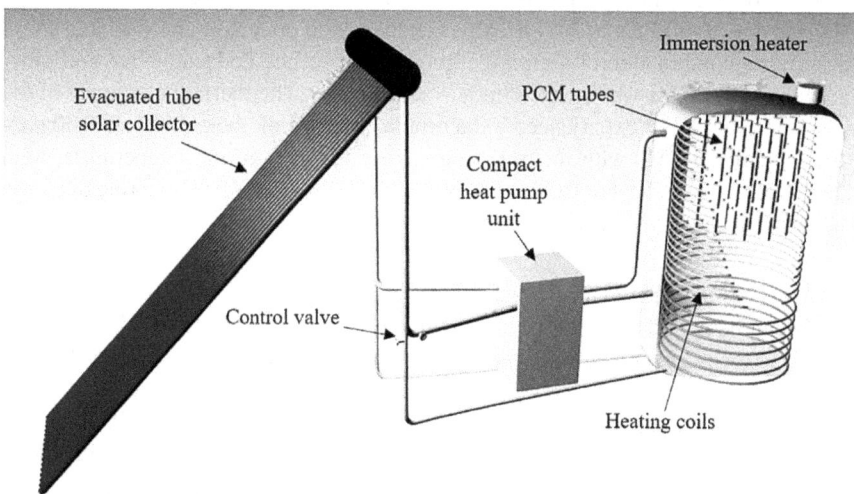

FIGURE 7.19   An evacuated tube collector integrated with a PCM storage tank [198].

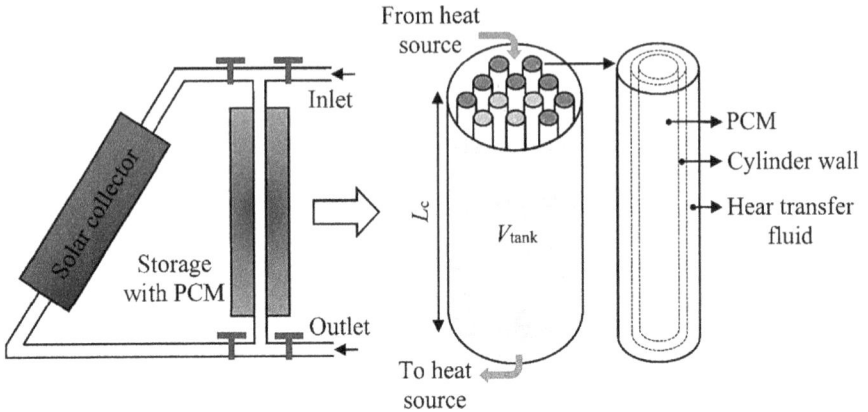

FIGURE 7.20 A solar collector integrated with a PCM storage tank [204].

paraffin and copper porous metal as a TES system. The collector's maximum daily efficiency rose from 36.91% to 85.64%. Furthermore, the cost of hot water production was reduced by 11.57% and 9.43% at lifetime of 5 and 10 years [205]. A hybrid material of polyaniline (PANI) and MXene used to form a porous aerogel that encapsulated polyethylene glycol (PEG) as the PCM with a high light absorption, a high heat storage capacity, and a high cycling stability was a promising material for solar energy utilization [206].

Among PCMs, organic PCMs are the best option due to their working temperature ranges between 46 °C and 60 °C. In addition, in major studies of PCMs, paraffin has been applied [207]. For instance, paraffin wax with a melting temperature of 49 °C was applied under an absorber plate of a flat plate collector with a volume of 1 m ×0.81 m×0.03 m. The thermal efficiency of the collector using PCM increased from 29% to 41% at 4:00 p.m. [208]. In an evacuated tube collector, encapsulated PCM including a combination of aluminum ammonium sulfate dodecahydrate and other inorganic salt PCMs was utilized in the vacuum tube. The thermal efficiency of the collector using PCM experienced a maximum increase of about 150% at 6:00 p.m. [209]. Finally, PCMs with melting points close to the working temperature, high density, large latent heat of fusion, and high specific heat play a crucial role to obtain an energy storage system with high capacity integrated with solar collectors [3].

7.6.3.4 Solar Dryers

PCMs can be applied in energy storage units that are integrated into solar dryers to overcome the mismatch between solar energy extraction and supply [210]. Solar dryers operate at temperatures between 45 °C and 75 °C; as a consequence, the melting temperature of PCMs should be under 80 °C [211]. PCM selection is a complex and challenging task, as it involves multiple criteria that may be conflicting or uncertain. The VIKOR method, which is a MCDM tool that can handle multiple criteria and provide a compromise that is closest to the ideal solution and farthest from the negative ideal solution can help to choose the best PCM. Using this method leads to choosing paraffin wax as the best option for solar dryers [212]. So, paraffin wax is extensively used in drying applications [213]. Al_2O_3 paraffin wax was applied in a concentrated PV/T dryer. The overall thermal efficiency and the exergy efficiency were 20% and 8%, respectively [214]. In a solar potato dryer, paraffin was utilized. The moisture of potatoes decreased from 81% in the case of the conventional dryer to 13.3% in that using PCMs [215]. At the bottom of the drying enclosure of a solar dryer for drying chili peppers, paraffin wax was utilized. The dehydrating time decreased by 86% [216]. To dry medicinal herbs, an air collector was coupled with a sensible heat storage system and a drying unit including paraffin RT-42. Using the PCM decreased the time of drying from 216 hours to 120 hours [217]. A PCM flat plate collector was applied for drying. The air passed through the heat exchanger pipes presented in Figure 7.21, which were under the ground. The thermal efficiency of the PCM flat plate collector was 20.5% more than the conventional one [218]. Al_2O_3 (4% vol.) nanofluids, water, glycerin, and engine oil were used as a working fluid passed through a parabolic trough collector. Then, heat was transferred between the working fluid and air using a heat exchanger. The warm air was utilized in a dryer. A PCM was applied in the heat storage tank. The maximum enhancement of the overall efficiency of the dryer was 20.2% for oil [219]. Paraffin wax with a melting temperature of 57 °C was utilized in thermal storage coupled with a solar dryer. Fans had the maximum exergy destruction cost of $0.2286/h. In addition, the payback period of energy was 6.82 years [220]. A PCM, namely paraffin wax including Al_2O_3 nanoparticles, was used in tubes under the chamber of a solar dryer to dry kiwi. The thermal energy absorption increased from 3393 kJ to 6109 kJ for PCMs with 0% and 0.5% wt nanoparticles, respectively [221]. In a solar dryer comprising a solar air heater and a thermal energy storage system to dry mushrooms, 14 kg of paraffin wax was applied. The electrical energy was saved by about 40–70% [222].

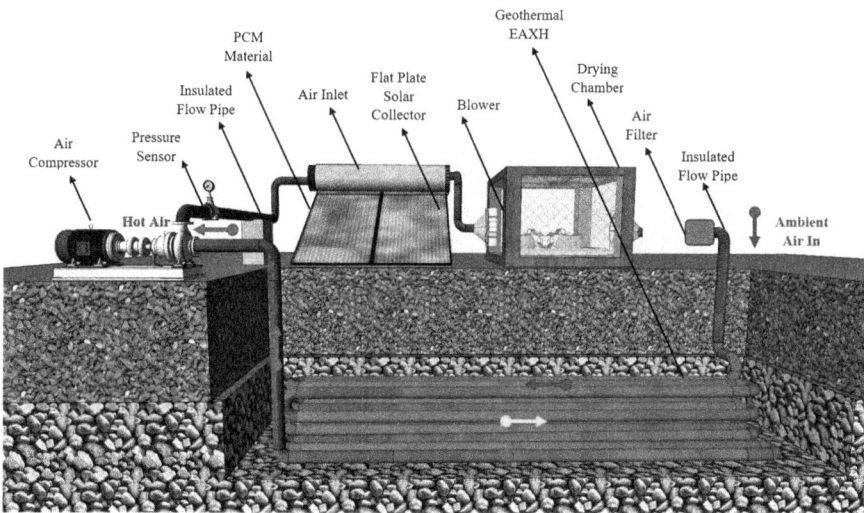

FIGURE 7.21   A PCM flat plate collector [218].

Drying bitter gourd slices with an indirect solar dryer setup that had fins inserted in a PCM energy storage unit was the most efficient and economical choice, as it reduced the drying time from 15 h to 11 h and improved the energy performance from 17.3% to 19.6% [223]. In addition, a novel design of a solar dryer that used a photovoltaic/thermal (PV/T) unit, a V-corrugated collector, a nano-PCM packed compartment, and a drying chamber resulted in increasing the rehydration capacity from 68.57% to 74.06% [224]. By mixing paraffin with n-docosane and kerosene in a 2:1 ratio, the solar dryers achieved a 50% increase in their thermal efficiency. More importantly, the melting temperature of paraffin rose from 38 °C to 75.9 °C and the latent heat varied between 205 kJ/kg and 269 kJ/kg as the number of carbon atoms increased from 15 to 34 [225].

7.7 CONCLUSIONS AND FUTURE RESEARCH DIRECTION

Phase change materials (PCMs) play a critical role as thermal storage and management media because of their high capacity for releasing and absorbing heat. PCMs can improve the efficiency of cooling and heating systems by decreasing their sizes and costs. But the thermal conductivity of PCMs is low; as a result, many studies have been conducted to improve it. In this chapter, methods of improving the thermophysical properties of PCMs were categorized into two main groups, including heat transfer enhancement methods and thermal conductivity enhancement methods. In addition, based on their characteristics, especially the melting temperature, they have wide and different applications. Highlights are as follows:

1. PCMs can not only reduce the temperature fluctuations but also decrease the thermal load. As a result, they can be applied to concrete. Moreover, the freezing and melting points of PCMs are selected based on their applications. Temperature ranges for cooling, human comfort, and hot water applications are 21 °C, 22 °C to 28 °C, and 29 °C to 60 °C, respectively.
2. To improve the thermal conductivity of PCMs, many techniques including nanoparticles, fins, metal foams, mixing PCMs, and surface modification of the channel in PV/T modules have been applied.
3. Graphite foils, aluminum, stainless steel, carbon steel, and copper are widely used as extended surfaces. Among extended surfaces, fins are widely used due to their efficiency and simplicity.
4. Inorganic PCMs have high thermal conductivity, high volumetric latent heat density, and stability in cycles.
5. The choice of PCMs depends on the aim and the requirements of the considered application. The materials which are applied in LHTES should have high latent heat in the phase change process, high thermal conductivity, high density, low vapor pressure, small volume changes, and no toxicity, and they should be non-flammable, non-corrosive, and cost effective.
6. The latent heat of fusion, mass fraction, and Stefan number are important parameters that show the performance of PCMs in heat transfer systems. When the Stefan number is below 1, PCMs have optimal impacts.

7. The volume changes in PCMs during the melting process should be considered in the design of their containers. Their flammability and toxicity should be checked for domestic applications.
8. Salt hydrates have high thermal conductivity and low costs, but their chemical and thermal stabilities are challenging, and controlling actions are required.
9. Paraffin has a low capacity for thermal energy storage. In addition, a large space is needed to contain paraffin due to its low density and thermal conductivity.
10. The increase in the number of fins and their length has a significant impact on the solidification process than the melting process.
11. The fractions of PCMs and their melting temperatures are crucial parameters in multiple PCMs.
12. The combination of RT-20 and RT-25 showed good potential for air conditioning units.
13. A biochar–PCM was applied in an LTES. The high thermal conductivity and porosity of biochar due to its high carbon content make it an appropriate option to use in PCM. When the ratio of PCM to the ratio of biochar was 6:4 (wt/wt %) it was the best mixing ratio.
14. Bio-based PCMs including cellulose fibers, clay powder, and graphite were used in the envelope of a building. The environmentally friendly and low-cost bio-composite PCMs resulted in applying them in buildings.
15. In the macroencapsulation method, PCMs are stored in tubes or spheres without any impact on the structure of buildings. Their low cost and direct usage in concrete make them a suitable option.
16. PCMs with high electrical conduction such as gallium have the potential to improve their convective energy transport properties under the magnetic field.
17. The speed of melting increased with a decrease in porosity due to easier penetration of heat via conductivity in the solid region. The conduction had a dominant impact compared to the convection heat transfer by increasing the Hartmann number. More importantly, porous media had more impact on the melting process compared to the using nanoparticles.
18. One of the promising active methods is the electrohydrodynamic technique due to its fast response, easy control, and no moving parts. Dielectric materials such as organic PCMs have a desirable potential to bear the electric field.
19. The low density and excellent stability of carbon-based nanoparticles have made carbon-based nanoparticles superior to metallic nanoparticles.
20. The high thermal conductivity (2000–4000 W/m·K) and high specific area (> 2000 m^2/g) of graphene make it an appropriate candidate for PCMs.
21. Metallic foams increase the performance of PCMs more than metallic nanoparticles due to lower density and larger aspect ratio. In heat transfer enhancement, metal foams with low porosities are utilized, while in heat storage capacity improvement, high porosities are preferred. Using

metallic foams decreased the time of charging and discharging due to increasing the thermal conductivity.

22. The stability of organic paraffin wax, including its suitable melting point, low cost, and high heat capacity, result in using it widely in concrete.

23. Paraffin, fatty acids, and alcohol are widely used in cooling electronic devices. In addition, paraffin wax has great potential to use as a material for storing energy in LHES. It is non-toxic, chemically stable, non-corrosive, and low in cost.

24. LHTS systems using PCMs have the potential to store heat 14 times more than sensible heat storage systems.

25. Solar dryers usually operate at temperatures between 45 °C and 75 °C; as a consequence, the melting temperature of PCMs should be under 80 °C. In addition, among PCMs, paraffin wax is extensively used in drying applications.

For future research in this field, the following topics are proposed:

- Using the combination of different types of nanoparticles, for instance graphene nanoparticles and MXene, can be the topic for more research in this field. MXene is a new ultrathin two-dimensional nanomaterial with a graphene-like structure. Because of its high thermal conductivity, MXene has attracted more attention in research on improving heat transfer efficiency. Furthermore, reaching an appropriate ratio in the mixture of nanoparticles to have a reasonable viscosity enhancement is another important parameter that should be analyzed.

- Metal foams can improve the thermal properties of PCMs. But there are not many studies on the impact of the material of foams and their porosities on the performance of PCMs used in different applications.

- The combination of different PCMs can improve their thermo-physical properties. In addition, their melting time can increase. As a result, it is so important to investigate the impact of mixing different PCMs as well as that of their thicknesses to reach an appropriate configurations of them.

- The combination of corrugated tubes or wavy absorbers and PCMs under PV panels to study their effects on the electrical and thermal efficiencies of PV systems can be the topic of new research.

- There are not enough exergy and economic analyses concerning heat transfer augmentation methods when PCMs are used. A study on exergy can help to identify useful energy. In addition, investigation of the cost of PCMs, which can have important effects on the payback period of systems, is one of the critical parameters to design economical units.

REFERENCES

[1] M. Dehghan, M. Ghasemizadeh, S. Rahgozar, A. Pourrajabian, A. Arabkoohsar, Chapter 4—Latent thermal energy storage, in *Future Grid-Scale Energy Storage Solutions*, Academic Press, 2023. https://doi.org/10.1016/B978-0-323-90786-6.00008-X.

[2] N.I. Ibrahim, F.A. Al-Sulaiman, S. Rahman, B.S. Yilbas, A.Z. Sahin, Heat transfer enhancement of phase change materials for thermal energy storage applications: A critical review, *Renew. Sustain. Energy Rev.* 74 (2017) 26–50. https://doi.org/10.1016/j.rser.2017.01.169.

[3] F.S. Javadi, H.S.C. Metselaar, P. Ganesan, Performance improvement of solar thermal systems integrated with phase change materials (PCM), a review, *Sol. Energy.* 206 (2020) 330–352. https://doi.org/10.1016/j.solener.2020.05.106.

[4] A Sharma, V.V Tyagi, C.R. Chen, D. Buddhi, Review on thermal energy storage with phase change materials and applications, *Renew Sustain Energy Rev.* 13 (2009) 318–345. https://doi.org/10.1016/j.rser.2007.10.005.

[5] J.L. Alvarado, C. Marsh, C. Sohn, G. Phetteplace, T. Newell, Thermal performance of microencapsulated phase change material slurry in turbulent flow under constant heat flux, *Int. J. Heat Mass Transf.* 50 (2007) 1938–1952. https://doi.org/10.1016/j.ijheatmasstransfer.2006.09.026.

[6] S. Rangarajan, C. Balaji, *Material-Based Heat Sinks a Multi-Objective Perspective*, Taylor & Francis Group, 2020.

[7] J.A. Noël, S. Kahwaji, L. Desgrosseilliers, D. Groulx, M.A. White, *Phase Change Materials*, Elsevier Inc., 2022. https://doi.org/10.1016/B978-0-12-824510-1.00005-2.

[8] C.R.E.S. Nóbrega, K.A.R. Ismail, F.A.M. Lino, Solidification around axial finned tube submersed in PCM: Modeling and experiments, *J. Energy Storage* 29 (2020). https://doi.org/10.1016/j.est.2020.101438.

[9] C. Nie, S. Deng, J. Liu, Numerical investigation of PCM in a thermal energy storage unit with fins: Consecutive charging and discharging, *J. Energy Storage* 29 (2020) 101319. https://doi.org/10.1016/j.est.2020.101319.

[10] M. Liu, W. Saman, F. Bruno, Review on storage materials and thermal performance enhancement techniques for high temperature phase change thermal storage systems, *Renew Sustain Energy Rev.* 16 (2012) 2118–2132. https://doi.org/10.1016/j.rser.2012.01.020.

[11] Y.B. Tao, Y.K. Liu, Y.L. He, Effects of PCM arrangement and natural convection on charging and discharging performance of shell-and-tube LHS unit, *Int. J. Heat Mass Transf.* 115 (2017) 99–107. https://doi.org/10.1016/j.ijheatmasstransfer.2017.07.098.

[12] L. Kasper, D. Pernsteiner, M. Koller, A. Schirrer, S. Jakubek, R. Hofmann, Numerical studies on the influence of natural convection under inclination on optimal aluminium proportions and fin spacings in a rectangular aluminium finned latent-heat thermal energy storage, *Appl. Therm. Eng.* 190 (2021). https://doi.org/10.1016/j.applthermaleng.2020.116448.

[13] A. Arshad, M. Jabbal, P.T. Sardari, M.A. Bashir, H. Faraji, Y. Yan, Transient simulation of finned heat sinks embedded with PCM for electronics cooling, *Therm. Sci. Eng. Prog.* (2020) 100520. https://doi.org/10.1016/j.tsep.2020.100520.

[14] T. Bouzennada, F. Mechighel, T. Ismail, L. Kolsi, K. Ghachem, Heat transfer and fluid flow in a PCM-filled enclosure : Effect of inclination angle and mid-separation fin, *Int. Commun. Heat Mass Transf.* 124 (2021) 105280. https://doi.org/10.1016/j.icheatmasstransfer.2021.105280.

[15] I. Sarani, S. Payan, S.A. Nada, A. Payan, Numerical investigation of an innovative discontinuous distribution of fins for solidification rate enhancement in PCM with and without nanoparticles, *Appl. Therm. Eng.* 176 (2020) 115017. https://doi.org/10.1016/j.applthermaleng.2020.115017.

[16] T. Sathe, A.S. Dhoble, Thermal analysis of an inclined heat sink with finned PCM container for solar applications, *Int. J. Heat Mass Transf.* 144 (2019) 118679. https://doi.org/10.1016/j.ijheatmasstransfer.2019.118679.

[17] M. Gürtürk, B. Kok, A new approach in the design of heat transfer fin for melting and solidification of PCM, *Int. J. Heat Mass Transf.* 153 (2020) 119671. https://doi.org/10.1016/j.ijheatmasstransfer.2020.119671.

[18] A.R. Mazhar, A. Shukla, S. Liu, Numerical analysis of rectangular fins in a PCM for low-grade heat harnessing, *Int. J. Therm. Sci.* 152 (2020) 106306. https://doi.org/10.1016/j.ijthermalsci.2020.106306.

[19] H. Xu, N. Wang, C. Zhang, Z. Qu, M. Cao, Optimization on the melting performance of triplex-layer PCMs in a horizontal finned shell and tube thermal energy storage unit, *Appl. Therm. Eng.* 176 (2020) 115409. https://doi.org/10.1016/j.applthermaleng.2020.115409.

[20] S. Ni, A. Shahsavar, Numerical investigation of natural convection behavior of molten PCM in an enclosure having rectangular and tree-like branching fins zeti, *Energy* 207 (2020). https://doi.org/10.1016/j.energy.2020.118223.

[21] L. Alberto, O. Rocha, C. Biserni, R. De C, G. Eduardo, S. Eberhardt, Design of fin structures for phase change material (PCM) melting process in rectangular cavities, *J. Energy Storage* 35 (2021). https://doi.org/10.1016/j.est.2021.102337.

[22] Y. Zhang, B. Sun, X. Zheng, P. Kumar, Case Studies in Thermal Engineering Investigation on effect of connection angle of "L" shaped fin on charging and discharging process of PCM in vertical enclosure, *Case Stud. Therm. Eng.* 33 (2022) 101908. https://doi.org/10.1016/j.csite.2022.101908.

[23] A.N. Desai, H. Shah, V.K. Singh, Novel inverted fin configurations for enhancing the thermal performance of PCM based thermal control unit: A numerical study, *Appl. Therm. Eng.* 195 (2021) 117155. https://doi.org/10.1016/j.applthermaleng.2021.117155.

[24] A.H.N. Al-Mudhafar, A.F. Nowakowski, F.C.G.A. Nicolleau, Enhancing the thermal performance of PCM in a shell and tube latent heat energy storage system by utilizing innovative fins, *Energy Rep.* 7 (2021) 120–126. https://doi.org/10.1016/j.egyr.2021.02.034.

[25] M.E. Nakhchi, J.A. Esfahani, Improving the melting performance of PCM thermal energy storage with novel stepped fins, *J. Energy Storage* 30 (2020) 101424. https://doi.org/10.1016/j.est.2020.101424.

[26] B. Debich, A. El Hami, A. Yaich, W. Gafsi, L. Walha, M. Haddar, Design optimization of PCM-based finned heat sinks for mechatronic components: A numerical investigation and parametric study, *J. Energy Storage.* 32 (2020) 101960. https://doi.org/10.1016/j.est.2020.101960.

[27] A.N. Desai, A. Gunjal, V.K. Singh, Numerical investigations of fin efficacy for phase change material (PCM) based thermal control module, *Int. J. Heat Mass Transf.* 147 (2020) 118855. https://doi.org/10.1016/j.ijheatmasstransfer.2019.118855.

[28] S. Yao, X. Huang, Study on solidification performance of PCM by longitudinal triangular fins in a triplex-tube thermal energy storage system, *Energy.* 227 (2021) 120527. https://doi.org/10.1016/j.energy.2021.120527.

[29] M.Y. Yazici, M. Avci, O. Aydin, Combined effects of inclination angle and fin number on thermal performance of a PCM-based heat sink, *Appl. Therm. Eng.* 159 (2019) 113956. https://doi.org/10.1016/j.applthermaleng.2019.113956.

[30] R. Triki, S. Chtourou, M. Baccar, Heat transfer enhancement of phase change materials PCMs using innovative fractal H-shaped fin configurations, *J. Energy Storage.* 73 (2023) 109020. https://doi.org/10.1016/j.est.2023.109020.

[31] M. Fang, G. Chen, Effects of different multiple PCMs on the performance of a latent thermal energy storage system, *Appl. Therm. Eng.* 27 (2007) 994–1000. https://doi.org/10.1016/j.applthermaleng.2006.08.001.

[32] O.S. Elsanusi, E.C. Nsofor, Melting of multiple PCMs with different arrangements inside a heat exchanger for energy storage, *Appl. Therm. Eng.* 185 (2021) 116046. https://doi.org/10.1016/j.applthermaleng.2020.116046.

[33] G.S. Sodhi, P. Muthukumar, Compound charging and discharging enhancement in multi-PCM system using non-uniform fin distribution, *Renew Energy*. 171 (2021).

[34] S. Liu, M. Iten, A. Shukla, Numerically study the performance of an air-multiple PCMs unit for free cooling and ventilation, *Energy Build*. (2017). https://doi.org/10.1016/j.enbuild.2017.07.005.

[35] A.H. Mosaffa, C.A.I. Ferreira, F. Talati, M.A. Rosen, Thermal performance of a multiple PCM thermal storage unit for free cooling, *Energy Convers. Manag*. 67 (2013) 1–7. https://doi.org/10.1016/j.enconman.2012.10.018.

[36] A.H. Mosaffa, L.G. Farshi, C.A.I. Ferreira, M.A. Rosen, Energy and exergy evaluation of a multiple-PCM thermal storage unit for free cooling applications, *Renew. Energy*. 68 (2014) 452–458. https://doi.org/10.1016/j.renene.2014.02.025.

[37] J. Gasia, L. Mir, L.F. Cabeza, G. Peir, Experimental evaluation at pilot plant scale of multiple PCMs (cascaded) vs. single PCM configuration for thermal energy storage, *Renew. Energy*. 83 (2015). https://doi.org/10.1016/j.renene.2015.05.029.

[38] V. Ranawade, K.S. Nalwa, Multilayered PCMs-based cooling solution for photovoltaic modules: Modelling and experimental study, *Renew. Energy*. 216 (2023) 119136. https://doi.org/10.1016/j.renene.2023.119136.

[39] J.M. Maldonado, A. de Gracia, L.F. Cabeza, Systematic review on the use of heat pipes in latent heat thermal energy storage tanks, *J. Energy Storage*. 32 (2020) 101733. https://doi.org/10.1016/j.est.2020.101733.

[40] N. Putra, A.F. Sandi, B. Ariantara, N. Abdullah, T.M. Indra Mahlia, Performance of beeswax phase change material (PCM) and heat pipe as passive battery cooling system for electric vehicles, *Case Stud. Therm. Eng*. 21 (2020) 100655. https://doi.org/10.1016/j.csite.2020.100655.

[41] J. Qu, A. Zuo, H. Liu, J. Zhao, Z. Rao, Three-dimensional oscillating heat pipes with novel structure for latent heat thermal energy storage application, *Appl. Therm. Eng*. 187 (2021) 116574. https://doi.org/10.1016/j.applthermaleng.2021.116574.

[42] H. Behi, D. Karimi, F.H. Gandoman, M. Akbarzadeh, S. Khaleghi, T. Kalogiannis, M.S. Hosen, J. Jaguemont, J. Van Mierlo, M. Berecibar, PCM assisted heat pipe cooling system for the thermal management of an LTO cell for high-current profiles, *Case Stud. Therm. Eng*. 25 (2021) 100920. https://doi.org/10.1016/j.csite.2021.100920.

[43] A. Ghanbarpour, M.J. Hosseini, A.A. Ranjbar, M. Rahimi, R. Bahrampoury, M. Ghanbarpour, Evaluation of heat sink performance using PCM and vapor chamber/heat pipe, *Renew. Energy*. 163 (2021) 698–719. https://doi.org/10.1016/j.renene.2020.08.154.

[44] D.G. Atinafu, Y.S. Ok, H.W. Kua, S. Kim, Thermal properties of composite organic phase change materials (PCMs): A critical review on their engineering chemistry, *Appl. Therm. Eng*. 181 (2020) 115960. https://doi.org/10.1016/j.applthermaleng.2020.115960.

[45] D. Das, U. Bordoloi, H.H. Muigai, P. Kalita, A novel form stable PCM based bio composite material for solar thermal energy storage applications, *J. Energy Storage*. 30 (2020) 101403. https://doi.org/10.1016/j.est.2020.101403.

[46] E. Meng, H. Yu, B. Zhou, Study of the thermal behavior of the composite phase change material (PCM) room in summer and winter, *Appl. Therm. Eng*. 126 (2017) 212–225. https://doi.org/10.1016/j.applthermaleng.2017.07.110.

[47] E. Meng, R. Cai, Z. Sun, J. Yang, J. Wang, Experimental study of the passive and active performance of real-scale composite PCM room in winter, *Appl. Therm. Eng*. 185 (2021) 116418. https://doi.org/10.1016/j.applthermaleng.2020.116418.

[48] L. Li, H. Yu, R. Liu, Research on composite-phase change materials (PCMs)-bricks in the west wall of room-scale cubicle: Mid-season and summer day cases, *Build. Environ*. 123 (2017) 494–503. https://doi.org/10.1016/j.buildenv.2017.07.019.

[49] J. Yoo, S.J. Chang, S. Wi, S. Kim, Spent coffee grounds as supporting materials to produce bio-composite PCM with natural waxes, *Chemosphere*. 235 (2019) 626–635. https://doi.org/10.1016/j.chemosphere.2019.06.195.

[50] L. Boussaba, A. Foufa, S. Makhlouf, G. Lefebvre, L. Royon, Elaboration and properties of a composite bio-based PCM for an application in building envelopes, *Constr. Build. Mater.* 185 (2018) 156–165. https://doi.org/10.1016/j.conbuildmat.2018.07.098.

[51] Y. Zhang, J. Liu, Z. Su, M. Lu, S. Liu, T. Jiang, Preparation of low-temperature composite phase change materials (C-PCMs) from modified blast furnace slag (MBFS), *Constr. Build. Mater.* 238 (2020) 117717. https://doi.org/10.1016/j.conbuildmat.2019.117717.

[52] Y. Yang, W. Wu, S. Fu, H. Zhang, Study of a novel ceramsite-based shape-stabilized composite phase change material (PCM) for energy conservation in buildings, *Constr. Build. Mater.* 246 (2020) 118479. https://doi.org/10.1016/j.conbuildmat.2020.118479.

[53] H.G. Kim, A. Qudoos, I.K. Jeon, B.H. Woo, J.S. Ryou, Assessment of PCM/SiC-based composite aggregate in concrete: Energy storage performance, *Constr. Build. Mater.* 258 (2020) 119637. https://doi.org/10.1016/j.conbuildmat.2020.119637.

[54] C. Xu, S. Xu, R.D. Eticha, Experimental investigation of thermal performance for pulsating flow in a microchannel heat sink filled with PCM (paraffin/CNT composite), *Energy Convers. Manag.* 236 (2021) 114071. https://doi.org/10.1016/j.enconman.2021.114071.

[55] L. Jiang, H. Zhang, J. Li, P. Xia, Thermal performance of a cylindrical battery module impregnated with PCM composite based on thermoelectric cooling, *Energy*. 188 (2019) 116048. https://doi.org/10.1016/j.energy.2019.116048.

[56] A. Raul, M. Jain, S. Gaikwad, S.K. Saha, Modelling and experimental study of latent heat thermal energy storage with encapsulated PCMs for solar thermal applications, *Appl. Therm. Eng.* 143 (2018) 415–428. https://doi.org/10.1016/j.applthermaleng.2018.07.123.

[57] S. Yu, X. Wang, D. Wu, Microencapsulation of n-octadecane phase change material with calcium carbonate shell for enhancement of thermal conductivity and serving durability: Synthesis, microstructure, and performance evaluation, *Appl. Energy*. 114 (2014) 632–643.

[58] X. Jin, M.A. Medina, X. Zhang, On the importance of the location of PCMs in building walls for enhanced thermal performance, *Appl. Energy*. 106 (2013) 72–78. https://doi.org/10.1016/j.apenergy.2012.12.079.

[59] H. Cui, S.A. Memon, R. Liu, Development, mechanical properties and numerical simulation of macro encapsulated thermal energy storage concrete, *Energy Build*. 96 (2015).

[60] H. Cui, W. Tang, Q. Qin, F. Xing, W. Liao, H. Wen, Development of structural-functional integrated energy storage concrete with innovative macro-encapsulated PCM by hollow steel ball, *Appl. Energy*. 185 (2017) 107–118. https://doi.org/10.1016/j.apenergy.2016.10.072.

[61] Q. Rao, Y. Xia, J. Li, J. McConnell, J. Sutherland, Z. Li, A modified many-body dissipative particle dynamics model for mesoscopic fluid simulation: Methodology, calibration, and application for hydrocarbon and water, *Mol. Simul.* 47 (2021) 363–375. https://doi.org/10.1080/08927022.2021.1876233.

[62] S. Aziz, N.A.M. Amin, M.S. Abdul Majid, M. Belusko, F. Bruno, CFD simulation of a TES tank comprising a PCM encapsulated in sphere with heat transfer enhancement, *Appl. Therm. Eng.* 143 (2018) 1085–1092. https://doi.org/10.1016/j.applthermaleng.2018.08.013.

[63] A.K. Raj, M. Srinivas, S. Jayaraj, A cost-effective method to improve the performance of solar air heaters using discrete macro-encapsulated PCM capsules for drying applications, *Appl. Therm. Eng.* 146 (2019) 910–920. https://doi.org/10.1016/j.applthermaleng.2018.10.055.

[64] B. Praveen, S. Suresh, V. Pethurajan, Heat transfer performance of graphene nano-platelets laden micro-encapsulated PCM with polymer shell for thermal energy storage based heat sink, *Appl. Therm. Eng.* 156 (2019) 237–249. https://doi.org/10.1016/j.applthermaleng.2019.04.072.

[65] Z.I. Djamai, F. Salvatore, A. Si Larbi, G. Cai, M. El Mankibi, Multiphysics analysis of effects of encapsulated phase change materials (PCMs) in cement mortars, *Cem. Concr. Res.* 119 (2019) 51–63. https://doi.org/10.1016/j.cemconres.2019.02.002.

[66] V.V. Tyagi, A.K. Pandey, D. Buddhi, R. Kothari, Thermal performance assessment of encapsulated PCM based thermal management system to reduce peak energy demand in buildings, *Energy Build.* 117 (2016) 44–52. https://doi.org/10.1016/j.enbuild.2016.01.042.

[67] W. Wu, N. Liu, W. Cheng, Y. Liu, Study on the effect of shape-stabilized phase change materials on spacecraft thermal control in extreme thermal environment, *Energy Convers. Manag.* 69 (2013) 174–180.

[68] Y. Wu, X. Zhang, X. Xu, X. Lin, L. Liu, A review on the effect of external fields on solidification, melting and heat transfer enhancement of phase change materials, *J. Energy Storage.* 31 (2020) 101567. https://doi.org/10.1016/j.est.2020.101567.

[69] N. Zhang, Y. Du, Ultrasonic enhancement on heat transfer of palmitic-stearic acid as PCM in unit by experimental study, *Sustain. Cities Soc.* 43 (2018) 532–537. https://doi.org/10.1016/j.scs.2018.08.040.

[70] Z. Yan, Z.J. Yu, T. Yang, S. Li, G. Zhang, Impact of ultrasound on the melting process and heat transfer of phase change material, *Energy Procedia.* 158 (2019) 5014–5019. https://doi.org/10.1016/j.egypro.2019.01.663.

[71] M. Legay, S. Le Person, N. Gondrexon, P. Boldo, A. Bontemps, Performances of two heat exchangers assisted by ultrasound, *Appl. Therm. Eng.* 37 (2012) 60–66. https://doi.org/10.1016/j.applthermaleng.2011.12.051.

[72] S. Han, S. Lyu, S. Wang, F. Fu, High-intensity ultrasound assisted manufacturing of melamine-urea-formaldehyde/paraffin nanocapsules, Colloids Surfaces A Physicochem. *Eng. Asp.* 568 (2019) 75–83. https://doi.org/10.1016/j.colsurfa.2019.01.054.

[73] M. Izadi, M. Sheremet, A. Hajjar, A.M. Galal, I. Mahariq, F. Jarad, M.B. Ben Hamida, Numerical investigation of magneto-thermal-convection impact on phase change phenomenon of Nano-PCM within a hexagonal shaped thermal energy storage, *Appl. Therm. Eng.* 223 (2023) 119984. https://doi.org/10.1016/j.applthermaleng.2023.119984.

[74] M. Izadi, A. Hajjar, H.M. Alshehri, M. Sheremet, A.M. Galal, Charging process of a partially heated trapezoidal thermal energy storage filled by nano-enhanced PCM using controlable uniform magnetic field, *Int. Commun. Heat Mass Transf.* 138 (2022) 106349. https://doi.org/10.1016/j.icheatmasstransfer.2022.106349.

[75] S.D. Farahani, A.D. Farahani, A.J. Mamoei, W.M. Yan, Enhancement of phase change material melting using nanoparticles and magnetic field in the thermal energy storage system with strip fins, *J. Energy Storage.* 57 (2023) 106282. https://doi.org/10.1016/j.est.2022.106282.

[76] W. Zhang, X. Li, W. Wu, J. Huang, Influence of mechanical vibration on composite phase change material based thermal management system for lithium-ion battery, *J. Energy Storage.* 54 (2022) 105237. https://doi.org/10.1016/j.est.2022.105237.

[77] M. Sheikholeslami, O. Mahian, Enhancement of PCM solidification using inorganic nanoparticles and an external magnetic field with application in energy storage systems, *J. Clean. Prod.* 215 (2019) 963–977. https://doi.org/10.1016/j.jclepro.2019.01.122.

[78] F. Selimefendigil, H.F. Öztop, F. Izadi, Non-uniform magnetic field effects on the phase transition dynamics for PCM-installed 3D conic cavity having ventilation ports under hybrid nanofluid convection, *J. Build. Eng.* 49 (2022). https://doi.org/10.1016/j.jobe.2022.104074.

[79] S. Lashgari, A.R. Mahdavian, H. Arabi, V. Ambrogi, V. Marturano, Preparation of acrylic PCM microcapsules with dual responsivity to temperature and magnetic field changes, *Eur. Polym. J.* 101 (2018) 18–28. https://doi.org/10.1016/j.eurpolymj.2018.02.011.

[80] M.T. Kohyani, B. Ghasemi, A. Raisi, S.M. Aminossadati, Melting of cyclohexane—Cu nano-phase change material (nano-PCM) in porous medium under magnetic field, *J. Taiwan Inst. Chem. Eng.* 77 (2017) 142–151. https://doi.org/10.1016/j.jtice.2017.04.037.

[81] D. Nakhla, J.S. Cotton, Effect of electrohydrodynamic (EHD) forces on charging of a vertical latent heat thermal storage module filled with octadecane, *Int. J. Heat Mass Transf.* 167 (2021) 120828. https://doi.org/10.1016/j.ijheatmasstransfer.2020.120828.

[82] Z. Sun, K. Luo, J. Wu, Experimental study on the melting characteristics of n-octadecane with passively installing fin and actively applying electric field, *Int. Commun. Heat Mass Transf.* 127 (2021) 105570. https://doi.org/10.1016/j.icheatmasstransfer.2021.105570.

[83] K. He, L. Wang, J. Huang, Electrohydrodynamic enhancement of phase change material melting in circular-elliptical annuli, *Energies.* 14 (2021) 1–20. https://doi.org/10.3390/en14238090.

[84] R. Pérez-Masiá, A. López-Rubio, M.J. Fabra, J.M. Lagaron, Use of electrohydrodynamic processing to develop nanostructured materials for the preservation of the cold chain, *Innov. Food Sci. Emerg. Technol.* 26 (2014) 415–423. https://doi.org/10.1016/j.ifset.2014.10.010.

[85] W. Zhou, H.I. Mohammed, S. Chen, M. Luo, Y. Wu, Effects of mechanical vibration on the heat transfer performance of shell-and-tube latent heat thermal storage units during charging process, *Appl. Therm. Eng.* 216 (2022) 119133. https://doi.org/10.1016/j.applthermaleng.2022.119133.

[86] W. Du, S. Chen, Effect of mechanical vibration on phase change material based thermal management module of a lithium-ion battery at high ambient temperature, *J. Energy Storage.* 59 (2023) 106465. https://doi.org/10.1016/j.est.2022.106465.

[87] Y. Yu, S. Chen, Utilize mechanical vibration energy for fast thermal responsive PCMs-based energy storage systems: Prototype research by numerical simulation, *Renew. Energy.* 187 (2022) 974–986. https://doi.org/10.1016/j.renene.2022.02.010.

[88] G. Wu, S. Chen, S. Zeng, Effects of mechanical vibration on melting behaviour of phase change material during charging process, *Appl. Therm. Eng.* 192 (2021) 116914. https://doi.org/10.1016/j.applthermaleng.2021.116914.

[89] N. Joshy, M. Hajiyan, A.R.M. Siddique, S. Tasnim, H. Simha, S. Mahmud, Experimental investigation of the effect of vibration on phase change material (PCM) based battery thermal management system, *J. Power Sources.* 450 (2020) 227717. https://doi.org/10.1016/j.jpowsour.2020.227717.

[90] J. Yang, Q. Yu, S. Chen, M. Luo, W. Du, Y. Yu, Y. Wu, W. Zhou, Z. Zhou, Effect of mechanical vibration on thermal performance of PCM-fin structure Li-ion battery thermal management system under high-rate discharge and high-temperature environment, *Int. J. Heat Mass Transf.* 217 (2023) 124722. https://doi.org/10.1016/j.ijheatmasstransfer.2023.124722.

[91] Z.A. Qureshi, H.M. Ali, S. Khushnood, Recent advances on thermal conductivity enhancement of phase change materials for energy storage system: A review, *Int. J. Heat Mass Transf.* 127 (2018) 838–856. https://doi.org/10.1016/j.ijheatmasstransfer.2018.08.049.

[92] S. Kashyap, S. Kabra, B. Kandasubramanian, Graphene aerogel-based phase changing composites for thermal energy storage systems, *J. Mater. Sci.* 55 (2020) 4127–4156. https://doi.org/10.1007/s10853-019-04325-7.

[93] E.B.S. Mettawee, G.M.R. Assassa, Thermal conductivity enhancement in a latent heat storage system, *Sol. Energy.* 81 (2007) 839–845. https://doi.org/10.1016/j.solener.2006.11.009.

[94] T. ur Rehman, H.M. Ali, M.M. Janjua, U. Sajjad, W.M. Yan, A critical review on heat transfer augmentation of phase change materials embedded with porous materials/foams, *Int. J. Heat Mass Transf.* 135 (2019) 649–673. https://doi.org/10.1016/j.ijheatmasstransfer.2019.02.001.

[95] T. Oya, T. Nomura, M. Tsubota, N. Okinaka, T. Akiyama, Thermal conductivity enhancement of erythritol as PCM by using graphite and nickel particles, *Appl. Therm. Eng.* 61 (2013) 825–828. https://doi.org/10.1016/j.applthermaleng.2012.05.033.

[96] J.D. Renteria, D.L. Nika, A.A. Balandin, Graphene thermal properties: Applications in thermal management and energy storage, *Appl. Sci.* 4 (2014) 525–547. https://doi.org/10.3390/app4040525.

[97] A. Sathishkumar, V. Kumaresan, R. Velraj, Solidification characteristics of water based graphene nanofluid PCM in a spherical capsule for cool thermal energy storage applications, *Int. J. Refrig.* 66 (2016) 73–83. https://doi.org/10.1016/j.ijrefrig.2016.01.014.

[98] P. Kumar, P. Kumar Singh, S. Nagar, K. Sharma, M. Saraswat, Effect of different concentration of functionalized graphene on charging time reduction in thermal energy storage system, *Mater. Today Proc.* 44 (2021) 146–152. https://doi.org/10.1016/j.matpr.2020.08.548.

[99] M. Amin, N. Putra, E.A. Kosasih, E. Prawiro, R.A. Luanto, T.M.I. Mahlia, Thermal properties of beeswax/graphene phase change material as energy storage for building applications, *Appl. Therm. Eng.* 112 (2017) 273–280. https://doi.org/10.1016/j.applthermaleng.2016.10.085.

[100] X. Xu, H. Cui, S.A. Memon, H. Yang, W. Tang, Development of novel composite PCM for thermal energy storage using CaCl2·6H2O with graphene oxide and SrCl2·6H2O, *Energy Build.* 156 (2017) 163–172. https://doi.org/10.1016/j.enbuild.2017.09.081.

[101] L. Liu, K. Zheng, Y. Yan, Z. Cai, S. Lin, X. Hu, Graphene Aerogels Enhanced Phase Change Materials prepared by one-pot method with high thermal conductivity and large latent energy storage, *Sol. Energy Mater. Sol. Cells.* 185 (2018) 487–493. https://doi.org/10.1016/j.solmat.2018.06.005.

[102] M. Vivekananthan, V.A. Amirtham, Characterisation and thermophysical properties of graphene nanoparticles dispersed erythritol PCM for medium temperature thermal energy storage applications, *Thermochim. Acta.* 676 (2019) 94–103. https://doi.org/10.1016/j.tca.2019.03.037.

[103] M.R. Safaei, H.R. Goshayeshi, I. Chaer, Solar still efficiency enhancement by using graphene oxide/paraffin nano-PCM, *Energies.* 12 (2019) 1–13. https://doi.org/10.3390/en12102002.

[104] M. Nitsas, I.P. Koronaki, Performance analysis of nanoparticles-enhanced PCM: An experimental approach, *Therm. Sci. Eng. Prog.* 25 (2021) 100963. https://doi.org/10.1016/j.tsep.2021.100963.

[105] C. Nie, J. Liu, S. Deng, Effect of geometric parameter and nanoparticles on PCM melting in a vertical shell-tube system, *Appl. Therm. Eng.* 184 (2021) 116290. https://doi.org/10.1016/j.applthermaleng.2020.116290.

[106] F. Li, N. Muhammad, E. Abohamzeh, A.K.A. Hakeem, M.R. Hajizadeh, Z. Li, Q.V. Bach, Finned unit solidification with use of nanoparticles improved PCM, *J. Mol. Liq.* 314 (2020). https://doi.org/10.1016/j.molliq.2020.113659.

[107] D. Li, Y. Wu, C. Liu, G. Zhang, M. Arıcı, Numerical investigation of thermal and optical performance of window units filled with nanoparticle enhanced PCM, *Int. J. Heat Mass Transf.* 125 (2018) 1321–1332. https://doi.org/10.1016/j.ijheatmasstransfer.2018.04.152.

[108] M. Sheikholeslami, R. ul Haq, A. Shafee, Z. Li, Heat transfer behavior of nanoparticle enhanced PCM solidification through an enclosure with V shaped fins, *Int. J. Heat Mass Transf.* 130 (2019) 1322–1342. https://doi.org/10.1016/j.ijheatmasstransfer.2018.11.020.

[109] P. Murugan, P. Ganesh Kumar, V. Kumaresan, M. Meikandan, K. Malar Mohan, R. Velraj, Thermal energy storage behaviour of nanoparticle enhanced PCM during freezing and melting, *Phase Transitions.* 91 (2018) 254–270. https://doi.org/10.1080/0141159 4.2017.1372760.

[110] J.M. Mahdi, E.C. Nsofor, Solidification enhancement of PCM in a triplex-tube thermal energy storage system with nanoparticles and fins, *Appl. Energy.* 211 (2018) 975–986. https://doi.org/10.1016/j.apenergy.2017.11.082.

[111] R. Du, W. Li, T. Xiong, X. Yang, Y. Wang, K.W. Shah, Numerical investigation on the melting of nanoparticle-enhanced PCM in latent heat energy storage unit with spiral coil heat exchanger, *Build. Simul.* 12 (2019) 869–879. https://doi.org/10.1007/s12273-019-0527-3.

[112] H. Senobar, M. Aramesh, B. Shabani, Nanoparticles and metal foams for heat transfer enhancement of phase change materials: A comparative experimental study, *J. Energy Storage.* 32 (2020) 101911. https://doi.org/10.1016/j.est.2020.101911.

[113] T. ur Rehman, H.M. Ali, Thermal performance analysis of metallic foam-based heat sinks embedded with RT-54HC paraffin: An experimental investigation for electronic cooling, *J. Therm. Anal. Calorim.* 140 (2020) 979–990. https://doi.org/10.1007/s10973-019-08961-8.

[114] D. Gowthami, R.K. Sharma, Influence of Hydrophilic and Hydrophobic modification of the porous matrix on the thermal performance of form stable phase change materials: A review, *Renew. Sustain. Energy Rev.* 185 (2023) 113642. https://doi.org/10.1016/j.rser.2023.113642.

[115] M. Esapour, A. Hamzehnezhad, A.A. Rabienataj Darzi, M. Jourabian, Melting and solidification of PCM embedded in porous metal foam in horizontal multi-tube heat storage system, *Energy Convers. Manag.* 171 (2018) 398–410. https://doi.org/10.1016/j.enconman.2018.05.086.

[116] M. Iasiello, M. Mameli, S. Filippeschi, N. Bianco, Metal foam/PCM melting evolution analysis: Orientation and morphology effects, *Appl. Therm. Eng.* 187 (2021) 116572. https://doi.org/10.1016/j.applthermaleng.2021.116572.

[117] A.M. Bassam, K. Sopian, A. Ibrahim, M.F. Fauzan, A.B. Al-Aasam, G.Y. Abusaibaa, Experimental analysis for the photovoltaic thermal collector (PVT) with nano PCM and micro-fins tube nanofluid, *Case Stud. Therm. Eng.* 41 (2023) 102579. https://doi.org/10.1016/j.csite.2022.102579.

[118] F. Afsharpanah, M. Izadi, F. Akbarzadeh Hamedani, S. S. Mousavi Ajarostaghi, W. Yaïci, Solidification of nano-enhanced PCM-porous composites in a cylindrical cold thermal energy storage enclosure, *Case Stud. Therm. Eng.* 39 (2022) 102421.

[119] P.T. Sardari, D. Grant, D. Giddings, G.S. Walker, M. Gillott, Composite metal foam/PCM energy store design for dwelling space air heating, *Energy Convers. Manag.* 201 (2019) 112151. https://doi.org/10.1016/j.enconman.2019.112151.

[120] C. Zhang, M. Yu, Y. Fan, X. Zhang, Y. Zhao, L. Qiu, Numerical study on heat transfer enhancement of PCM using three combined methods based on heat pipe, *Energy.* 195 (2020) 116809. https://doi.org/10.1016/j.energy.2019.116809.

[121] H. Soltani, M. Soltani, H. Karimi, J. Nathwani, Heat transfer enhancement in latent heat thermal energy storage unit using a combination of fins and rotational mechanisms, Int. *J. Heat Mass Transf.* 179 (2021) 121667. https://doi.org/10.1016/j.ijheatmasstransfer.2021.121667.

[122] M. Jourabian, M. Farhadi, A.A. Rabienataj Darzi, Constrained ice melting around one cylinder in horizontal cavity accelerated using three heat transfer enhancement techniques, *Int. J. Therm. Sci.* 125 (2018) 231–247. https://doi.org/10.1016/j.ijthermalsci.2017.12.001.

[123] *Retrofitted buildings are key to the energy transition. 5 ways to unlock progress,* (2024), https://www.weforum.org/agenda/2024/01/retrofitted-buildings-energy-transition/ (Accessed March 2024).

[124] A. D'Alessandro, A.L. Pisello, C. Fabiani, F. Ubertini, L.F. Cabeza, F. Cotana, Multifunctional smart concretes with novel phase change materials: Mechanical and thermo-energy investigation, *Appl. Energy.* 212 (2018) 1448–1461.

[125] L. Navarro, A. De Gracia, A. Castell, L.F. Cabeza, Experimental evaluation of a concrete core slab with phase change materials for cooling purposes, *Energy Build.* 116 (2016) 411–419.

[126] L.F.C.L. Navarro, A. Solé, M. Martín, C. Barreneche, L. Olivieri, J. Tenorio, Benchmarking of useful phase change materials for a building application, *Energy Build.* 182 (2018) 45–50.

[127] L.F. Cabeza, A. Castell, C. Barreneche, A. De Gracia, A.I. Fernández, Materials used as PCM in thermal energy storage in buildings: A review, *Renew. Sustain. Energy Rev.* 15 (2011) 1675–1695.

[128] T.C. Ling, C.S. Poon, Use of phase change materials for thermal energy storage in concrete: An overview, *Constr. Build. Mater.* 46 (2013) 55–62.

[129] F.O. Cedeño, M.M. Prieto, A. Espina, J.R. García, Measurements of temperature and melting heat of some pure fatty acids and their binary and ternary mixtures by differential scanning calorimetry, *Thermochim. Acta 3.* 69 (2001) 39–50.

[130] V.V. Rao, R. Parameshwaran, V.V. Ram, PCM-mortar based construction materials for energy efficient buildings: A review on research trends, *Energy Build.* 158 (2018) 95–122.

[131] L. Navarro, A. de Gracia, S. Colclough, M. Browne, S.J. McCormack, P. Griffiths, L.F. Cabeza, Thermal energy storage in building integrated thermal systems: A review. Part 1. active storage systems, *Renew. Energy.* 88 (2016) 526–547.

[132] D.P. Bentz, R. Turpin, Potential applications of phase change materials in concrete technology, *Cem. Concr. Compos.* 29 (2007) 527–532.

[133] D.W. Hawes, Latent heat storage in building materials, *Energy Build.* (1993) 77–86.

[134] U. Berardi, A.A. Gallardo, Properties of concretes enhanced with phase change materials for building applications, *Energy Build.* 199 (2019) 402–414. https://doi.org/10.1016/j.enbuild.2019.07.014.

[135] H. Mehling, M. Brütting, T. Haussmann, PCM products and their fields of application—An overview of the state in 2020/2021, *J. Energy Storage.* 51 (2022) 104354. https://doi.org/10.1016/j.est.2022.104354.

[136] PCM Products Ltd (n.d.). www.pcmproducts.net/files/underfloor_pcm_heating.pdf (Accessed 22 September 2021).

[137] Datum Phase Change Ltd (2021). https://therma.cool/products/thermacool-tile/.

[138] PCM Technology (2021). https://doi.org/www.pcmtechnology.eu/en/applications/for-office-cooling.

[139] H. Mehling, Estimation of the worldwide installed capacity of cold storage with ice and its effect in the electricity grid, *GREENSTOCK 2015—13th Int. Conf. Energy Storage.* (2015) 1–6.

[140] www.pcm-ral.org/pcm/en/links/ (n.d.).

[141] Y. Khetib, A. Gari, R. Kalbasi, Introducing two scenarios to reduce building energy usage: PCM installation and integrating nanofluid solar collectors with DHW system, *J. Taiwan Inst. Chem. Eng.* 128 (2021) 327–337. https://doi.org/10.1016/j.jtice.2021.06.013.

[142] F.L. Tan, C.P. Tso, Cooling of mobile electronic devices using phase change materials, *Appl. Therm. Eng.* 24 (2004) 159–169. https://doi.org/10.1016/j.applthermaleng.2003.09.005.

[143] Y.L. Tua, R.C. Chu, W.S. Janna, Thermal management of micro-electronic equipment: Heat transfer theory, analysis methods, and design practices, *Appl. Mech. Rev.* 56 (2003) B46–B48.

[144] W. Hua, L. Zhang, X. Zhang, Research on passive cooling of electronic chips based on PCM: A review, *J. Mol. Liq.* 340 (2021) 117183. https://doi.org/10.1016/j.molliq.2021.117183.

[145] Y. Konuklu, et al. Cellulose-based myristic acid composites for thermal energy storage applications, *Sol. Energy Mater. Sol. Cells.* 193 (2019) 85–91.

[146] C. Alkan, A. Sari, Fatty acid/poly(methyl methacrylate) (PMMA) blends as form-stable phase change materials for latent heat thermal energy storage, *Sol. Energy.* 82 (2008) 118–124.

[147] C. Zhu, Y. Chen, R. Cong, F. Ran, G. Fang, Improved thermal properties of stearic acid/ high density polyethylene/carbon fiber composite heat storage materials, *Sol. Energy Mater. Sol. Cells.* 219 (2021) 110782. https://doi.org/10.1016/j.solmat.2020.110782.

[148] X. Yang, Y. Liu, Z. Lv, Q. Hua, L. Liu, B. Wang, J. Tang, Synthesis of high latent heat lauric acid/silica microcapsules by interfacial polymerization method for thermal energy storage, *J. Storage Mater.* 33 (2021) 102059. https://doi.org/https://doi.org/10.1016/j.est.2020.102059.

[149] X. Huang, C. Zhu, Y. Lin, G. Fang, Thermal properties and applications of microencapsulated PCM for thermal energy storage: A review, *Appl. Therm. Eng.* 147 (2019) 841–855.

[150] D. Zou, X. Ma, X. Liu, P. Zheng, Y. Hu, Thermal performance enhancement of composite phase change materials (PCM) using graphene and carbon nanotubes as additives for the potential application in lithium-ion power battery, *Int. J. Heat Mass Transf.* 120 (2018) 33–41.

[151] A.N. Keshteli, M. Sheikholeslami, Nanoparticle enhanced PCM applications for intensification of thermal performance in building: A review, *J. Mol. Liq.* 274 (2019) 516–533.

[152] K.Y. Leong, M.R. Abdul, B. Rahman, A. Gurunathan, Nano-enhanced phase change materials: A review of thermophysical properties, applications and challenges, *J. Storage Mater.* 21 (2019) 18–31.

[153] A. Novikov, D. Lexow, M. Nowottnick, Cooling of electronic assemblies through PCM containing coatings, *ESTC 2014—5th Electron. Syst. Technol. Conf.* (2014). https://doi.org/10.1109/ESTC.2014.6962787.

[154] X.Z. Xuelai, L.T. Weiwen, L. Xiaoyang, D. Jinhong, Preparation and properties of lauric acid-decanoic / tetradecyl alcoholdodecane composite as PCMs for thermal energy storage, *J. Refrig.* 37 (2016) 60–64.

[155] Z. Ling, J. Chen, T. Xu, X. Fang, X. Gao, Z. Zhang, Thermal conductivity of an organic PCM/expanded graphite composite across the phase change temperature range and a novel thermal conductivity model, *Energy Convers. Manag.* 102 (2015) 202–208.

[156] G. KumarMarri, C. Balaji, Experimental and numerical investigations on the effect of porosity and PPI gradients of metal foams on the thermal performance of a composite phase change material heat sink, *Int. J. Heat Mass Transf.* 164 (2021) 120454. https://doi.org/10.1016/j.ijheatmasstransfer.2020.120454.

[157] R. Kalbasi, Introducing a novel heat sink comprising PCM and air—Adapted to electronic device thermal management, *Int. J. Heat Mass Transf.* 169 (2021) 120914. https://doi.org/10.1016/j.ijheatmasstransfer.2021.120914.

[158] A. Arshad, I. Alabdullatif, M., M. Jabbal, Y. Yan, Towards the thermal management of electronic devices: A parametric investigation of finned heat sink filled with PCM, *Transf. Int. Commun. Heat Mass.* 129 (2021) 105643. https://doi.org/10.1016/j.icheatmasstransfer.2021.105643.

[159] A. Kurhade, V. Talele, T. Venkateswara Rao, A. Chandak, V.K. Mathew, Computational study of PCM cooling for electronic circuit of smart-phone, *Mater. Today Proc.* 47 (2021) 3171–3176. https://doi.org/10.1016/j.matpr.2021.06.284.

[160] I. Zahid, M. Farhan, M. Farooq, M. Asim, M. Imran, Experimental investigation for thermal performance enhancement of various heat sinks using Al2O3NePCM for cooling of electronic devices, *Case Stud. Therm. Eng.* 41 (2023) 102553. https://doi.org/10.1016/j.csite.2022.102553.

[161] M. Mozafari, A. Lee, J. Mohammadpour, Thermal management of single and multiple PCMs based heat sinks for electronics cooling, *Therm. Sci. Eng. Prog.* 23 (2021) 100919. https://doi.org/10.1016/j.tsep.2021.100919.

[162] A. Sadeghian, M.R. Zargarabadi, M. Dehghan, Effects of rib on cooling performance of photovoltaic modules (PV/PCM-Rib), *J. Cent. South Univ.* 28 (2021) 3449–3465. https://doi.org/10.1007/s11771-021-4867-7.

[163] M. Dehghan, M. Ghasemizadeh, S. Rashidi, Solar-driven water treatment: Generation II technologies, INC, 2021. https://doi.org/10.1016/B978-0-323-90991-4.00006-2.

[164] M. Faegh, P. Behnam, M.B. Shafii, A review on recent advances in humidification-dehumidification (HDH) desalination systems integrated with refrigeration, power and desalination technologies, *Energy Convers. Manag.* 196 (2019) 1002–1036. https://doi.org/10.1016/j.enconman.2019.06.063.

[165] E.K. Summers, M.A. Antar, J.H. Lienhard, Design and optimization of an air heating solar collector with integrated phase change material energy storage for use in humid-ification-dehumidification desalination, *Sol. Energy.* 86 (2012) 3417–3429. https://doi.org/10.1016/j.solener.2012.07.017.

[166] M. Jafaripour, F.A. Roghabadi, S. Soleimanpour, S.M. Sadrameli, Barriers to implementation of phase change materials within solar desalination units: Exergy, thermal conductivity, economic, and environmental aspects review, *Desalination.* 546 (2023) 116191. https://doi.org/10.1016/j.desal.2022.116191.

[167] R. Kumar R, A.K. Pandey, M. Samykano, B. Aljafari, Z. Ma, S. Bhattacharyya, V. Goel, I. Ali, R. Kothari, V.V. Tyagi, Phase change materials integrated solar desalination system: An innovative approach for sustainable and clean water production and storage, *Renew. Sustain. Energy Rev.* 165 (2022) 112611. https://doi.org/10.1016/j.rser.2022.112611.

[168] V.S. Vigneswaran, G. Kumaresan, B.V. Dinakar, K.K. Kamal, R. Velraj, Augmenting the productivity of solar still using multiple PCMs as heat energy storage, *J. Energy Storage.* 26 (2019). https://doi.org/10.1016/j.est.2019.101019.

[169] M. Abdelgaied, A.E. Kabeel, Performance improvement of pyramid solar distillers using a novel combination of absorber surface coated with CuO nano black paint, reflective mirrors, and PCM with pin fins, *Renew. Energy.* 180 (2021) 494–501. https://doi.org/10.1016/j.renene.2021.08.071.

[170] B. Ghorbani, M. Mehrpooya, A. Dadak, Thermo-economic analysis of a solar-driven multi-stage desalination unit equipped with a phase change material storage system to provide heating and fresh water for a residential complex, *J. Energy Storage.* 30 (2020) 101555. https://doi.org/10.1016/j.est.2020.101555.

[171] M.S. Yousef, H. Hassan, Energy payback time, exergoeconomic and enviroeconomic analyses of using thermal energy storage system with a solar desalination system: An experimental study, *J. Clean. Prod.* 270 (2020) 122082. https://doi.org/10.1016/j.jclepro.2020.122082.

[172] A.W. Kandeal, N.M. El-Shafai, M.R. Abdo, A.K. Thakur, I.M. El-Mehasseb, I. Maher, M. Rashad, A.E. Kabeel, N. Yang, S.W. Sharshir, Improved thermo-economic performance of solar desalination via copper chips, nanofluid, and nano-based phase change material, *Sol. Energy.* 224 (2021) 1313–1325. https://doi.org/10.1016/j.solener.2021.06.085.

[173] H. Mousa, A.M. Gujarathi, Modeling and analysis the productivity of solar desalination units with phase change materials, *Renew. Energy.* 95 (2016) 225–232. https://doi.org/10.1016/j.renene.2016.04.013.

[174] Q. Chen, G. Xu, P. Xia, The performance of a solar-driven spray flash evaporation desalination system enhanced by microencapsulated phase change material, *Case Stud. Therm. Eng.* 27 (2021) 101267. https://doi.org/10.1016/j.csite.2021.101267.

[175] M.M. Khairat Dawood, T. Nabil, A.E. Kabeel, A.I. Shehata, A.M. Abdalla, B.E. Elnaghi, Experimental study of productivity progress for a solar still integrated with parabolic trough collectors with a phase change material in the receiver evacuated tubes and in the still, *J. Energy Storage.* 32 (2020) 102007. https://doi.org/10.1016/j.est.2020.102007.

[176] M.E.A.E. Ahmed, S. Abdo, M.A. Abdelrahman, O.A. Gaheen, Finned-encapsulated PCM pyramid solar still—Experimental study with economic analysis, *J. Energy Storage.* 73 (2023) 108908. https://doi.org/10.1016/j.est.2023.108908.

[177] M.A. Fikri, M. Samykano, A.K. Pandey, K. Kadirgama, R.R. Kumar, J. Selvaraj, N. Abd, V.V. Tyagi, K. Sharma, R. Saidur, Solar Energy Materials and Solar Cells Recent progresses and challenges in cooling techniques of concentrated photovoltaic thermal system: A review with special treatment on phase change materials (PCMs) based cooling Power Conversion efficiency, *Sol. Energy Mater. Sol. Cells.* 241 (2022) 111739. https://doi.org/10.1016/j.solmat.2022.111739.

[178] M. Xiao, L. Tang, X. Zhang, I.Y.F. Lun, Y. Yuan, A review on recent development of cooling technologies for concentrated photovoltaics (CPV) systems, *Energies.* 11 (2018) 12. https://doi.org/https://doi.org/10.3390/en11123416.

[179] S. Khanna, K.S. Reddy, T.K. Mallick, Optimization of finned solar photovoltaic phase change material (finned pv pcm) system, *Int. J. Therm. Sci.* 130 (2018) 313–322. https://doi.org/10.1016/j.ijthermalsci.2018.04.033.

[180] M.T. Chaichan, H.A. Kazem, A.H.A. Al-Waeli, K. Sopian, Controlling the melting and solidification points temperature of PCMs on the performance and economic return of the water-cooled photovoltaic thermal system, *Sol. Energy.* 224 (2021) 1344–1357. https://doi.org/10.1016/j.solener.2021.07.003.

[181] M. Eisapour, A.H. Eisapour, M.J. Hosseini, P. Talebizadehsardari, Exergy and energy analysis of wavy tubes photovoltaic-thermal systems using microencapsulated PCM nano-slurry coolant fluid, *Appl. Energy.* 266 (2020) 114849. https://doi.org/10.1016/j.apenergy.2020.114849.

[182] M. Sardarabadi, M. Passandideh-Fard, M.J. Maghrebi, M. Ghazikhani, Experimental study of using both ZnO/ water nanofluid and phase change material (PCM) in photovoltaic thermal systems, *Sol. Energy Mater. Sol. Cells.* 161 (2017) 62–69. https://doi.org/10.1016/j.solmat.2016.11.032.

[183] S. Preet, B. Bhushan, T. Mahajan, Experimental investigation of water based photovoltaic/thermal (PV/T) system with and without phase change material (PCM), *Sol. Energy.* 155 (2017) 1104–1120. https://doi.org/10.1016/j.solener.2017.07.040.

[184] A.H.A. Al-waeli, H.A. Kazem, M.T. Chaichan, K. Sopian, Experimental investigation of using nano-PCM / nanofluid on a photovoltaic thermal system (PVT): Technical and economic study, *Therm. Sci. Eng. Prog.* (2019). https://doi.org/10.1016/j.tsep.2019.04.002.

[185] M. Moein-Jahromi, H. Rahmanian-Koushkaki, S. Rahmanian, S. Pilban Jahromi, Evaluation of nanostructured GNP and CuO compositions in PCM-based heat sinks for photovoltaic systems, *J. Energy Storage.* 53 (2022) 105240. https://doi.org/10.1016/j.est.2022.105240.

[186] J.M. Mahdi, H.I. Mohammed, P. Talebizadehsardari, A new approach for employing multiple PCMs in the passive thermal management of photovoltaic modules, *Sol. Energy.* 222 (2021) 160–174. https://doi.org/10.1016/j.solener.2021.04.044.

[187] A.S. Soliman, L. Xu, J. Dong, P. Cheng, A novel heat sink for cooling photovoltaic systems using convex/concave dimples and multiple PCMs, *Appl. Therm. Eng.* 215 (2022). https://doi.org/10.1016/j.applthermaleng.2022.119001.

[188] Z. Fu, Y. Li, X. Liang, S. Lou, Z. Qiu, Z. Cheng, Q. Zhu, Experimental investigation on the enhanced performance of a solar PVT system using micro-encapsulated PCMs, *Energy.* 228 (2021) 120509. https://doi.org/10.1016/j.energy.2021.120509.

[189] N. Savvakis, T. Tsoutsos, Theoretical design and experimental evaluation of a PV+PCM system in the mediterranean climate, *Energy.* 220 (2021) 119690. https://doi.org/10.1016/j.energy.2020.119690.

[190] J. Feng, J. Huang, Z. Ling, X. Fang, Z. Zhang, Performance enhancement of a photovoltaic module using phase change material nanoemulsion as a novel cooling fluid, *Sol. Energy Mater. Sol. Cells.* 225 (2021) 111060. https://doi.org/10.1016/j.solmat.2021.111060.

[191] M.M. Islam, M. Hasanuzzaman, N.A. Rahim, A.K. Pandey, M. Rawa, L. Kumar, Real time experimental performance investigation of a NePCM based photovoltaic thermal system: An energetic and exergetic approach, *Renew. Energy.* 172 (2021) 71–87. https://doi.org/10.1016/j.renene.2021.02.169.

[192] A. Kazemian, M. Khatibi, S. Reza Maadi, T. Ma, Performance optimization of a nanofluid-based photovoltaic thermal system integrated with nano-enhanced phase change material, *Appl. Energy.* 295 (2021) 116859. https://doi.org/10.1016/j.apenergy.2021.116859.

[193] F. Abo-elnour, E.B. Zeidan, A.A. Sultan, E. El-negiry, A.S. Soliman, Enhancing the bifacial PV system by using dimples and multiple PCMs, *J. Energy Storage.* 70 (2023) 108079. https://doi.org/10.1016/j.est.2023.108079.

[194] A. Mourad, A. Aissa, Z. Said, O. Younis, M. Iqbal, A. Alazzam, Recent advances on the applications of phase change materials for solar collectors, practical limitations, and challenges: A critical review, *J. Energy Storage.* 49 (2022) 104186. https://doi.org/10.1016/j.est.2022.104186.

[195] M. Carmona, M. Palacio, Thermal modelling of a flat plate solar collector with latent heat storage validated with experimental data in outdoor conditions, *Sol. Energy.* 177 (2019) 620–633. https://doi.org/10.1016/j.solener.2018.11.056.

[196] F. Zhou, J. Ji, W. Yuan, X. Zhao, S. Huang, Study on the PCM flat-plate solar collector system with antifreeze characteristics, *Int. J. Heat Mass Transf.* 129 (2019) 357–366. https://doi.org/10.1016/j.ijheatmasstransfer.2018.09.114.

[197] M.J. Alshukri, A.A. Eidan, S.I. Najim, Thermal performance of heat pipe evacuated tube solar collector integrated with different types of phase change materials at various location, *Renew. Energy.* 171 (2021) 635–646. https://doi.org/10.1016/j.renene.2021.02.143.

[198] C. Kutlu, Y. Zhang, T. Elmer, Y. Su, S. Riffat, A simulation study on performance improvement of solar assisted heat pump hot water system by novel controllable crystallization of supercooled PCMs, *Renew. Energy.* 152 (2020) 601–612. https://doi.org/10.1016/j.renene.2020.01.090.

[199] H. Huang, Y. Xiao, J. Lin, T. Zhou, Y. Liu, Q. Zhao, Improvement of the efficiency of solar thermal energy storage systems by cascading a PCM unit with a water tank, *J. Clean. Prod.* 245 (2020) 118864. https://doi.org/10.1016/j.jclepro.2019.118864.

[200] R. Elarem, T. Alqahtani, S. Mellouli, W. Aich, N. Ben Khedher, L. Kolsi, A. Jemni, Numerical study of an Evacuated Tube Solar Collector incorporating a Nano-PCM as a latent heat storage system, *Case Stud. Therm. Eng.* 24 (2021) 100859. https://doi.org/10.1016/j.csite.2021.100859.

[201] W. Ke, J. Ji, L. Xu, B. Yu, X. Tian, J. Wang, Numerical study and experimental validation of a multi-functional dual-air-channel solar wall system with PCM, *Energy.* 227 (2021) 120434. https://doi.org/10.1016/j.energy.2021.120434.

[202] X. Kong, L. Wang, H. Li, G. Yuan, C. Yao, Experimental study on a novel hybrid system of active composite PCM wall and solar thermal system for clean heating supply in winter, *Sol. Energy.* 195 (2020) 259–270. https://doi.org/10.1016/j.solener.2019.11.081.

[203] P. Pourmoghadam, M. Farighi, F. Pourfayaz, A. Kasaeian, Annual transient analysis of energetic, exergetic, and economic performances of solar cascade organic Rankine cycles integrated with PCM-based thermal energy storage systems, *Case Stud. Therm. Eng.* 28 (2021) 101388. https://doi.org/10.1016/j.csite.2021.101388.

[204] G. Wheatley, R.I. Rubel, Design improvement of a laboratory prototype for efficiency evaluation of solar thermal water heating system using phase change material (PCMs), *Results Eng.* 12 (2021) 100301. https://doi.org/10.1016/j.rineng.2021.100301.

[205] V.R. Pawar, S. Sobhansarbandi, Heat transfer enhancement of a PCM-porous metal based heat pipe evacuated tube solar collector: An experimental study, *Sol. Energy.* 251 (2023) 106–118. https://doi.org/10.1016/j.solener.2022.10.054.

[206] M. Weng, J. Su, J. Lin, J. Huang, Y. Min, Solar Energy Materials and Solar Cells Intrinsically lighting absorptive PANI / MXene aerogel encapsulated PEG to construct PCMs with efficient photothermal energy storage and stable reusability, *Sol. Energy Mater. Sol. Cells.* 254 (2023) 112282. https://doi.org/10.1016/j.solmat.2023.112282.

[207] H.M. Teamah, M. Teamah, Integration of phase change material in flat plate solar water collector: A state of the art, opportunities, and challenges, *J. Energy Storage.* 54 (2022) 105357. https://doi.org/10.1016/j.est.2022.105357.

[208] M.F.I. Al Imam, R.A. Beg, M.J. Haque, M.S. Rahman, Effect of novel phase change material (PCM) encapsulated design on thermal performance of solar collector, *Results Mater.* 18 (2023). https://doi.org/10.1016/j.rinma.2023.100388.

[209] Q. Luo, B. Li, Z. Wang, S. Su, H. Xiao, C. Zhu, Thermal modeling of air-type double-pass solar collector with PCM-rod embedded in vacuum tube, *Energy Convers. Manag.* 235 (2021) 113952. https://doi.org/10.1016/j.enconman.2021.113952.

[210] S. Madhankumar, K. Viswanathan, W. Wu, M. Ikhsan Taipabu, Analysis of indirect solar dryer with PCM energy storage material: Energy, economic, drying and optimization, *Sol. Energy.* 249 (2023) 667–683. https://doi.org/10.1016/j.solener.2022.12.009.

[211] G. Srinivasan, D.K. Rabha, P. Muthukumar, A review on solar dryers integrated with thermal energy storage units for drying agricultural and food products, *Sol. Energy.* 229 (2021) 22–38. https://doi.org/10.1016/j.solener.2021.07.075.

[212] P. Balasundaram, B. Baranidharan, N.M. Sivaram, Materials today: Proceedings A VIKOR based selection of phase change material for thermal energy storage in solar dryer system, *Mater. Today Proc.* 90 (2023) 245–249. https://doi.org/10.1016/j.matpr.2023.06.174.

[213] M. Sharma, D. Atheaya, A. Kumar, Recent advancements of PCM based indirect type solar drying systems: A state of art, *Mater. Today Proc.* 47 (2021) 5852–5855. https://doi.org/10.1016/j.matpr.2021.04.280.

[214] M. Everts, J.P. Meyer, Flow regime maps for smooth horizontal tubes at a constant heat flux, *Int. J. Heat Mass Transf.* 117 (2018) 1274–1290. https://doi.org/10.1016/j.ijheatmasstransfer.2017.10.073.

[215] N. Vigneshkumar, M. Venkatasudhahar, P. Manoj Kumar, A. Ramesh, R. Subbiah, P. Michael Joseph Stalin, V. Suresh, M. Naresh Kumar, S. Monith, R. Manoj Kumar, M. Kriuthikeswaran, Investigation on indirect solar dryer for drying sliced potatoes using phase change materials (PCM), *Mater. Today Proc.* 47 (2021) 5233–5238. https://doi.org/10.1016/j.matpr.2021.05.562.

[216] A.K. Bhardwaj, R. Kumar, R. Chauhan, S. Kumar, Experimental investigation and performance evaluation of a novel solar dryer integrated with a combination of SHS and PCM for drying chilli in the Himalayan region, *Therm. Sci. Eng. Prog.* 20 (2020) 100713. https://doi.org/10.1016/j.tsep.2020.100713.

[217] A.K. Bhardwaj, R. Kumar, S. Kumar, B. Goel, R. Chauhan, Energy and exergy analyses of drying medicinal herb in a novel forced convection solar dryer integrated with SHSM and PCM, *Sustain. Energy Technol. Assessments.* 45 (2021) 101119. https://doi.org/10.1016/j.seta.2021.101119.

[218] A.A. Ananno, M.H. Masud, P. Dabnichki, A. Ahmed, Design and numerical analysis of a hybrid geothermal PCM flat plate solar collector dryer for developing countries, *Sol. Energy.* 196 (2020) 270–286. https://doi.org/10.1016/j.solener.2019.11.069.

[219] Z. Alimohammadi, H. Samimi Akhijahani, P. Salami, Thermal analysis of a solar dryer equipped with PTSC and PCM using experimental and numerical methods, *Sol. Energy.* 201 (2020) 157–177. https://doi.org/10.1016/j.solener.2020.02.079.

[220] H. Atalay, E. Cankurtaran, Energy, exergy, exergoeconomic and exergo-environmental analyses of a large scale solar dryer with PCM energy storage medium, *Energy.* 216 (2021) 119221. https://doi.org/10.1016/j.energy.2020.119221.

[221] M. Bahari, B. Najafi, A. Babapoor, Evaluation of α-AL2O3-PW nanocomposites for thermal energy storage in the agro-products solar dryer, *J. Energy Storage.* 28 (2020). https://doi.org/10.1016/j.est.2019.101181.

[222] A. Reyes, A. Mahn, F. Cubillos, P. Huenulaf, Mushroom dehydration in a hybrid-solar dryer, *Energy Convers. Manag.* 70 (2013) 31–39. https://doi.org/10.1016/j.enconman.2013.01.032.

[223] S. Madhankumar, K. Viswanathan, W. Wu, M. Ikhsan, Analysis of indirect solar dryer with PCM energy storage material: Energy, economic, drying and optimization, *Sol. Energy.* 249 (2023) 667–683. https://doi.org/10.1016/j.solener.2022.12.009.

[224] M. Reza, M.H. Abbaspour-fard, M. Hedayatizadeh, Design, thermal simulation and experimental study of a hybrid solar dryer with heat storage capability, *Sol. Energy.* 258 (2023) 232–243. https://doi.org/10.1016/j.solener.2023.05.003.

[225] A. Lingayat, P. Das, M.C. Gilago, V.P. Chandramohan, A detailed assessment of paraffin waxed thermal energy storage medium for solar dryers, *Sol. Energy.* 261 (2023) 14–27. https://doi.org/10.1016/j.solener.2023.05.047.

8 Nanomaterial-Enhanced Phase Change Materials

P. Sivasamy, S. Harikrishnan, and Hafiz Muhammad Ali

HIGHLIGHTS

- Improved thermal properties were observed in nanocomposite phase change materials.
- Higher thermal conductivity was found in nanocomposite phase change materials using various volume concentrations of nanomaterials.
- The use of nanocomposite phase change materials in buildings diminished the energy consumption and enhanced the system performance.

LIST OF SYMBOLS AND ABBREVIATIONS

CA	Capric acid
CLF	Cavity liquid fraction
DSC	Differential scanning calorimetry
Exo	Exothermic process
FESEM	Field emission scanning electron microscope
HE	Heat exchanger
HTF	Heat transfer fluid
LA	Lauric acid
LHTES	Latent heat thermal energy storage
MA	Myristic acid
NCPCM	Nanocomposite phase change material
NePCM	Nanomaterial-enhanced phase change material
nm	Nanometer
NP	Nanoparticle
PA	Palmitic acid
PCM	Phase change material
RTD	Resistance temperature detector
SA	Stearic acid
SDBS	Sodium dodecyl benzene sulfonate
STHE	Shell-and-tube heat exchanger
TEM	Transmission electron microscope
TES	Thermal energy storage
TGA	Thermogravimetric analysis

DOI: 10.1201/9781003331957-8

| wt% | Weight percentage |
| XRD | X-ray diffraction |

NOMENCLATURE

Heat flow	mW
m	Meter
mins	Minutes
Nu	Nusselt number
Nu$_{avg}$	Average Nusselt number
Ra	Rayleigh number
sec	Seconds
T	Temperature ($^\circ$C)
θ_{avg}	Average temperature
θ_{max}	Maximum temperature
τ	Dimensionless time

8.1 INTRODUCTION

Energy requirements for a number of different applications in various forms and phases are time-dependent due to the limited availability of energy sources. This suggests that the existing energy sources and applications ought to be supplemented by an effective energy storage system. Regarding renewable energy sources, thermal energy storage (TES) technologies are a rapidly developing area of innovation. Researchers' attention has recently been drawn to TES technologies because of their unique behavior in storing energy for later use, which would lower total energy consumption. The latent heat thermal energy storage (LHTES) systems using phase change materials (PCMs) are the most advantageous, due to their high energy density storage and minimal temperature changes, among the many TES technologies that have been explored over the past two decades [1]. Despite these good benefits, most PCMs have a unique disadvantage, namely an extremely low thermal conductivity [2]. Enhancing the thermal conductivity of PCMs is one technique to increase the efficiency of PCM-based TES systems by reducing the thermal energy phase change processing time and temperature difference. As a result of earlier studies, PCM has recently become the most important research topic in order to improve the performance of TES [3]. Numerous studies have been undertaken to enhance the thermal conductivity of PCM by inserting highly conductive metal fins and fibers in diverse forms such as fins, honeycomb, metallic wool, and foam [4]. Metal fins and fibers, however, add weight and cost to TES systems. Moreover, there are difficulties in identifying the optimal structures of these fixed garnishes and their connections with conduction or convection heat transfer associated with phase transition. The novel idea of employing ultrafine nanoparticles (NPs), which typically have a diameter on the order of 10 to 50 nm, has, on the other hand, quickly advanced in the field of nanotechnology, and these NPs are now commercially available in a variety of metals and metal oxides. Advanced heat transfer fluids known as nanofluids have been created using these extremely conductive NPs. These nanofluids have far more thermal conductivity than base fluids [5–8].

As a result, the use of manufactured NPs creates several prospects for new technological advancements in the synthesis of materials, leading to the identification of nanocomposite phase change materials (NCPCMs) that have been functionally evaluated. The thermo-physical characteristics of pure PCMs, such as thermal conductivity, latent heat, viscosity, supercooling, density, etc., might be significantly altered by adding NPs. Khodadadi et al. [9] and Ramm Dheep and Sreekumar [10] recently published a review article demonstrating the utility of nanoparticles in improving the thermal conductivity of PCMs for TES. In the published research, a variety of works have been preceded by an investigation into the thermos-physical characteristics of PCM with the presence of NPs, in which various reports focus on various characteristics of NCPCM. The literature is included with these investigations. A detailed literature study has been conducted in this area to show how the dispersion of various NPs has an impact on the thermo-physical properties of PCMs.

8.2   PREPARATION METHODS

The manufacture of NCPCMs does not include the mixing of NPs directly into the base PCM. Instead, in order to achieve a homogenous dispersion of NPs, the preparation of NCPCMs requires the use of specific mixing and stabilizing processes. NCPCMs may be prepared using a one- or two-step process. NPs are manufactured and distributed in the pure PCM in the one-step technique, but in the two-step process, NPs are produced first and then distributed in the pure PCMs. Moreover, the two-step approach is more cost-effective, and it is widely employed nowadays to obtain NCPCMs in mass production [11,12].

NPs are first produced in dry powder form utilizing a variety of mechanical and chemical techniques, including the sol–gel process, ball milling, and chemical reduction. The produced NPs are then combined with the base PCM, and various techniques, including high-shear mixing, ultrasonic bathing, and magnetic stirring,

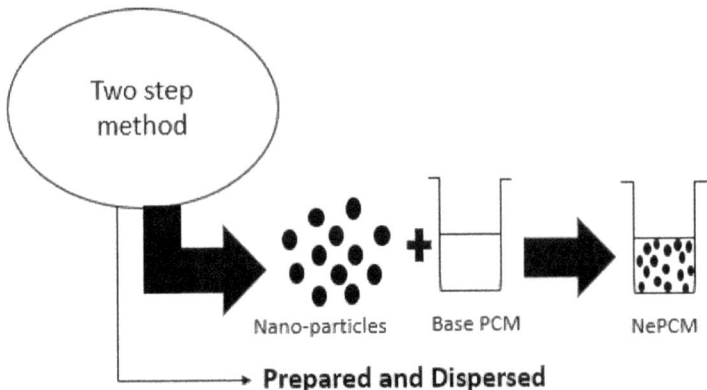

FIGURE 8.1   Schematic of a two-step process [13].

are utilized to ensure that the particles are homogeneously dispersed. A two-step process is schematically represented in Figure 8.1 [13].

NCPCM preparation process was influenced by the PCM's requirements. The procedure for preparing NCPCM is heavily influenced by the PCM's melting point. Kumar et al. [14] stated that NCPCM with methyl ester having a phase transition temperature range from 20–40°C was suitable for preparation, as shown in Figure 8.2. For the preparation of the NCPCM, a novel two-step process is used. The homogenous solution was obtained by dispersing copper-titania in methyl ester for 30 minutes using a stirrer.

In recent investigations, paraffin was employed as a PCM to prepare CuO, Al_2O_3, and Fe_3O_4 NCPCMs, as reported by Jin Wang et al. [15]. These NPs were dispersed in the paraffin using a two-step process, and sodium dodecylbenzene sulfonate (SDBS) was added as a capping agent to ensure their homogenous dispersion. After 50 minutes of sonication, the NPs were uniformly dispersed in the pure PCM. The preparation process for composite PCMs is shown in Figure 8.3. Accordingly, prominent researchers [16–20] presented different NCPCMs using this method at various sonication times, as shown in Table 8.1.

NPs are prepared and dispersed concurrently in the one-step technique. This approach improves NP dispersion in the corresponding PCM and lessens the issue of particle agglomeration because it does not involve drying, storing, moving, or mixing NPs in the pure PCM [21]. Because it costs more than the two-step process, this process is only used on a small scale. Figure 8.4 illustrates a diagram of a one-step process.

By employing a direct synthesis approach, Teng et al. [22] obtained an NCPCM by combining paraffin wax with Al_2O_3, TiO_2, SiO_2, and ZnO NPs in mass fractions

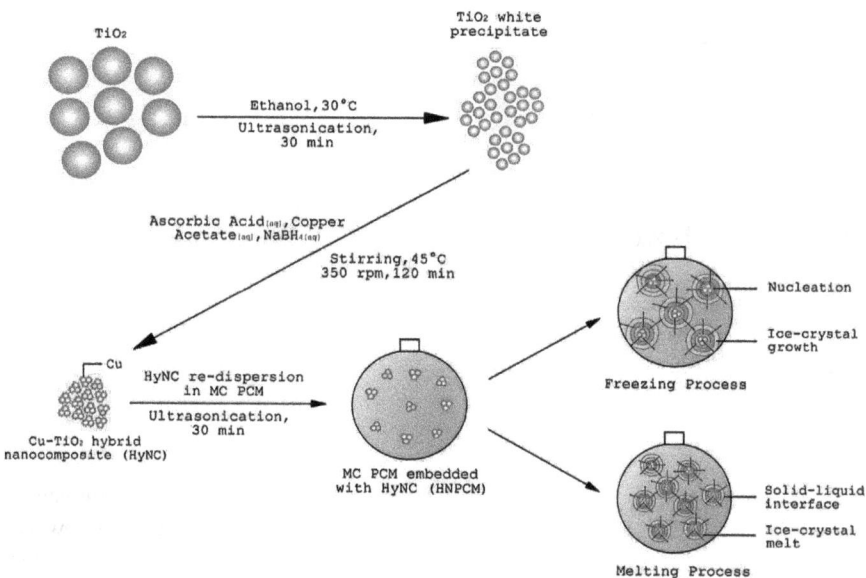

FIGURE 8.2 Schematic diagram of NCPCM preparation method [14].

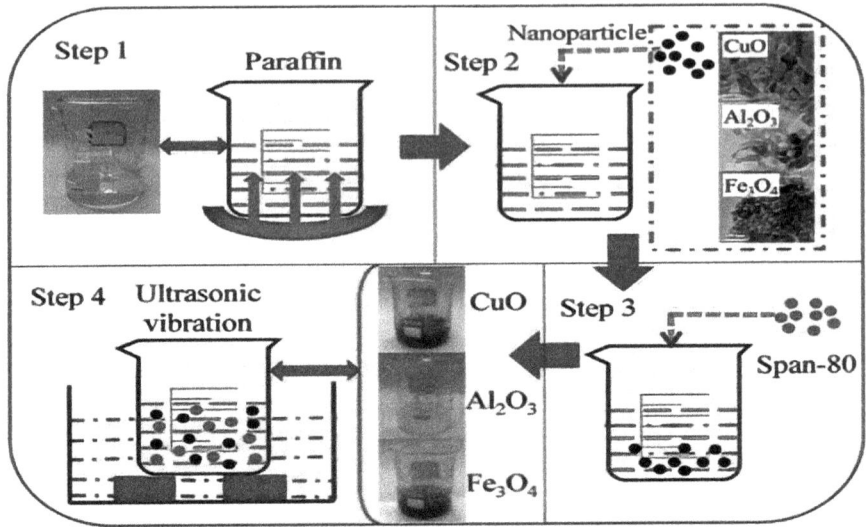

FIGURE 8.3   Schematic preparation of nanocomposite PCMs [15].

TABLE 8.1
An Overview of NCPCMs Made Using the Sonication Process

Reference	Nanoparticles	Pure PCM	Preparation method	Time for sonication (mins)	Surfactant
[14]	Cu–TiO$_2$	Methyl ester	Two-step	30	Ethanol
[15]	CuO, Al$_2$O$_3$, and Fe$_3$O$_4$	Paraffin	Two-step	50	-
[19]	MWCNT	Paraffin	Two-step	60	-
[20]	γ-Al$_2$O$_3$	CaCl$_2$·6H$_2$O	Two-step	30	SrCl$_2$·6H$_2$O
[18]	Ag-TiO$_2$	Ethyl transcinnamate	Two-step	30	Ethanol
[17]	TiO$_2$, CuO, ZnO	LA and SA mixture	Two-step	60	SDBS
[16]	TiO$_2$	Poly carboxylic acid	Two-step	30	-
[22]	α-Nano alumina	Petroleum wax	One-step	30	-
[23]	Al2O$_3$, TiO$_2$, SiO$_2$, and ZnO	Paraffin wax	One-step	40	-

of 1, 2, and 3. The mixture was then agitated for 45 mins to obtain a homogenous dispersion. NCPCM was examined by Mohamed et al. [25] after being microwaved for 4–6 minutes. Petroleum wax was then mixed with this solution, and the mixture was sonicated for 35 minutes to produce the necessary NCPCM. Table 8.1 shows the details of NCPCMs prepared using this method.

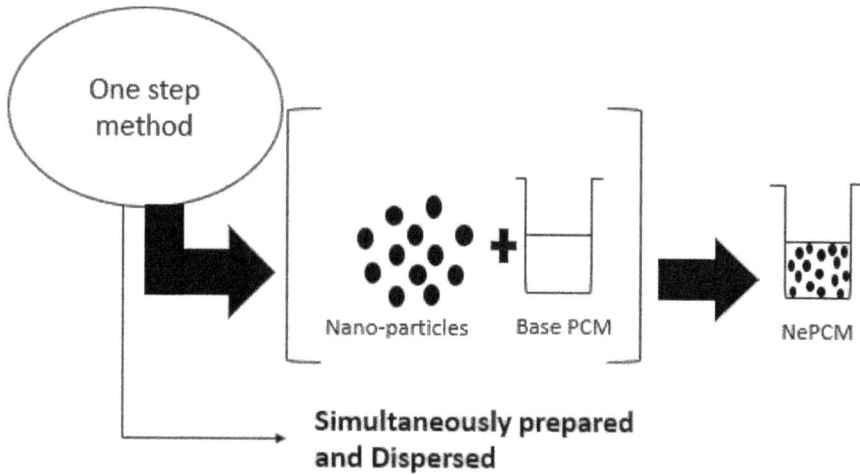

One step
method

Nano-particles   Base PCM

NePCM

**Simultaneously prepared
and Dispersed**

FIGURE 8.4   Schematic of a one-step process [13].

8.3   NANOPARTICLE-ENHANCED PCMS

The majority of studies in the literature focused on enhancing the thermal characteristics of PCMs like paraffin and fatty acids. Researchers are very interested in paraffin because of its good heat storage density, ability to melt or solidify with little or no subcooling, lack of reactivity with certain common chemical reagents, and low price [24]. The characteristics of fatty acids are quite similar to those of paraffin [25]. The main disadvantage of these PCMs is their poor thermal conductivity [26]. The thermal conductivity of these PCMs could be improved by dispersing NPs. The heat transfer rates during melting/freezing operations in the heat storage system are constrained by the poor thermal conductivity of PCMs. To address this issue, a variety of experiments have aimed to increase the thermal conductivity of organic PCMs or enhance their heat transfer ability. A few of the studies discussed in this chapter focused on the thermal properties of these PCMs with various NPs.

According to the authors' knowledge, Elgafy and Lafdi were among the first to investigate NCPCMs in 2005. They investigated paraffin wax improved with carbon nanofibers both analytically and through experiments. Different carbon nanofiber mass fractions' thermo-physical characteristics and freezing temperatures were recorded. Thermal diffusivities were evaluated with a laser flash method. Moreover, as the number of carbon fibers was improved, the specific heat was diminished. It was found that increasing the number of carbon nanofibers in the NCPCM shortened the freezing time and enhanced the output power. There was a good correlation between the analytical solution and the experimental findings. The rate of freezing was also sped up by changing the nanofibers' areas [27].

One of the first researcher teams to explore NCPCMs was Khoddadi and Hosseinizadeh, whose numerical model for base PCM was augmented with copper nanoparticles. They examined how nanoparticle dispersion affected the latent heat of fusion, freezing rate, and thermal conductivity. The dispersion of nanoparticles accelerated the speed of heat release, improved thermal conductivity, and lowered the latent heat of fusion. NCPCM was found to be promising for TES applications [28]. Another water PCM study [29] numerically examined copper nanoparticles for improvement. The effect of nanoparticle mass fraction and Grashof number on heat transmission in NCPCM was investigated using a lattice Boltzmann method. Depending on the nanoparticle mass fraction, larger Grashof numbers resulted in various asymmetric convection patterns. Due to the enhanced thermal conductivity, the local Nusselt number fell as the nanoparticle mass fraction increased. The dispersion of nanoparticles increased the efficiency of heat transfer, and a large volume proportion accelerated heating and temperature change. Mass fraction also led to an increase in energy storage. The PCM heating process was definitely improved by the copper nanoparticles [29].

Paraffin-based PCMs are a good option, and they are very crucial to investigate due to their diverse spectrum of minimal temperature heating points and suitability with various kinds of nanoparticles. In order to quantitatively analyze different LHTES systems employing different paraffin wax PCMs with low and high heating and cooling processes, Akhmetov et al. [30] employed Comsol multi-physics. To improve the system's ability to transport heat, Al_2O_3 nanoparticles were utilized. The three-dimensional analyses made use of temperature-dependent characteristics. They produced pure wax, 2-mass-fraction NP, and 4-mass-fraction NP mixes for each type of wax. For usage in the analyses, such combinations have been employed to test and compute the thermo-physical characteristics. The high LHTES received heat first, followed by the low LHTES in the order in which the LHTES systems were set up. The poor thermal conductivities of both kinds of paraffin were improved by the Al_2O_3 nanoparticles. The PCM containers were found to receive energy efficiently and equally from the LHTES system (represented in Figure 8.5). The two LHTES systems' sequential design was likewise successful [30].

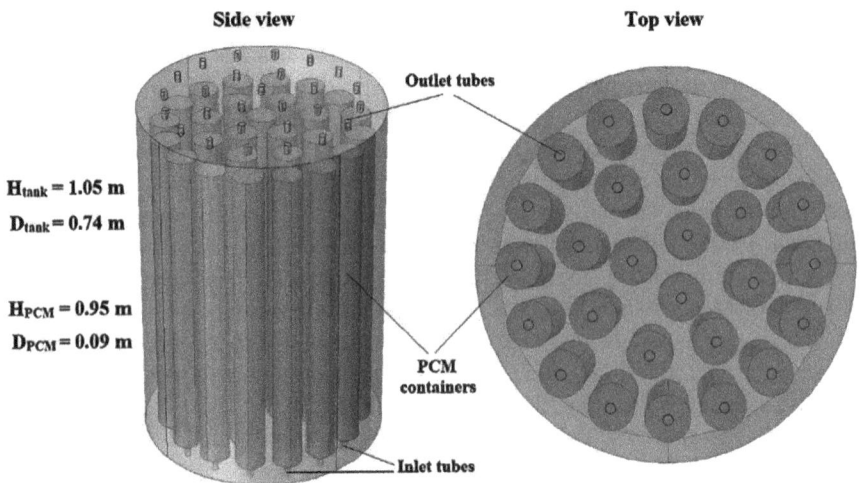

FIGURE 8.5 Configuration of the LHTES system [30].

In past studies, alumina was used as a nanoparticle improvement for paraffin in a variety of studies. In order to study the LHTES system with pure PCM and alumina nanoparticles, Abdulateef et al. [31] conducted mathematical and experimental studies. The heat exchanger (HE) had eight equally sized PCM fins in the central tube and heat transfer fluid in the inlet and outlet. In the same HE, longitudinal and triangular fins were examined numerically. The experimental findings demonstrated that the exterior melting method completely melted the pure PCM at a lower HTF temperature and time, while the internal melting process did not completely charge the base PCM. The calculated optimal heat transfer fluid mass flow rate was 30 kg/min. The triangular fins sped up the heating process, according to the numerical analysis, and the dispersion of nanoparticles made the system even better [31]. Two-dimensional mathematical research was carried out by Farsani et al. [32] using base PCM and Al_2O_3 nanoparticles. The impacts of the rate of heat transfer and phase transition were first investigated, followed by investigations into the effects of nanoparticles on the heating process. This revealed that the convection vortex produced while heating was greater when greater Rayleigh number (Ra) was used. This was discovered when the Rayleigh values were modified to indicate the rate of heat generation. After heating of the NCPCM, Figure 8.6 displays the liquid percent, average non-dimensional temperature, maximum non-dimensional temperature, and Nu_{avg} for Rayleigh values. As a result, the researchers concluded that nanoparticles

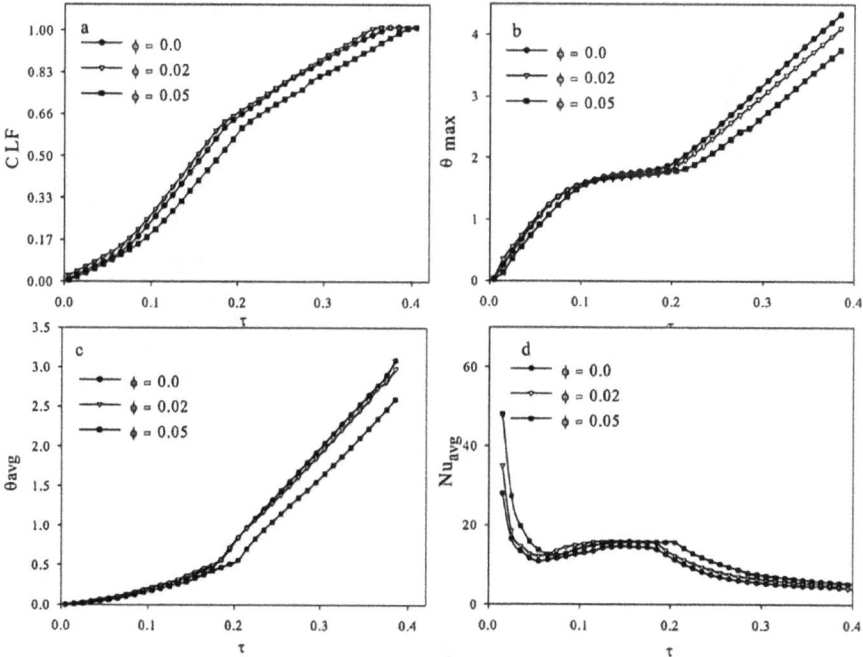

FIGURE 8.6 Macroscopic characteristics of the composite PCM during the melting process (a) CLF, (b) θ_{max}, (c) θ_{avg}, and (d) Nu_{avg} [32].

had a negligible impact on the Nu and temperature during the heating process (as demonstrated in Figure8.6) [32].

A two-dimensional simulation model was created in another investigation by Elbahjaoui and El Qarnia to analyze the freezing of base PCM with Cu NPs, another well-liked NP type. The TES system was made up of rectangular PCM slabs with heat transfer fluid positioned in the middle. The impacts of the NP wt%, PCM proportions, and heat transfer fluid inlet temperature on the thermal properties of the LHTES system during the freezing process were investigated using the enthalpy–porosity approach to the melting and freezing process. The system's functioning was enhanced by raising the aspect ratio and bringing down the HTF inlet temperature. The freezing rate was also increased by the NPs [33].

Carbon-based nanoparticles are suggested to be more efficient than metallic NPs, which are already quite successful. Aqib et al. looked into the performance of nonmetallic MWCNTs and base PCM augmented with Al_2O_3 NPs at three different mass fractions. Mixtures had 2, 4, or 6 wt% in them. The temperatures of the combinations were monitored using thermocouples as they went through the phase change process. The heat transfer improved along with the nanoparticle dispersion. The MWCNTs were a better choice for improvement since they had higher potential temperatures compared to Al_2O_3 NPs [34]. Even very low concentrations of carbon NPs can make systems better. Murugan et al. [35] studied the phase transition temperature of the composite PCM comprising paraffin and MWCNTs and it was determined as 59–61°C. DSC measurements of the NCPCMs' specific heats were followed by temperature measurements using RTDs after the sample had a phase change. Heating times decreased by 32% at 0.3 wt% MWCNTs, possibly as a result of increased viscosity beyond this point. The lowest cooling periods, with a reduction of 43%, were at 0.9 wt% [35]. Moreover, Harikrishnan et al. [17] studied TiO_2, ZnO, and CuO in a lauric acid (LA) and stearic acid (SA) mixture using a two-step method. Despite the fact that all NPs improved the system, CuO was proven to be the dominant NP type and TiO_2 to be the weakest, as seen in Figure 8.7. TGA results

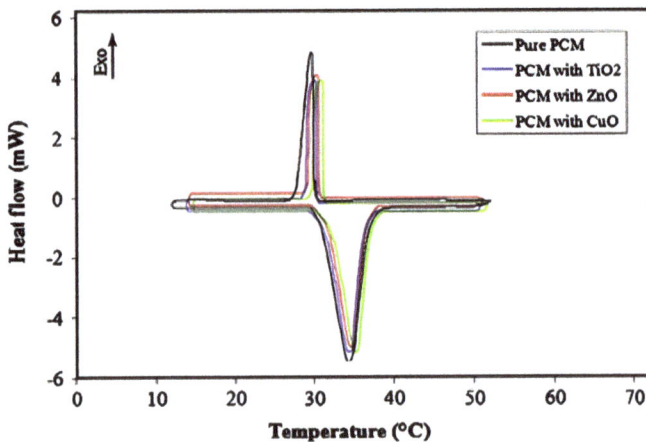

FIGURE 8.7 DSC measurements of the pure PCM and composite PCMs [17].

showed that adding NPs increased the thermal stability of the NCPCMs. However, ZnO NPs, as opposed to TiO_2 and CuO NPs, significantly increased the thermal stability of the NCPCM. In comparison to the 1.0 wt% TiO_2 and ZnO NPs, the thermal conductivity of the NCPCM was much higher for CuO NPs [17].

By employing mathematical research, Khatibi et al. [36] employed a variety of metal oxide NPs, including a few of the ones mentioned earlier, to improve the base PCM. To explore the freezing process and test various heat transfer fluid temperatures in three separate LHTES system, Al_2O_3, ZnO, CuO, and SiO_2 were dispersed in the PCM. The systems included a triplex tube and STHE, one for internal cooling and the other for external cooling. For all NP comparison experiments, the triplextube system was modeled. Both with and without NPs, the various systems were then compared. For a mass fraction of 0.02 wt%, all of the NPs showed decreased freezing time, with Al_2O_3 having the greatest effect. It was preceded by ZnO, CuO, and SiO_2 in decreasing order. CuO was the most effective at a mass fraction of 0.04 wt%, followed in that order by ZnO, $Al2O_3$, and SiO_2. The improvement in thermal conductivity and reduction in latent heat were the two factors that contributed to the shorter freezing time. The greatest result was provided by the triplextube system, with external cooling coming in second. The kind of LHTES system had an impact on the impact of the NPs as well as how much the heat transfer fluid temperature dropped [36]. In order to assess the heat transfer efficiency for solar dryer application, Harikrishnan et al. [37] studied the newly prepared NCPCMs employing 0.2, 0.5, 0.8, and 1 wt% SiO_2 NPs. Each NCPCM was shown to have higher thermal conductivities than the base PCM alone. According to the experimental results, NCPCMs with 0.2, 0.5, 0.8, and 1.0 wt% SiO_2 NPs took 7.07, 15.12, 22.21, and 31.96% less time to melt completely than base PCM, respectively. This is shown in Figure 8.8 [37].

To increase thermal conductivity, Maher et al. [38] tested two distinct nanomaterials used in a base PCM. After the NCPCMs were created, they were examined using a FESEM (seen in Figure 8.9), a DSC, and a laser flash device to measure their

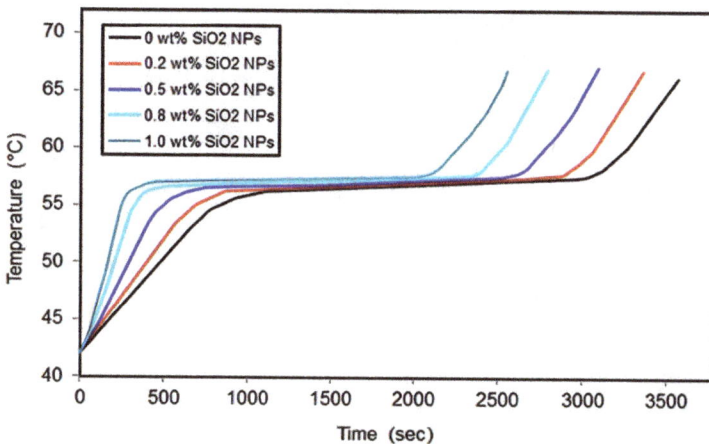

FIGURE 8.8 Heating curve of MA and composite PCMs [37].

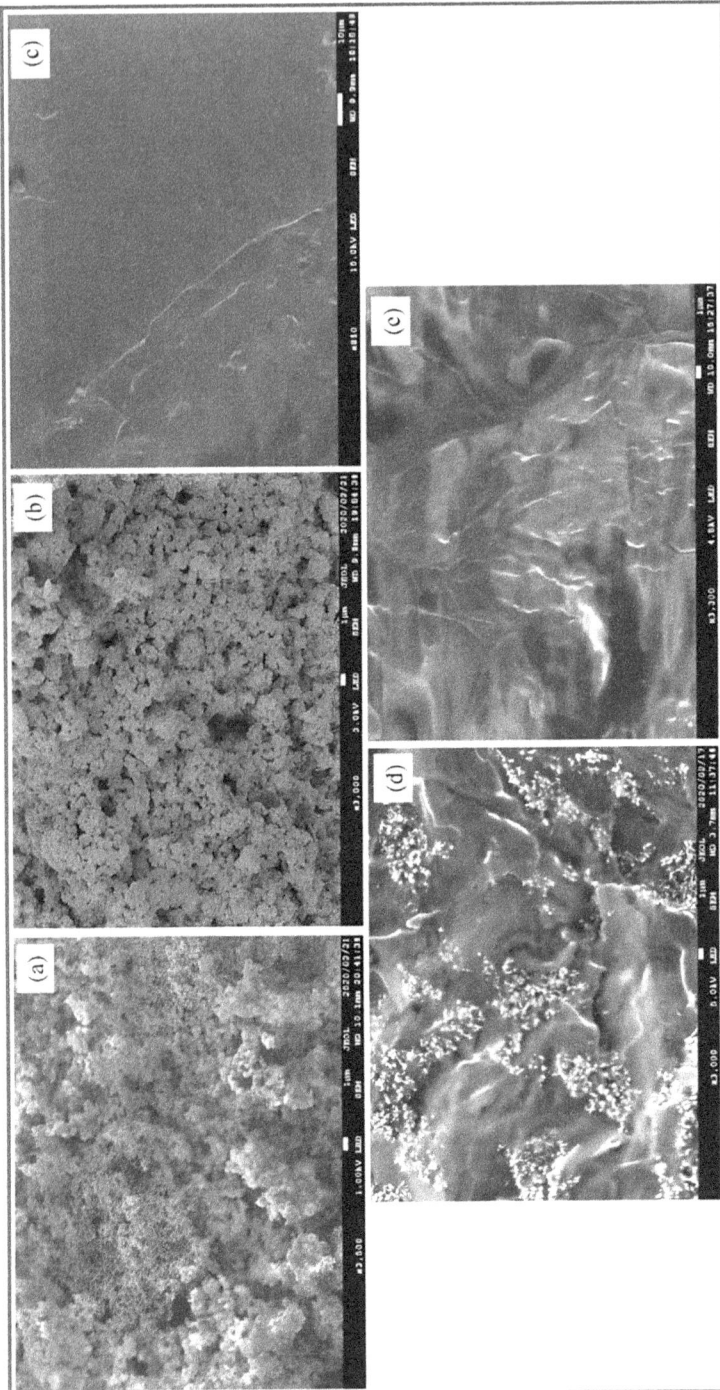

FIGURE 8.9   SEM images of (a) SiC, (b) Ag, (c) paraffin wax, (d) composite PCM (paraffin/SiC), and (e) composite PCM (paraffin/Ag) [38].

thermal conductivity. The maximum amount, 15 wt%, of the SiC NCPCM produced the best results, outperforming the Ag NCPCM in terms of thermal conductivity. However, the NPs decreased the latent heat, heating temperature, and specific heat capacity. As a result, the largest mass fraction is not constantly the best choice, and the number of NPs must be tuned individually [38].

For low-temperature applications, Marcos et al. [39] studied PVP-treated Ag NPs in a base PCM. The mass fraction of the NP dispersion ranged from 0.1 to 1 wt%. They examined the sample density, latent heat, viscosity, and thermal conductivity. TEM, TGA, and DSC were used to examine properties. The 1 wt% NPs enhanced thermal conductivity by up to 4 wt%. The inclusion of NPs significantly decreased inappropriate subcooling [39].

PA and CA PCMs improved by copper oxide NPs are two other fatty acid NCPCMs that were characterized by Barreneche et al. [40]. The NCPCMs were characterized by SEM, XRD, FTIR and TGA. Thermal conductivity and LH were assessed after the samples had been melted. Thermal conductivity increased for CA up to 1.5 wt% NPs and for PA up to 3 wt%. For both, maximum thermal storage was attained at 1 wt% [40].

A high-temperature PCM was the subject of a different investigation by Chieruzzi et al. [41], who tested pure PCM with 1 wt% SiO_2, Al_2O_3, and hybrid NPs. With these ingredients, the researchers intended to create a PCM with a heating point between 280 and 360°C. NPs raised the specific heat in the solidified phase by around 4–11% and in the melted phase by about 7%. SiO_2 emerged as the most attractive of the NPs, decreasing the pure PCM phase change temperature by 3°C and raising the latent heat by up to 13%. The SiO_2 NPs were much more evenly disseminated compared to the other NPs, according to SEM. Tt was determined thatthe materials might be advantageous for CSP systems [41].

Similar experiments were conducted by Han et al. [42] using a base PCM augmented with Al_2O_3, CuO, and ZnO NPs as different high-temperature PCM mixtures. The dispersion of NPs resulted in a small drop in latent heat but a rise in sensible heat, with little apparent change in the TES. Al_2O_3 nanoparticles had the highest impact on thermal conductivity improvements, followed by CuO and ZnO. Al_2O_3 again showed the greatest improvement in the thermal stability according to TGA, which was good. Al_2O_3 was found to be the best NP for improving this base PCM [42].

A variety of PCMs and NPs, including a significant number of pure PCMs, have been studied using both numerical and experimental methods. NPs were discovered to be an efficient improvement over the base PCM in almost every instance. The most efficient NP varied depending on the investigation and the types examined when they were compared. It appears to be influenced by the heating temperature and the PCM being used. Future studies might involve testing various heating temperatures with various NP kinds in order to establish a link. Based on the NCPCM application, pure PCM, and PCM heating temperature, the "best" NP varies greatly based on the circumstances. The studies discussed in this section are included in Table 8.2.

As was already mentioned, NCPCMsare frequently used with other techniques for improving heat transfer to further improve the LHTES system. Due to the size

TABLE 8.2
A Summary of Experiments that Used NPs to Improve PCM Thermal Characteristics

Reference	Nanoparticles	Pure PCM	Properties	Result
[27]	Carbon nanofibers	Paraffin	• Thermal diffusivity • Heat capacity • Thermal conductivity	The carbon nanofibers improved thebase PCM.
[28]	Cu	Water	• Thermal conductivity • Freezing Process	The NPs improved the thermal conductivity and freezing.
[29]	Cu	Water	• Melting • Freezing	The Cu nanoparticles improved the phase change properties.
[30]	Aluminum oxide	Paraffin wax	• Thermal diffusivity • Thermal conductivity • Phase change properties	The dispersion of Al_2O_3 NPs enhanced the pure PCM.
[31]	Aluminum oxide	Paraffin RT-82	• Thermal conductivity • Melting process	Improved the performance of paraffin due to addition of NPs.
[32]	Al_2O_3	Paraffin	• Thermal conductivity • Viscosity • Specific heat capacity	NPs offered little to no enhancement.
[33]	Cu	n-Octadecane	• Freezing Process	Enhanced the performance of PCM owing to dispersion of NPs.
[34]	Cu and Al_2O_3	Paraffin RT-50	• Heating • Cooling	Al_2O_3 was less effective than MWCNT.
[35]	MWCNT	Paraffin	• DSC • Melting • Freezing	Phase change properties were enhanced due to 0.3 wt%. MWCNT NPs.
[17]	TiO_2, ZnO, and CuO NPs	Lauric acid/ stearic acid mixture	• Thermal stability • Thermal reliability • Thermal conductivity • Phase change properties	CuO NPs offered the best enhancement.
[36]	Al_2O_3, CuO, SiO_2, ZnO	Paraffin RT-82	• Freezing process	Solidification time was decreased by using CuO-PCM 4%.
[37]	SiO_2	Myristic acid	• Thermal Conductivity • Phase change properties	Melting and solidification time was decreased by using 1 wt% SiO_2.
[38]	SiC and Ag	Paraffin wax	• DSC • Thermal Conductivity	SiC addition at 15 wt % showed better improvement.

Reference	Nanoparticles	Pure PCM	Properties	Result
[39]	Silver	PEG	• Thermal conductivity • Density • Diffusivity • Viscosity	Ag NPs improved thermal conductivity.
[40]	CuO	Capric acid Palmitic acid	• Viscosity • Density • Thermal conductivity	Palmitic acid was improved with up to 3.0 wt% NPs.
[41]	SiO_2, Al_2O_3, and SiO_2-Al_2O_3	Potassium nitrate	• DSC • Morphology studies	Silica improved substantially more than any other NCPCM.
[42]	Al2O3, CuO, and ZnO	KCl:MgCl$_2$: NaCl	• Thermal diffusivity • Thermal conductivity	Al_2O_3 exhibited the greatest improvement.

of the particles, NCPCMs, unlike certain other improvement techniques, can be coupled with almost any kind of improvement. An NCPCM with fins was evaluated in the following studies. The improvement of an LHTES system with branched fins and hybrid NPs in water as the NCPCM was statistically explored by Hosseinzadeh et al. [43]. Both individually and in combination with NPs and fins, the system was investigated. Compared to rectangular fins and no fins, the outcomes from the tree-like fins were good. The addition of nanomaterials without fins demonstrated that the nanomaterials did enhance the process, but not to the same extent as the fins. The maximum weight percentage of NPs and the branching fins were the optimum combination for enhancing performance. This was greatly superior to NPs alone and marginally superior to branching fins alone [43]. Branching fins and NCPCM were also examined by Hajizadeh et al. [44]. The freezing of a base PCM boosted by Y-fins and copper oxide nanomaterials in three separate situations was the main topic of the mathematical analysis. As illustrated in Figure 8.10, the first example was finless, the second had lengthy Y-fins on top and V-fins on bottom, and the third example had V-fins on top and Y-fins on top. While both of the following cases were evaluated both with and without nanomaterials, the first example was studied using two mass fractions of nanofluid. The least effective system was the one without fins, whereas case two was more effective than case three. Due to buoyancy effects, example two was more affected by the nanomaterials added than example three. Despite the fact that example two conserved more energy than case three [44], the quantity of energy fell as the number of NPs increased.

Li et al. [45] investigated the melting process of base PCM with various sizes and shapes of Y-fins and also used CuO NPs. There were six different fin cases examined. The best scenario is shown in Figure 8.11 for the last example, which had long, thin fins that alternately adhered the outside and inner wall. The worst scenario included short fins that were only linked to the inner wall in instance one. Natural convection was made possible by longer, thinner fins, which sped up the heating [45].

For fin–NP-improved LTES systems, Cu and CuO appear to be widely used and efficient materials. Combinations of different fins and NP materials are what are missing. Additionally, several researchers concluded that the NPs were overkill as

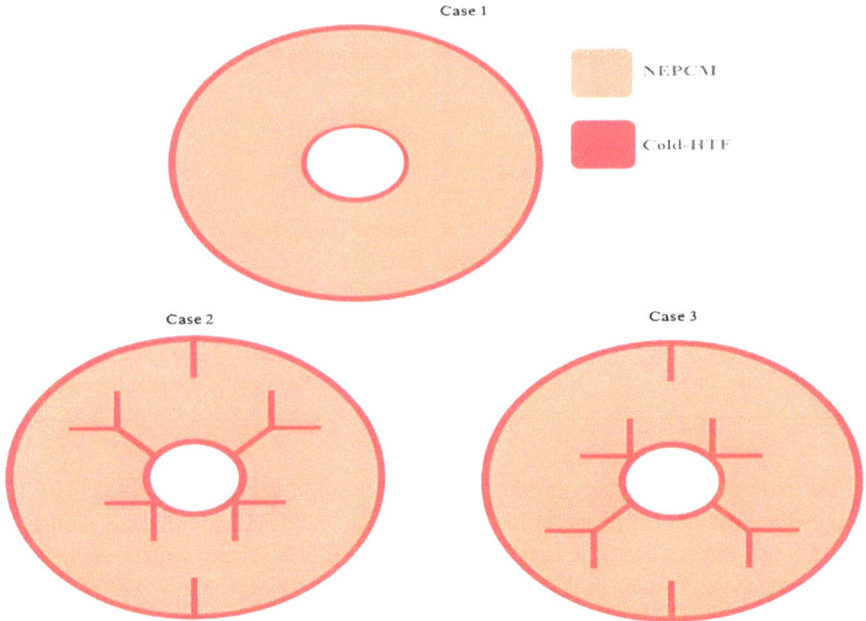

FIGURE 8.10 Three cases of composite PCM fin configurations [44].

compared to fins. This was not the case in every study, and it is possible that this was because of how the NP was employed in the system along with the PCM and heating temperature. It might be possible to perform additional research on those systems using alternative NP materials. Furthermore, it appears that both experiments with NPs and fins at moderate and high temperatures are lacking. Most of these NP and fin studies, if not all of them, use mathematical analysis. Future research should involve experimental methods. The experiments that combined NCPCMs with fins to improve LTES systems are compiled in Table 8.3.

Fins are frequently added to heat pipes even though they can be employed on their own. Koukou et al. [46] explored this triple arrangement of fins, heat pipes, and nanomaterials. In a tiny LTES unit, they conducted an experimental study on the heating and cooling of a base PCM augmented with GNPs and finned heat pipes. The impacts of the NPs and heat transfer fluid flow rate were also studied. The temperature during heating and cooling was monitored using thermocouples, and DSC was used to determine the thermal characteristics. Tests on heat transfer fluid flow rates of 30, 45, and 60 l/hr showed that faster flow rates caused heating to occur more quickly. Thermistors were used in place of the thermocouples in the NCPCM because the ultrasonication that was used to maintain the nanomaterials dispersed with the thermocouples. Nanomaterials shortened the heating and cooling times and improved efficiency. NCPCM covered the heat pipes and caused a problem during cooling, but this was readily resolved by first slowing the heat transfer fluid and then accelerating it throughout cooling [46]. Using a two times deionized

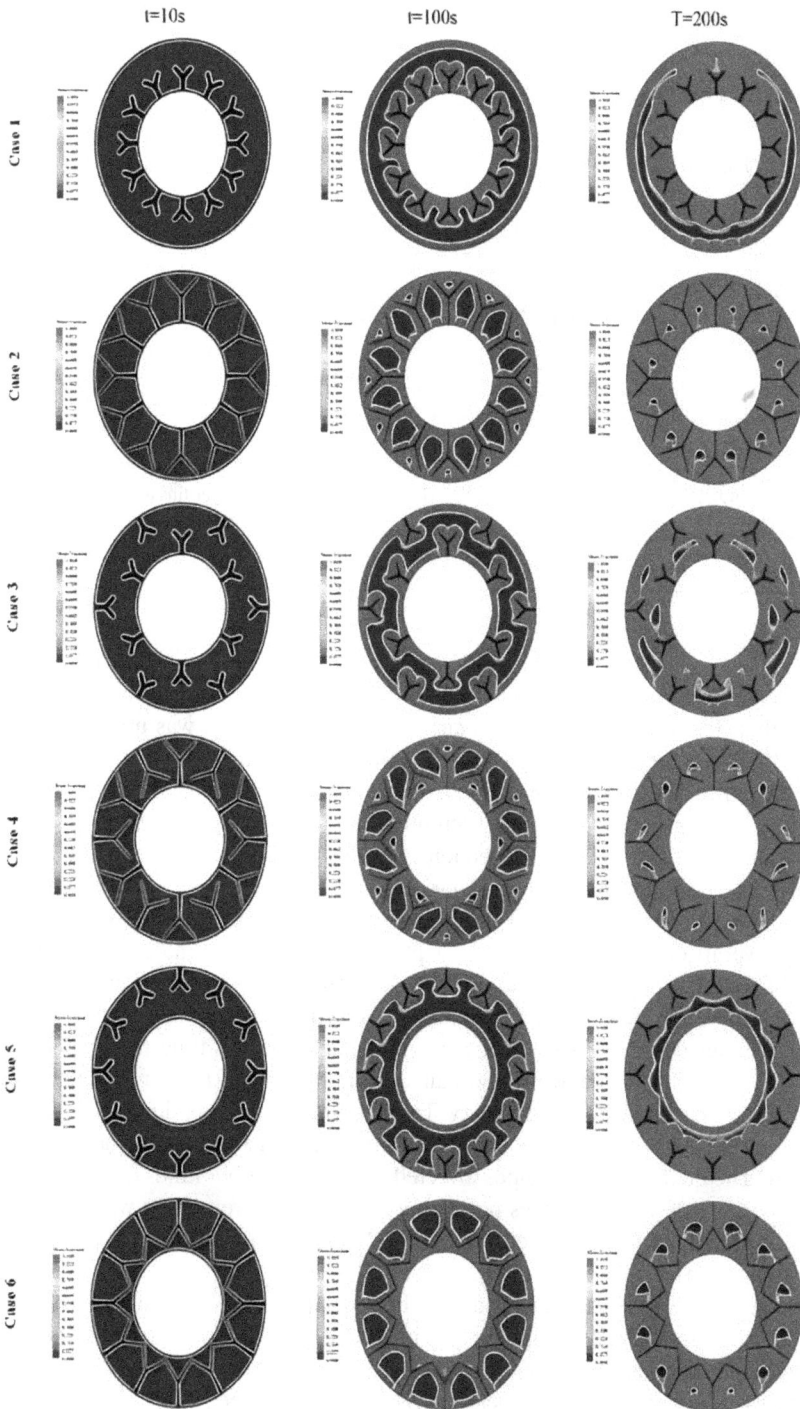

FIGURE 8.11 The liquid portion in each of the six cases at various intervals [45].

TABLE 8.3

A Summary of Experiments Employing NCPCMs Combined with Fins to Improve the Energy Storage System

Reference	Nanoparticles	Pure PCM	Properties/ enhancement method	Result
[43]	MoS_2–TiO_2	Water	• Solidification Process • Branched fins	The optimum augmentation method was high NP concentration with branching fins.
[44]	CuO	Paraffin RT-35	• Solidification process • Branched fins	The freezing periods for the second and third examples with composite PCM were decreased by roughly 69.92 and 68.84%, compared to the first example.
[45]	CuO	n-Octadecane	• Melting • Fins	Longer fins are more efficient and result in a lower melting time.

water axisymmetric analysis in ANSYS, Mahdavi et al. [47] used a mathematical analysis that involved only heat pipes and NPs. The base PCM was utilized in the model's shell-and-tube HE storage system. The impact of various improvements on the functionality of the system as a whole, particularly the heating and cooling periods and the capacity of the system to accumulate and release energy, was tested. The improvements incorporated the dispersion of various kinds of nanomaterials, varying nanomaterial mass fractions, and horizontal heat pipes. Figure 8.12 illustrates how the heat pipes dramatically reduced the time needed for heating and cooling, boosting overall efficiency. Without heat pipes, it took 2100 minutes to complete a full charge; with four heat pipes, it took 350 minutes. Since 95% of the PCM melted for four heat pipes in 170 minutes, melting time can be halved if just 95% melting is required. Using four heat pipes reduced discharge time in comparison to none by 96%. These noteworthy enhancements more than makeup for the heat pipes' 15% reduction in storage capacity. The heating and cooling durations and the quantity of energy that might be accumulated both reduced as nanomaterial volumes increased. The quantity of heat pipes affected how different NP kinds differed from one another, with fewer heat pipes producing a greater difference. Ag was the ideal NP type in the scenario with fewer heat pipes. Overall, the inclusion of NPs had less impact than the use of heat pipes [47].

A moderate-temperature shell-and-tube heat exchanger (STHE) storage system with pure PCM reinforced with heat pipes and Cu, CuO, Al_2O_3, or Ag NPs was the subject of another study by Mahdavi et al. [48]. Comparisons were made between the quantity of heat pipes, nanomaterial wt%, and different NP kinds. A more even dispersion of heat was produced and the heating time was sped up by adding more

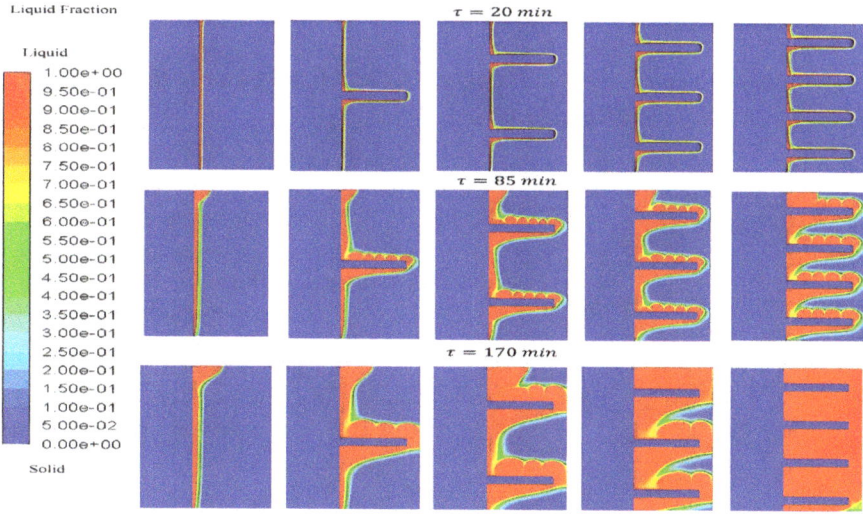

FIGURE 8.12   The impact of the number of heat pipes on the base PCM liquid fraction [47].

heat pipes. The heating time was sped up as the mass fraction of NPs rose, although the viscosity and overall heat storage were also increased. The best results were achieved by Ag, but ultimately, the variety of particles had no bearing [48].

The effects of NPs were shown in certain experiments to be minor in contrast to the improvement of the heat pipes, much like the fins were. Others, moreover, discovered that they continued to work well with the heat pipes, and those research findings of the kind or heating temperature reported some sort of improvement when using the NPs. Overall, it would be beneficial to investigate this arrangement in the future, possibly taking various NP types into account for each unique circumstance. The research that integrated NCPCMs with heat pipes to improve the latent heat storage system is compiled in Table 8.4.

By using ultrasonication and a dispersing agent, Yu et al. [49] aimed to address the problem of nanomaterial agglomeration. To improve the base PCM, CGF was combined with AlN, 1D tubular CNTs, or two times deionized water GnPs. The NPs were dispersed using OA, and the enhanced dispersion also enhanced the thermal conductivity. Ultrasound was employed to separate the particles because minute aggregations still formed, even though the OA prevented them from sinking. In order to determine volumes, the relationship between the duration of the ultrasound and the size of the nanoparticles was investigated. After 95 minutes, the particles had been evenly scattered, and the ultrasound had little effect. Additionally, utilizing GnPs and acetamide, a two times deionized water mathematical study was carried out using ANSYS. The GnPs exhibited a considerably stronger enhancing effect than the fundamental particles because they generated thermally conductive channels. In order to compare the mathematical results to the experimental findings, carbon foam was also examined, and the results showed a similar rising trend in thermal conductivity [49].

TABLE 8.4

A Summary of Experiments that Employ NCPCMs Together with Heat Pipes to Improve the Energy Storage System

Reference	Nanoparticles	Pure PCM	Properties/ enhancement method	Result
[46]	GNP	Paraffin	• Heating • Cooling • Finned heat pipes	According to the review, there are difficulties in determining the optimal concentration of NCPCM to define the lack of consistency in the LH behavior of NCPCM.
[47]	Al$_2$O$_3$, Ag, Cu, and CuO	Rubitherm 55	• Heating • Cooling • Horizontal heat pipes	More quantity of NPs and heat pipes reduced thermal conductivity.
[48]	Cu, CuO, Al$_2$O$_3$, and Ag	Rubitherm 55	• Heating • Heat pipes	Heating time will be reduced by 25% as the NP mass fraction rises from 0 to 10.

Kim et al. [50] studied experimental hybrid PCMs and various NPs in a low-temperature application. Silver, alumina, CNTs, and GNPs were among the materials studied. The foam was vacuumed with PCM, and the PCM temperature was checked with a thermocouple while it was kept in a sealed glass box immersed in hot oil. The thermal conductivity of all nanomaterial-foam mixtures was improved as contrasted with carbon foam without nanomaterials. Thermal conductivity was enhanced by 155–230% with increases in silver, alumina, CNTs, and GNPs, in that sequence. The rate of heating slowed down as well, with GNPs having the quickest rate. Furthermore, the LH drop during thermal cycling was lesser with the foam [50]. LH was lowered with the inclusion of foam but remained almost the same with the dispersion of the NPs.

Magnesium chloride, sodium chloride, and potassium chloride were used to create a high-temperature PCM by Yu et al. [51], which was then improved using enlarged graphite and SiO$_2$ NPs. It was demonstrated that enlarged graphite both significantly reduced PCM leaks and improved thermal conductivity. The PCM's specific heat and thermal conductivity rose with the dispersion of SiO$_2$ NPs, increasing by 23.4 and 9.3 times, respectively, in the solidification and melting states, and thermal cycling demonstrated excellent thermal stability [51].

The aforementioned research frequently tended to conclude that metal foam offered a better improvement than NPs. Although this is not to rule out the possibility of using NPs as an enhancer, it is crucial to try to comprehend why foams could have performed better in these trials. NPs take up much less space than metal foams; therefore, there is less of a drop in TES capacity, even though foams could have done better in heat transfer in these trials. According to the theory in [49], the conductive channel that the GNPs created increased their efficiency. It is likely that the paths created by the foam facilitate the passage of heat more effectively than isolated NPs do. Changing the forms of NPs to better establish these channels is one possible remedy for this. Additionally, it is possible to argue that the transfer of more heat storage capacity is sufficient in some circumstances to favor NPs over foams. A promising area for future studies to improve the effects of NCPCM is NP shapes. The research using NCPCMs in conjunction with a high-conductivity porous material to improve the latent heat system is compiled in Table 8.5.

The incorporation of nanomaterials into a multiple-PCM latent heat storage system is a possible alternative, despite the fact that it is not the most preferred improvement technique. In mathematical research, Mahdi et al. [52] investigated the freezing of multiple PCMs improved with cascaded metal foam and nanomaterials. The first and third PCMs with various heating points were stored in the STHE system in various portions. Nine cases were present, each having a distinct set of improvements. No improvement, NPs, or metal foam were present in three pairs of instances. The first PCM, second PCMs, and third PCM cases made up each of these three pairs. The PCMs were selected to maintain a constant mean phase transition temperature for every system. They employed conduction as the means of heat transfer since they were testing the freezing process. Without any improvements, the many PCMs freeze faster and with greater uniformity. For all three situations, freezing was enhanced and made faster with the dispersion of Al NPs. The metal foam accelerated the process of freezing. The most efficient scenario was the one with numerous PCMs and metal foam [52].

As mentioned by Mahdi et al. [52] and to the best of their knowledge, there is currently no other research that integrates various PCMs and nanoparticle improvement strategies. Further research is necessary to fill this gap in the research. Both improvements were successful and should be further researched for heating, with different nanoparticle kinds and for various applications. The investigation that used numerous PCMs to improve the latent heat storage system are summarized in Table 8.6.

TABLE 8.5

A Summary of Experiments that Employed the Combination of NCPCMs and Highly Conductive Porous Materials to Improve the Thermal Conductivity of TES Systems

Reference	Nanoparticles	Pure PCM	Properties	Result
[49]	CNTs, graphene, and carbon graphite foam	Acetamide	• Thermal conductivity	Thermal conductivity was increased through dispersion techniques.
[50]	Ag, Al, CNTs, graphene NPs, and carbon foam	Erythritol	• Melting • Solidification	Graphene improved substantially more than any other NCPCMs.
[51]	Expanded graphite and SiO_2	$MgCl_2$–NaCl–KCl	• Melting • Solidification	NCPCMs improved the phase change properties.

TABLE 8.6

A Summary of Experiments that Employed the Combination of NPs, Metal Foam, and Multiple PCMs (NCPCM) to Improve the Thermal Performance of Energy Storage Systems

Reference	Nanoparticles	Pure PCM	Properties	Result
[52]	Al_2O_3 NPs, metal foam, and multiple PCMs	RT-55, 60, 65	• Solidification	According to the number of multiple PCMs, the whole freezing time was reduced by up to 94% compared to the module of a single PCM without NPs or cascaded foam.

8.4 CONCLUSION

The idea of adding nanoparticles to enhance PCM's thermal capabilities is novel. Due to PCM's higher performance in thermal energy storage systems, there is currently considerable interest in this topic. Although adding nanoparticles affects PCM's thermal properties in certain good ways, it is impossible to ignore the detrimental effects. The information provided in the NCPCM open literature enables us to make the following conclusions.

- NCPCM comparison of NP types can be challenging. The specific scenario, considering heating temperature, PCM type, and some other applied improvement techniques, strongly influences the "best" NP type. The top NPs in each study do not seem to follow any clear patterns. Al_2O_3, for instance, is typically seen as a good NP for NCPCM; nevertheless, tests have revealed that sporadically, particular Al_2O_3 NPs performed better, and other times, these Al_2O_3 NPs sank. An unexplored possibility for future research is the processing of these NPs to produce a more constant performance in thermal energy storage systems.
- The systems have been proven to benefit from the inclusion of various improvements like metal foam, fins, heat pipes, and numerous base PCMs to NCPCMs. Numerous studies have shown that NPs improve these systems even further. Moreover, several studies did discover that NPs' effects were inferior to those of the other improvements. This was true in some circumstances, but not necessarily all, including metal foams, heat pipes, and fins. Further research on many PCMs is required to more thoroughly assess the impact of nanoparticle addition.
- The fact that other methods create conductive routes that transmit heat more quickly than spherical nanoparticles is one possible explanation for the circumstances in which NPs were found to have a lower impact. Spherical nanoparticles are unable to generate conductive pathways because the aggregation of particles from the dispersion of many NPs limits their capacity for improvement.

- Testing more NPs with various shapes, as some of the research presented here has done, is one relevant answer. The potential of longer and thinner particles to create favorable paths appears to be a phenomenon that needs to be further investigated. Another problem was the occurrence of hollow PCM parts occasionally caused by metal foams and NPs. A further possible future task is identifying the cause of the problem and fixing it.
- The investigation of various NPs shapes is another interesting area of future work. Investigations into this have already started. For instance, carbon is a fantastic enhancer, and it appears that the form influences how effectively it functions. Compared to standard graphite NPs, MWCNTs appear to be more efficient. Using low- and high-temperature NCPCMs and combining them with other upgrades are two further areas where the literature is lacking.
- Future research should focus on high temperatures because the majority of recent experiments used moderate-temperature NCPCMs. Additionally, there is less research on NPs alone than there is on NPs combined with other improvements. Developing the efficiency of various NPs at various temperatures to look for patterns might be another topic for future study.

REFERENCES

1. Mehling H, Cabeza L. Phase change materials and their basic properties. In: Paksoy HO, editor. *Thermal Energy Storage for Sustainable Energy Consumption*. Dordrecht: Springer; 2007. pp. 257–277.
2. Humphries, WR, Griggs, EI. *A design handbook for phase change thermal control and energy storage devices*. United States: N. p. 1977. https://www.osti.gov/biblio/6899545
3. Fernandes D, Pitie F, Caceres G, Baeyens J. Thermal energy storage: 'How previous findings determine current research priorities'. *Energy* 2012;39:246–257.
4. Fan L, Khodadadi JM. Thermal conductivity enhancement of phase change materials for thermal energy storage: A review. *Renew Sustain Energy Rev* 2011;15:24–46.
5. Yu W, France DM, Routbort JL, Choi SUS. Review and comparison of nanofluid thermal conductivity and heat transfer enhancements. *Heat Transfer Eng* 2008;29:432–460.
6. Sivasamy P, HarikrishnanS, DevarajuA. Experimental investigation of improved thermal characteristics of Al2O3/barium hydroxide octa hydrate as phase change materials (PCMs). *Mater Today Proc* 2018;5(6):14440–14447.
7. Muzhanje AT, HassanMA, OokawaraS, HassanH. An overview of the preparation and characteristics of phase change materials with nanomaterials. *J Energy Storage* 2022;51:104353.
8. Kumar R, MukhtarA, Md YasirASH, EldinSM, MusaDAR, RochaCMM, LeBN, GhalandariM. Simultaneous applications of fins and nanomaterials in phase change materials: A comprehensive review. *Energy Rep* 2023;10:1028–1040.
9. Khodadadi JM, Fan L, Babaei H. Thermal conductivity enhancement of nanostructure-based colloidal suspensions utilized as phase change materials for thermal energy storage: A review. *Renew Sustain Energy Rev* 2013;24:418–444.
10. Raam Dheep G, Sreekumar A. Influence of nanomaterials on properties of latent heat solar thermal energy storage materials—a review. *Energy Convers Manag* 2014;83:133–148.

11. Khan Z, Khan Z, Ghafoor A. A review of performance enhancement of PCM based latent heat storage system within the context of materials, thermal stability and compatibility. *Energy Convers Manag* 2016;115:132–158.

12. Chon CH, Kihm KD, Lee SP, Choi SU. Empirical correlation finding the role of temperature and particle size for nanofluid (Al 2 O 3) thermal conductivity enhancement. *Appl Phys Lett* 2005;87(15):153107.

13. Tariq SL, Ali HM, Akram MA, Janjua MM, Ahmadlouydarab M. Nanoparticles enhanced phase change materials (NePCMs)-A recent review. *Appl Therm Eng* 2020; 176:115305.

14. Kumar KS, Parameshwaran R, Kalaiselvam S. Preparation and characterization of hybrid nanocomposite embedded organic methyl ester as phase change material. *Sol Energy Mater Sol Cells* 2017;171:148–160.

15. Wang J, Li Y, Zheng D, Mikulčić H, Vujanović M, Sundén B. Preparation and thermophysical property analysis of nanocomposite phase change materials for energy storage. *Renew Sustain Energy Rev* 2021;151:111541.

16. Bozzi A, Yuranova T, Kiwi J. Self-cleaning of wool-polyamide and polyester textiles by TiO2-rutile modification under daylight irradiation at ambient temperature. *J Photochem Photobiol A Chem* 2005;172(1):27–34.

17. HarikrishnanS, DeenadhayalanM, KalaiselvamS. Experimental investigation of solidification and melting characteristics of composite PCMs for building heating application. *Energy Convers Manag* 2014;86:864–872.

18. ParameshwaranR, DeepakK, SaravananR, KalaiselvamS. Preparation, thermal and rheological properties of hybrid nanocomposite phase change material for thermal energy storage. *Appl Energy* 2014;115:320–330.

19. Murugan P, Ganesh Kumar P, Kumaresan V, Meikandan M, Malar Mohan K, Velraj R. Thermal energy storage behaviour of nanoparticle enhanced PCM during freezing and melting. *Ph Transit* 2018;91:254–270.

20. Li X, Zhou Y, Nian H, Zhang X, Dong O, Ren X, Zeng J, Hai C, Shen Y. Advanced nanocomposite phase change material based on calcium chloride hexahydrate with aluminum oxide nanoparticles for thermal energy storage. *Energy Fuels* 2017;31(6):6560–6567.

21. AliH, BabarH, ShahT, SajidM, QasimM, JavedS. Preparation techniques of TiO2 nanofluids and challenges: A review. *Appl Sci* 2018;8(4):587.

22. Teng T-P, Yu C-C. Characteristics of phase-change materials containing oxide nanoadditives for thermal storage. *Nanoscale Res Lett.* 2012;7(1):611.

23. MohamedNH, SolimanFS, El MaghrabyH, MoustfaYM. Thermal conductivity enhancement of treated petroleum waxes, as phase change material, by α nano alumina: Energy storage. *Renew Sustain Energy Rev* 2017;70:1052–1058.

24. Abhat A. Low temperature latent heat thermal energy storage: Heat storage materials. *Sol Energy* 1983;30:313–332.

25. Mehling H, Cabeza LF. *Solid-Liquid Phase Change Materials Heat and Cold Storage with PCM*. Berlin and Heidelberg: Springer; 2008. pp. 11–55.

26. Zalba B, Marín JM, Cabeza LF, Mehling H. Review on thermal energy storage with phase change: Materials, heat transfer analysis and applications. *Appl Therm Eng* 2003;23:251–283.

27. Elgafy A, Lafdi K. Effect of carbon nanofiber additives on thermal behavior of phase change materials. *Carbon* 2005;43:3067–3074.

28. Khodadadi JM, Hosseinizadeh SF. Nanoparticle-enhanced phase change materials (NEPCM) with great potential for improved thermal energy storage. *Int Commun Heat Mass Transf* 2007;34:534–543.

29. Feng Y, Li H, Li L, Bu L, Wang T. Numerical investigation on the melting of nanoparticle-enhanced phase change materials (NEPCM) in a bottom-heated rectangular cavity using lattice Boltzmann method. *Int J Heat Mass Transf* 2015;81:415–425.

30. Akhmetov B, Navarro ME, Seitov A, Kaltayev A, Bakenov Z, Ding Y. Numerical study of integrated latent heat thermal energy storage devices using nanoparticle-enhanced phase change materials. *Sol Energy* 2019;194:724–741.

31. Abdulateef AM, Jaszczur M, Hassan Q, Anish R, Niyas H, Sopian K, Abdulateef J. Enhancing the melting of phase change material using a fins–nanoparticle combination in a triplex tube heat exchanger. *J Energy Storage* 2021;35:102227.

32. Farsani RY, Raisi A, Nadooshan AA, Vanapalli S. Does nanoparticles dispersed in a phase change material improve melting characteristics? *Int Commun Heat Mass Transf* 2017;89:219–229.

33. Elbahjaoui R, el Qarnia H. Thermal analysis of nanoparticle-enhanced phase change material solidification in a rectangular latent heat storage unit including natural convection. *Energy Build.* 2017;153:1–17.

34. Aqib M, Hussain A, Ali HM, Naseer A, Jamil F. Experimental case studies of the effect of Al2O3 and MWCNTs nanoparticles on heating and cooling of PCM. *Case Stud Therm Eng.* 2020;22:100753.

35. Murugan P, Ganesh Kumar P, Kumaresan V, Meikandan M, Malar Mohan K, Velraj R. Thermal energy storage behaviour of nanoparticle enhanced PCM during freezing and melting. *Ph Transit.* 2018;91:254–270.

36. Khatibi M, Nemati-Farouji R, Taheri A, Kazemian A, Ma T, Niazmand H. Optimization and performance investigation of the solidification behavior of nano-enhanced phase change materials in triplex-tube and shell-and-tube energy storage units. *J Energy Storage* 2021;33:102055.

37. Harikrishnan S, Devaraju A, Sivasamy P, Kalaiselvam S. Experimental investigation of improved thermal characteristics of SiO2/myristic acid nanofluid as phase change material (PCM). *Mater Today Proc* 2019;9:397–409.

38. Maher H, Rocky KA, Bassiouny R, Saha BB. Synthesis and thermal characterization of paraffin-based nanocomposites for thermal energy storage applications. *Therm Sci Eng Prog* 2021;22:100797.

39. Marcos MA, Cabaleiro D, Hamze S, Fedele L, Bobbo S, Estellé P, Lugo L. NePCM based on silver dispersions in poly(ethylene glycol) as a stable solution for thermal storage. *Nanomaterials* 2019;10:19.

40. Barreneche C, Martín M, la Calvo-de Rosa J, Majó M, Fernández AI. Own-synthetize nanoparticles to develop nanoenhanced phase change materials (NEPCM) to improve the energy efficiency in buildings. *Molecules* 2019;24:1232.

41. Chieruzzi M, Miliozzi A, Crescenzi T, Torre L, Kenny JM. A new phase change material based on potassium nitrate with silica and alumina nanoparticles for thermal energy storage. *Nanoscale Res Lett* 2015;10:984.

42. Han D, Lougou BG, Xu Y, Shuai Y, Huang X. Thermal properties characterization of chloride salts/nanoparticles composite phase change material for high-temperature thermal energy storage. *Appl Energy* 2020;264:114674.

43. Hosseinzadeh K, Erfani Moghaddam MA, Asadi A, Mogharrebi AR, Jafari B, Hasani MR, Ganji DD. Effect of two different fins (longitudinal-tree like) and hybrid nanoparticles (MoS2–TiO2) on solidification process in triplex latent heat thermal energy storage system. *Alex Eng J* 2021;60:1967–1979.

44. Hajizadeh MR, Keshteli AN, Bach Q-V. Solidification of PCM within a tank with longitudinal-Y shape fins and CuO nanoparticle. *J Mol Liq* 2020;317:114188.

45. Li F, Almarashi A, Jafaryar M, Hajizadeh MR, Chu Y-M. Melting process of nanoparticle enhanced PCM through storage cylinder incorporating fins. *Powder Technol* 2021;381:551–560.

46. Koukou MK, Dogkas G, Vrachopoulos MG, Konstantaras J, Pagkalos C, Lymperis K, Stathopoulos V, Evangelakis G, Prouskas C, Coelho L, et al. Performance evaluation of a small-scale latent heat thermal energy storage unit for heating applications based on a nanocomposite organic PCM. *ChemEngineering* 2019;3:88.

47. Mahdavi M, Tiari S, Pawar V. A numerical study on the combined effect of dispersed nanoparticles and embedded heat pipes on melting and solidification of a shell and tube latent heat thermal energy storage system. *J Energy Storage* 2020;27:101086.

48. Tiari S, Mahdavi M, Pawar V. Heat transfer analysis of a low-temperature heat pipe-assisted latent heat thermal energy storage system with nano-enhanced PCM. In *Proceedings of the ASME 2018 International Mechanical Engineering Congress and Exposition*, Pittsburgh, PA, 9–15 November 2018. pp. 1–10.

49. Yu J, Yu ZC, Tang CL, Chen X, Song QF, Kong L. Preparation and characterization of composite phase change materials containing nanoparticles. *Kemija u industriji: Časopis kemičara i kemijskih inženjera Hrvatske* 2016;65:605–612.

50. Kim HG, Kim Y-S, Kwac LK, Shin HJ, Lee SO, Lee US, Shin HK. Latent heat storage and thermal efficacy of carboxymethyl cellulose carbon foams containing Ag, Al, carbon nanotubes, and graphene in a phase change material. *Nanomaterials* 2019;9:158.

51. Yu Q, Zhang C, Lu Y, Kong Q, Wei H, Yang Y, Gao Q, Wu Y, Sciacovelli A. Comprehensive performance of composite phase change materials based on eutectic chloride with SiO2 nanoparticles and expanded graphite for thermal energy storage system. *Renew Energy* 2021;172:1120–1132.

52. Mahdi JM, Mohammed HI, Hashim ET, Talebizadehsardari P, Nsofor EC. Solidification enhancement with multiple PCMs, cascaded metal foam and nanoparticles in the shell-and-tube energy storage system. *Appl Energy* 2020;257:113993.

9 Phase Change Material Applications in Thermal Management of Electronics and Electrical Systems

Tehmina Ambreen, Arslan Saleem, Paula Ruiz-Hincapie, Anirudh Kulkarni, Hafiz Muhammad Ali, and Cheol Woo Park

HIGHLIGHTS

- Escalating power densities in electronics drive the need for effective thermal management.
- Phase change materials (PCMs) offer promise in passive and hybrid cooling methodologies.
- PCM-enhanced heat sinks and heat pipes effectively absorb and dissipate heat in devices.
- Hybrid systems using PCMs offer precise temperature control and adaptability.
- Future research should focus on solid–solid PCMs, IoT integration, and environmental impact.

NOMENCLATURE

A	Area
EPCM	Encapsulated PCM
iNEMI	International Electronics Manufacturing Initiative
IoT	Internet of Things
PCM	Phase change material
PEG	Polyethylene glycol
Ra	Rayleigh number
SEM	Scanning electron microscope
TEC	Thermoelectric cooler
TIMs	Thermal interface materials

DOI: 10.1201/9781003331957-9

9.1 INTRODUCTION

The swift progression of semiconductor technology, coupled with the downsizing of electronic gadgets, has led to escalating power densities, particularly within high-end chips. This progression has introduced substantial hurdles in effectively controlling the heat produced by these devices. As the electronics continue to abide by Moore's law, marked by diminishing feature dimensions and a surge in transistor density, there has been a corresponding ascent in power densities and operational temperatures, which, in turn, compromises the performance and lifespan of these devices. [1,2]. The failure rates of electronic devices escalate steeply with rising operational temperatures.

As per the 2004 iNEMI (International Electronics Manufacturing Initiative) technological blueprint, it was expected that the maximum power dissipation from superior microprocessors would reach approximately 1900 kW/m² by 2020 [2,3]. Nonetheless, the actual heat flux production in several high-end electronic devices has surpassed these estimations. Therefore, it is crucial to implement effective cooling strategies to dissipate heat at a pace equal to or faster than its generation. This is necessary to avert temperature spikes that could impair reliability and performance, potentially causing device malfunctions.

Electronic cooling techniques can be categorized based on various factors, including the cooling medium, direct or indirect interfacing with the device, energy consumption, configuration and size of the cooling system, type and layout of heat transfer surfaces, thermal management plan, and control plus monitoring systems. Conventional cooling methodologies encompass natural convection, forced air cooling, forced liquid cooling, and liquid evaporation. Natural convection is apt for low-heat-flux scenarios due to its simplicity, affordability, and reliability. Among the methods, liquid evaporation stands out as the most potent heat elimination technique, followed by forced liquid convection. Forced air convection, although less effective in heat removal compared to liquid-based techniques, outperforms free convection.

Newer entrants to the realm of electronic cooling include nanofluids, heat pipes, metallic foams, microchannels, spray cooling, heat pumps, thermoelectric cooling, and phase change materials for centric cooling. These can be classified into passive or active cooling systems. The passive variants operate on capillary or gravitational forces for fluid movement, while active ones employ pumps or compressors to amplify cooling capacity and performance. The selection of a cooling approach is influenced by the specific demands and limitations of the given application.

This chapter offers an extensive examination of PCMs (phase change materials) as a viable solution for adept thermal management in electronic devices. It delves into various PCM enhancement strategies aimed at bolstering their thermal efficiency and stability. The discussion extends to both passive and hybrid cooling techniques incorporating PCMs. The passive cooling methodologies discussed encompass PCM-amplified heat sinks, heat pipes, and PCM-integrated thermal interface materials. Additionally, the section explores hybrid cooling systems that meld PCMs with air cooling, liquid cooling, and thermoelectric cooling.

9.2 PCMS FOR ELECTRONIC COOLING

Ideal characteristics for PCMs employed in electronic cooling applications encompass a substantial heat of fusion per unit volume, consistent melting and freezing attributes, superior thermal conductivity, negligible supercooling, and restrained thermal expansion [4–12]. Moreover, the melting temperature of the PCM should be suitably high, while its chemical and physical attributes should harmonize with the adjacent materials. Elevated thermal durability ensures the PCM's stability, preventing degradation or decomposition under high-temperature exposure or through repeated thermal cycles. Low flammability mitigates fire hazards in the event of device failure. PCMs characterized by low viscosity and robust surface tension are beneficial for microscale or nanoscale electronic gadgets since they can circulate effortlessly and distribute uniformly without the need for external pumps or pressure sources. Additionally, the orientation of these PCMs is not a concern, making them suitable for fulfilling various cooling needs while maintaining cost and size efficiency [13]. Aligning with existing manufacturing processes is another sought-after characteristic, facilitating effortless integration of the PCM without substantial alterations, thereby ensuring the scalability and cost efficiency of the cooling solution. Lastly, PCMs are not only cost effective but also eco-friendly, non-toxic, and biodegradable, making their environmental and health impacts minimal [7,14].

PCMs showcase a diverse array of phase transitions, including transitions from solid to liquid, liquid to vapor, and solid to solid, leading to changes from crystalline to amorphous formations. Among these, the transition from solid to liquid has attracted considerable interest and practical deployment for energy storage through phase changes in electronic thermal regulation. This type of transition brings forth advantages like minimal volume alterations and tolerable expansion rates, rendering it suitable for proficient and dependable thermal energy management in electronic devices. Conversely, PCMs undergoing transitions from solid to gas or liquid to gas possess high latent heat but are challenged by excessive expansion rates, restricting their applicability to electronic thermal management, as they could potentially cause structural damage and performance setbacks. On a different note, transitions from solid to solid, although having less latent heat than solid–liquid transitions, present a viable alternative for phase change energy storage. Solid–solid transitions offer perks such as minimal volume alterations, low subcooling (temperature variance during phase change), and a lack of container requirements, making them attractive for specific electronic thermal management applications [15]. However, research on solid–solid PCMs for electronic cooling is relatively limited. Consequently, this book chapter mainly focuses on solid–liquid PCMs to offer a thorough insight into their role in electronic thermal management.

The PCMs utilized for managing thermal aspects in electronics can be broadly segregated into three primary categories: organic PCMs, inorganic PCMs, and eutectic PCMs.

9.2.1 ORGANIC PCMS

PCMs bring to the table numerous benefits in the domain of electronic cooling applications. Primarily, they showcase a wide spectrum of phase change temperatures, lending flexibility to cater to the distinct cooling demands of electronic gadgets. This adaptability paves the way for prime thermal management across a multitude of electronic applications with varying temperature outlines. Secondly, organic PCMs are endowed with substantial latent heat capacities, empowering them to absorb and release notable quantities of thermal energy amidst phase transitions. This trait amplifies their heat storage and dispersion capacities, proficiently modulating the temperature of electronic devices. In addition, organic PCMs exhibit commendable compatibility with an extensive array of electronic materials, making their assimilation into prevailing electronic architectures a breeze without necessitating significant alterations. They also boast low viscosity and elevated flowability, facilitating effortless distribution and even spread within the cooling infrastructure, thereby fostering efficient heat transfer. Moreover, organic PCMs are typically non-corrosive and non-toxic, guaranteeing safe manipulation while minimizing environmental and health concerns [16]. Their scalability, alignment with manufacturing protocols, and cost-effectiveness render organic PCMs a compelling choice for electronic cooling applications [17].

Organic PCMs for electronic cooling applications include paraffin wax [18–20], fatty acids (for instance, stearic acid, palmitic acid) [21,22], polyethylene glycol (PEG) [23,24], eutectic mixtures [23,24], sugar alcohols (such as erythritol, mannitol) [25], and bio-based PCMs [26].

9.2.2 Inorganic PCMs

Inorganic PCMs, fabricated from metals, salts, or alloys, embody unique thermal attributes that bolster their efficacy in heat storage and dispersion for electronic cooling purposes. They boast high thermal conductivity, which expedites heat movement within the cooling apparatus. This heightened thermal conductivity fosters proficient dissipation of thermal energy and effective thermal regulation of electronic gadgets. Inorganic PCMs also illustrate exceptional thermal steadiness at raised temperatures, rendering them apt for scenarios involving heat sources that generate high operating temperatures. They display compatibility with electronic components and heat sinks, streamlining their incorporation into electronic blueprints. Furthermore, inorganic PCMs can be custom modified to possess specific melting thresholds, enabling precise temperature governance in electronic devices. Additionally, they boast notable durability and longevity, showing resilience to degradation, chemical interactions, and phase change cycling. This robustness ensures dependable and uniform thermal performance through prolonged operational durations [17,27,28].

In the sphere of inorganic PCMs, various materials offer specific advantages for electronic cooling. Metal-based PCMs exhibit high thermal conductivity and compatibility with electronic components [29,30]. Meanwhile, salt-based PCMs showcase high thermal storage capabilities and conductivity [31]. Additionally, alloy-based PCMs, such as eutectic blends of metals, can be engineered to have specific melting points. They also exhibit excellent thermal stability, low supercooling, and high latent heat of fusion for efficient thermal governance. [32,33].

Salt hydrates, a variant of inorganic PCMs, undergo reversible phase transitions and are frequently used for thermal energy storage. They result from the combination of salts with water molecules and have distinct melting points and high latent heat values. Sodium sulfate decahydrate, calcium chloride hexahydrate, and magnesium nitrate hexahydrate exemplify salt hydrates employed as PCMs. These materials exhibit suitable properties for various thermal management applications [34].

9.2.3 ENHANCED PCMs

Enhanced PCMs, also known as advanced or modified PCMs, are PCMs that have been modified or engineered to improve their thermal properties and performance characteristics. These modifications aim to overcome the limitations of traditional PCMs such as supercooling, leakage, and volume changes. Enhanced PCMs exhibit improved thermal conductivity, compatibility with surrounding materials, stability, and reliability. They offer tailored solutions for specific thermal management requirements in electronic devices, ensuring efficient heat storage, transfer, and dissipation. Some methods used to enhance PCMs include:

9.2.3.1 Nanostructuring

Nanostructuring involves incorporating nanoparticles or nanostructured materials into the PCM matrix. This approach increases the surface area available for heat transfer, allowing for faster heat absorption and release during the phase change process. Nanoparticles, such as metal oxides, metallic nanoparticles, carbon nanotubes, carbon nanofibers, and graphene, can significantly enhance the thermal conductivity of the PCMs, leading to improved heat transfer efficiency [35,36].

9.2.3.2 Encapsulation

Encapsulation involves enveloping the PCM core with a protective coating or shell material, as depicted in Figure 9.1. The primary purpose of encapsulation is to isolate the PCM from its surroundings, preventing any undesired interactions and maintaining the PCM's composition. Encapsulation offers several advantages, including reduced reactivity, enhanced flexibility in phase change operations, improved heat transfer rates, and increased thermal and mechanical stability [37].

Encapsulated PCMs can be classified based on their size. Macroencapsulation involves enclosing the PCM within containers that exceed 1 mm in size. These

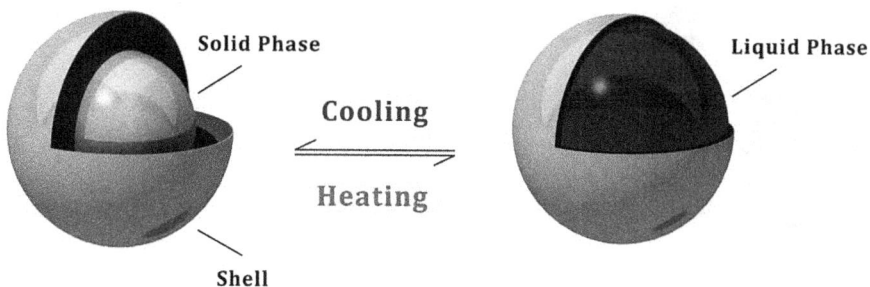

Solid Phase

Cooling

Heating

Liquid Phase

Shell

FIGURE 9.1 Illustration of an encapsulated PCM [38].

containers, which can be shaped as spheres, tubes, cylinders, or rectangles, are commonly used in thermal energy storage applications. However, macroencapsulation may suffer from temperature differentials between the core and the boundary of the PCM, leading to incomplete phase change and inefficient heat transfer within the system [37,39,40].

On the other hand, microencapsulation involves encapsulating PCMs with particle sizes ranging from 0 to 1000 μm. The manufacturing process for microencapsulated PCMs is more complex due to the smaller size scale. However, microencapsulation offers significant benefits, including improved heat transfer rates. The higher surface-area-to-volume ratio of microencapsulated PCMs enables faster melting and solidification, leading to more efficient heat transfer. Additionally, microencapsulation provides better control over PCM behavior and minimizes chemical reactivity between the PCM and the shell material, ensuring enhanced stability [37,38,41,42].

Another emerging area is nanoencapsulation, which involves encapsulating PCMs at the nanoscale (0 to 1000 nm). Nanoencapsulation holds promise for further enhancing the thermal properties and performance of PCMs, although it is an area of ongoing research [43] (Figure 9.2).

9.2.3.3   EPCM Slurries

Encapsulated PCM (EPCM) slurry is a specialized form of encapsulation where PCM particles are dispersed in a carrier liquid, forming a stable suspension. This approach combines the advantages of PCM encapsulation with the benefits of a fluid medium. The PCM particles are typically encapsulated at the microscale or nanoscale (Figure 9.3), providing a large surface area for efficient heat transfer. The carrier liquid facilitates the flow and distribution of the PCM within the system, enabling enhanced heat absorption and dissipation. Nano/microencapsulated PCM slurry offers advantages such as improved thermal conductivity, reduced viscosity, and the ability to conform to complex shapes and surfaces. This technology holds great promise for applications requiring compact and efficient thermal management, such as electronics cooling and heat exchangers in the automotive and aerospace industries.

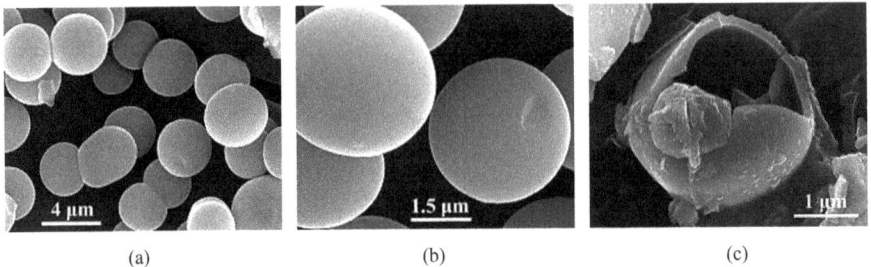

| (a) | (b) | (c) |

FIGURE 9.2   SEM images of microencapsulated n-eicosane synthesized with n-eicosane/ TBT (a); (b) intact and (c) damaged microcapsules [43]

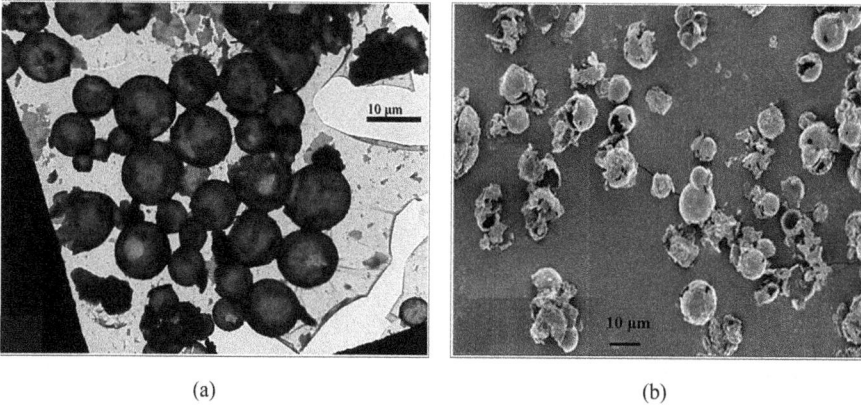

(a)   (b)

FIGURE 9.3 (a) SEM and (b) TEM images of water-based EPCM slurry [44].

9.2.3.4 PCM Composites

Enhanced PCMs can also be created by combining PCMs with other materials to form composites. This approach often involves embedding the PCM within a matrix material, such as metals (Figure 9.4), polymers, ceramics, or other suitable materials. The matrix material serves as a host for the PCM, providing structural support and acting as a conduit for heat transfer.

The composite structure enhances the overall performance of the PCM by improving its thermal conductivity, mechanical strength, or heat transfer capabilities, reduced energy consumption, and compact design. Examples include PCM-enhanced concrete for temperature regulation in buildings, PCM-infused textiles for thermal comfort, PCM-embedded thermal storage panels for solar thermal systems, and PCM-enhanced heat sinks for electronics cooling [45–48].

9.2.3.5 Chemical Functionalization

PCMs can be chemically modified by introducing functional groups or additives to improve their performance. For example, surfactants or dispersants can be added to enhance the dispersion and stability of PCMs within a matrix. Chemical modifications can also alter the melting point, latent heat, or thermal conductivity of the PCM, tailoring it to specific temperature ranges or applications [50,51].

9.2.3.6 PCM Blends

Blending different PCMs with complementary properties can create hybrid materials with enhanced performance (Figure 9.5). By combining PCMs with different melting points or latent heat capacities, the resulting blend can exhibit a broader operating temperature range and improved heat storage capacity [52,53]. This is particularly essential, as it seen that operating conditions can hamper the efficiency of the system [54].

FIGURE 9.4   SEM images of nickel foam before (a-1, a-2) and after (b-1, b-2) vacuum impregnation with erythritol PCM [49].

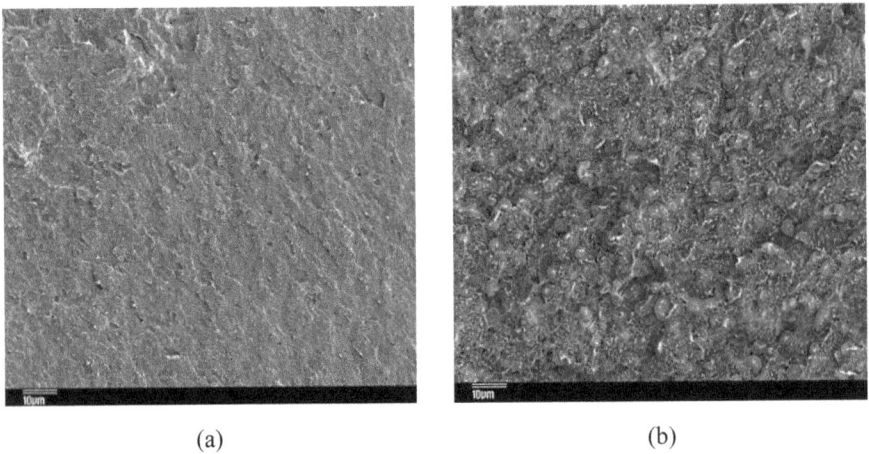

(a)   (b)

FIGURE 9.5   SEM images of (a) hard and (b) soft paraffin wax blends within an LDPE (low-density polyethylene) matrix [55].

9.3 PCM-BASED PASSIVE ELECTRONIC COOLING

A large portion of PCM-based electronic cooling techniques are passive in nature. These passive methods can be classified into multiple kinds based on their setup and usage.

9.3.1 PCM-ENHANCED HEAT SINKS

PCM heat sinks employ PCMs to bolster the heat dispersal capabilities of traditional heat sinks. This usually entails integrating PCMs within the structure of standard heat sinks or including PCM layers within the heat sink assembly. PCM-aided heat sinks enable efficient heat transfer by tapping into the high latent heat capacity of the PCM. During heightened thermal loads, the PCM soaks up heat and undergoes a phase change, effectively absorbing surplus thermal energy. As the device tempers, the PCM re-solidifies, relinquishing the stored heat to the heat sink, which then radiates it into the surrounding environment.

Research in this field focuses on choosing and improving PCMs with beneficial thermal properties like high latent heat capacity and suitable phase change temperatures. Techniques such as nanoparticle doping, hybrid PCM composites, and encapsulation are investigated to enhance thermal conductivity and stability. Enclosure designs are fine-tuned with internal fins, channels, enhanced surface area structures, PCM cores, and PCM capsules to amplify contact between the PCM and the heat-generating component. The selection of enclosure material encompasses evaluating materials with superior thermal conductivity and compatibility with the PCM, including advanced composites and metal alloys.

PCM-aided heat sinks span various types, including enclosure heat sinks, finned heat sinks, metal foam heat sinks, PCM core heat sinks, and PCM capsule heat sinks.

9.3.1.1 Enclosure Heat Sinks

Enclosure heat sinks are a simplified type of PCM-based heat sinks that use a dedicated enclosure filled with PCM. This enclosure directly connects to the heat-producing component, making it effective at absorbing heat. When the PCM reaches its phase change temperature, it changes from a solid to a liquid, absorbing a significant amount of heat. The PCM then releases this stored heat to the surrounding heat sink structure, which efficiently disperses it into the environment [56–58].

9.3.1.2 Finned Heat Sinks

Finned heat sinks are designed to enhance heat dissipation by incorporating PCM layers within the fins (see Figure 9.6). These PCM layers, with their high latent heat capacity, efficiently absorb and store excess heat when thermal loads are high. When the PCM changes from a solid to a liquid state, it absorbs heat and subsequently releases the stored heat to the heat sink fins. The presence of fins increases the surface area available for heat transfer, facilitating effective dissipation into the surrounding environment [59,61,62].

Continuous explorations in this domain focus on examining various fin types, fine-tuning their parameters (as shown in Figure 9.7), selecting suitable heat sink materials, and delving into diverse PCM enhancement techniques to further augment

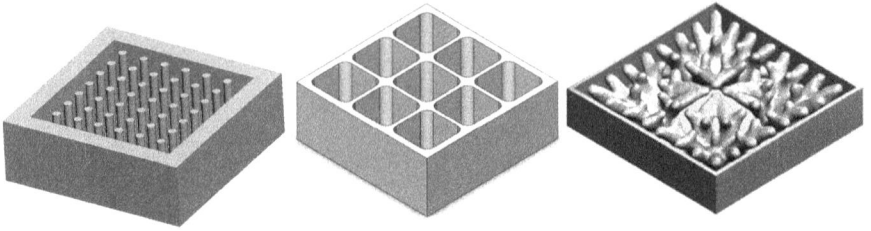

FIGURE 9.6   PCM-enhanced heat sinks with different fin configurations [59–61].

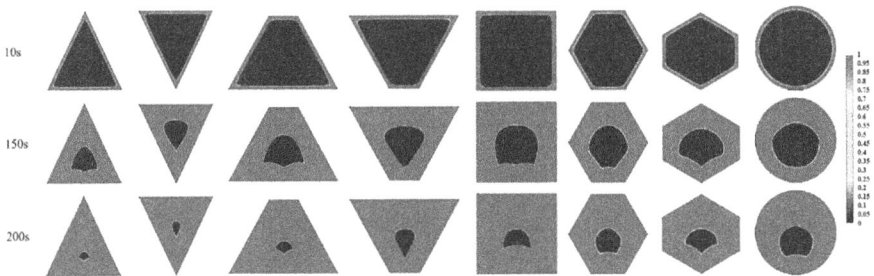

FIGURE 9.7   Liquid fraction contours at an initial, intermediate, and final melting time in different honeycomb structures at Ra = 3.095×10^5 (A = 100 mm^2) [64].

the performance of finned heat sinks. The convectional fin type design has also been modified more recently [63]. The proposed hybrid heat sink design utilizes finned structures for heat dissipation during high convective cooling periods, while employing PCM at the fin tips to absorb heat during reduced cooling rates. Unlike designs where the entire finned structure is immersed in PCM, this hybrid approach avoids superheating issues and maintains continuous heat dissipation as the PCM re-solidifies during high convective cooling periods. This design's thermal performance can be optimized through the right choice of governing parameters, ensuring sustained heat dissipation, even as the heat source continuously emits heat.

9.3.1.3   Metal Foam Heat Sinks

Metal foam heat sinks employ a porous framework made of metal foam as a heat transfer medium (Figure 9.8). The metal foam comprises interconnected metal cells or pores, offering a large surface area for heat exchange. PCM is infused into the metal foam structure, allowing for effective heat absorption and storage. The PCM transitions phase, discharging the stored heat and facilitating efficient heat dissipation. Metal foam heat sinks provide better thermal conductivity and superior cooling performance compared to conventional heat sinks [65–68].

The recent research trajectories regarding PCM-based metal foam heat sinks encompass optimization of metal foam design parameters like cell size and porosity, amalgamation of metal foam and fins, selection of suitable PCMs with enhanced thermal properties, and multi-physics modeling to grasp heat transfer mechanisms. Experimental validation and application-centric studies are conducted to gauge performance in real-world scenarios and to fine tune the design for specific thermal management requisites.

9.3.2 PCM-ENHANCED HEAT PIPES

PCM heat pipes are passive heat transfer devices that use PCMs to enhance their thermal efficiency. Heat pipes consist of a sealed tube with a wick structure and a working fluid, which is often a PCM. When heat is applied to one end of the heat pipe, the PCM absorbs the heat, vaporizes, and transports it to the other end of the pipe, where it releases the heat to the surrounding environment. The condensed PCM then returns to the heat source through capillary action, completing the heat transfer cycle [70–72]. The schematic of a low-melting-point PCM-based finned heat pipe is depicted in Figure 9.9 [73].

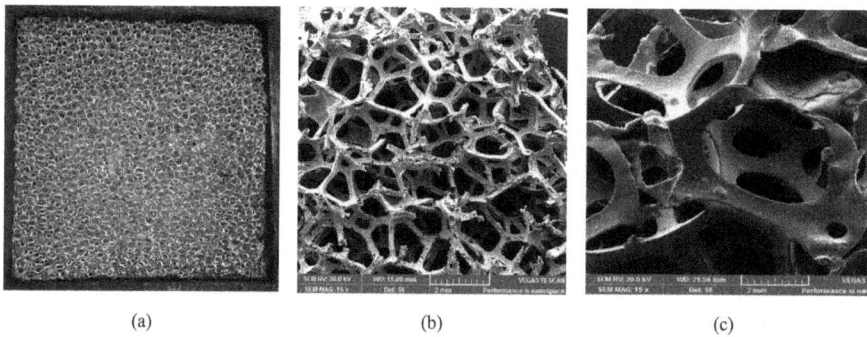

(a)   (b)   (c)

FIGURE 9.8   (a) Schematic of a paraffin-based metal foam (copper) heat sink and SEM images of (b) copper and (c) nickel metal foams investigated by Ref. [69].

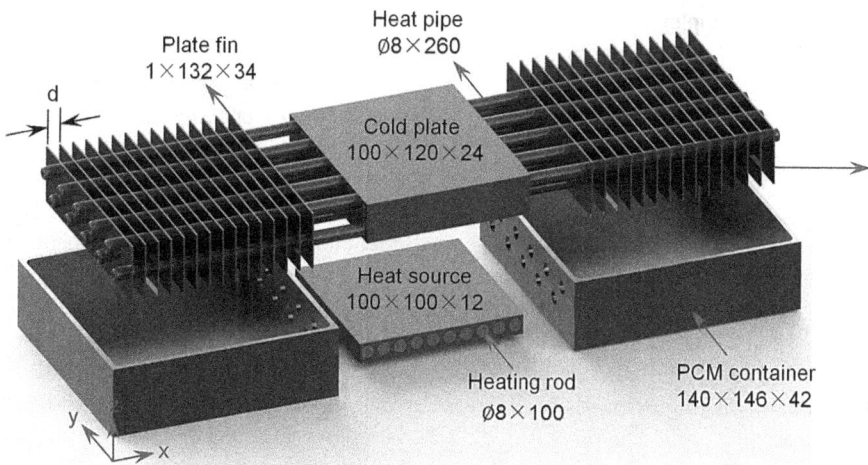

FIGURE 9.9   Schematic of a low-melting-point PCM-based finned heat pipe [73].

PCM-based heat pipes find applications in various electronic cooling scenarios, including computers, mobile devices, and other heat-generating electronic components. They are particularly useful in situations where traditional heat pipes may experience limitations in heat dissipation, such as during high thermal loads or transient heat events.

Ongoing research in this field focuses on optimizing the selection of PCM materials, designing heat pipe configurations suitable for PCM integration, pulsating heat pipes [70,71], and improving the overall thermal performance of PCM-based heat pipes through the introduction of fins [73] and metal foams in heat pipes [74].

9.3.3 PCM-Integrated Thermal Interface Materials (TIMs)

TIMs serve the purpose of bridging the voids between electronic components and heat sinks, facilitating proficient heat transfer. TIMs fortified with PCMs amalgamate the benefits of PCMs to augment their thermal conductivity and heat storage prowess. These specialized TIMs soak up heat emanating from the electronic component and dispatch it to the heat sink, thereby amplifying the overall heat dissipation efficacy [75–77].

9.4 PCM-BASED HYBRID ELECTRONIC COOLING

PCM-based hybrid cooling systems meld the distinctive attributes of PCMs with alternate cooling strategies to amplify cooling capacity, accelerate heat dissipation, fine-tune temperature control, and enhance flexibility and adaptability. Typically, these systems embed a PCM within a heat sink that is in direct contact with the heat-generating components, ensuring adept heat transfer from the electronic gadgets to the PCM. During operation, the PCM absorbs the thermal energy emitted by the electronic components. Upon reaching its phase change temperature, the PCM transitions phase, absorbing a notable amount of heat in the process. This absorbed heat is then disseminated through the additional cooling method employed in the system. This supplemental cooling method takes on a part of the heat removal duty when the PCM is in its melting phase. However, at temperatures below the melting temperature of the PCM, it becomes the primary mechanism of heat removal [78].

The hallmark advantages of PCM-based hybrid cooling encompass precise temperature control and defense against thermal runaways. PCMs operate within a confined temperature range during phase transitions, making them superb temperature regulators. By marrying PCMs with active cooling methodologies, precise temperature stewardship can be realized, dynamically reacting to variances in heat loads or ambient conditions. This meticulous control ensures that electronic systems function within ideal temperature bounds, safeguarding their performance, reliability, and longevity. Furthermore, PCM-based hybrid cooling methods display unmatched design flexibility and adaptability. These methods can be custom-fitted to address the specific cooling demands of a diverse array of electronic devices and complex system architectures. The versatility of PCM-based hybrid cooling facilitates smooth integration into intricate electronic designs, ensuring efficient and effective thermal management.

9.4.1 PCM AIR COOLING

A PCM–air hybrid cooling systems combines PCMs with forced or natural air cooling techniques. The system integrates a PCM into the heat sink to absorb and store excess heat generated by the electronics. During operation, the PCM modules undergo a phase change, utilizing their latent heat storage capacity. Simultaneously, air cooling mechanisms, such as fans or natural convection, facilitate the transfer of heat from the PCM modules to the surrounding environment. The PCM acts as a thermal energy buffer, preventing the overheating of electronic components and maintaining stable operating temperatures. In such systems, air flow rate plays a significant role in the heat dissipation capacity of the system. Forced convection, which utilizes fans to enhance airflow, generally provides higher cooling efficiency compared to natural convection alone. However, it is important to consider that forced convection requires additional components like fans and consumes power.

Optimizing the configuration of PCM modules within the system further enhances thermal performance. Researchers are focused on designing the layout and arrangement of PCM containers or modules to maximize heat transfer and improve overall cooling efficiency. This includes maximizing the surface area of PCM exposed to airflow, optimizing the spacing between modules, and incorporating heat transfer enhancement techniques such as fins, metal foam, or microstructures [79]. Figures 9.10 and 9.11 depict the schematic of a PCM–air hybrid cooling system incorporating natural [80] and forced air cooling [78], respectively.

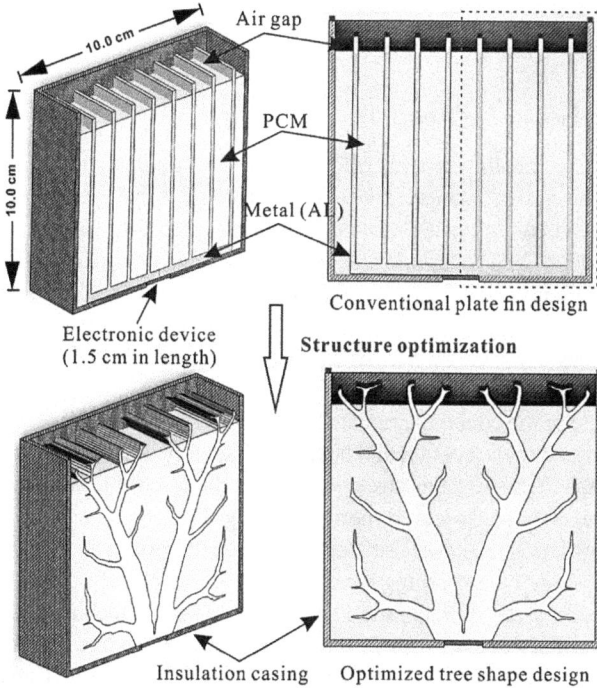

FIGURE 9.10 PCM–air (natural convection) hybrid cooling-based heat sink [80].

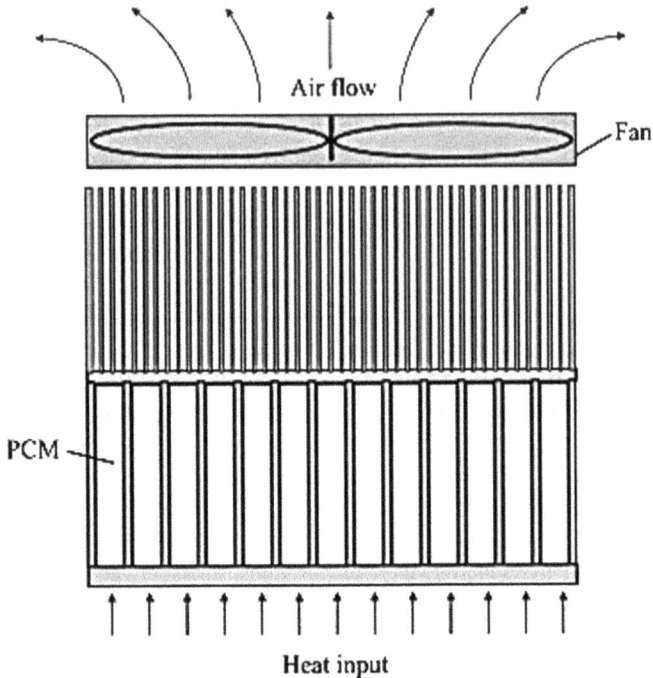

FIGURE 9.11   PCM–air (forced convection) hybrid cooling-based heat sink [78].

9.4.2   PCM LIQUID COOLING

PCM–liquid hybrid cooling systems offer efficient heat dissipation through the integration of PCMs and liquid coolants. There are two approaches in these systems: PCM slurry cooling and liquid cooling of PCM-containing heat sinks. The liquid coolant includes conventional options like water or oil, as well as advanced fluids such as nanofluids.

In the PCM slurry-based approach, micro or nanosized PCM particles are encapsulated and dispersed within the liquid coolant, forming an EPCM slurry. The liquid coolant, acting as the primary heat transfer medium, facilitates the circulation of the slurry mixture through the cooling loop. This loop typically comprises a pump to control the flow rate of the coolant, a radiator to dissipate the heat to the environment, and a container for storing the coolant and heat sink assembly. As the slurry passes through the heat sink, the PCM particles absorb heat from the electronic components, undergo a phase change, and release stored thermal energy. The liquid coolant, along with the dispersed PCM particles, then dissipates the heat to the external environment through a heat exchanger or other cooling mechanisms. Research efforts in this area focus on various aspects such as preparation and characterization techniques, hydrothermal performance evaluation, and simulation methods. These studies aim to optimize the combination of EPCM slurries with different types of heat sinks (including finned, metal foam, and micro/minichannel) under varying operational conditions [81–85]. The schematic of a PCM slurry cooling-based heat sink is depicted in Figure 9.12.

In the liquid cooling of PCM-containing heat sinks, the PCM is integrated into the heat sink structure itself, rather than being dispersed in a cooling loop. The heat sink comes into direct contact with the heat-generating electronic components. During operation, the PCM absorbs heat from the electronics, undergoes a phase change, and stores thermal energy. The liquid coolant, on the other hand, circulates through a separate cooling loop. This loop typically includes components such as a pump, radiator, and liquid container. As the liquid coolant flows through the heat sink, it absorbs the heat from the PCM-containing heat sink and dissipates it into the surrounding environment. The performance of such systems is influenced by various factors, including the thermophysical properties of the liquid coolant, flow rate, heat sink design, and operational conditions, as well as the thermophysical properties, distribution, and concentration of the PCM. Figure 9.13 depicts the schematic of a PCM–liquid cooling-based heat sink.

Recent studies have explored the use of jet impingement cooling techniques on PCM-containing heat sinks to enhance their thermal performance. These techniques involve directing high-velocity liquid jets onto the heat sink surface, improving the heat transfer between the PCM and the liquid coolant (Figure 9.14) [86].

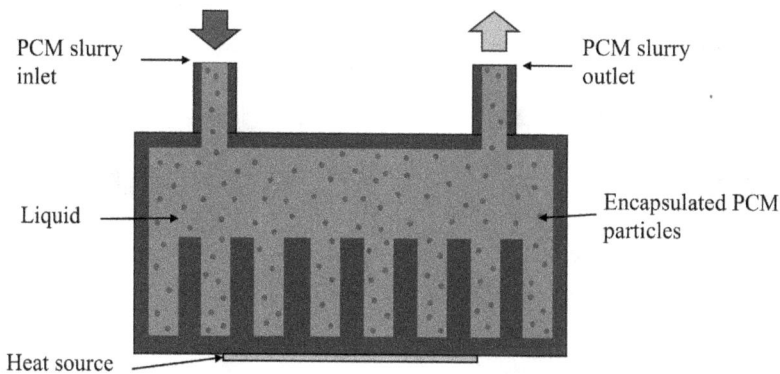

FIGURE 9.12 Illustration of a PCM slurry cooling-based heat sink.

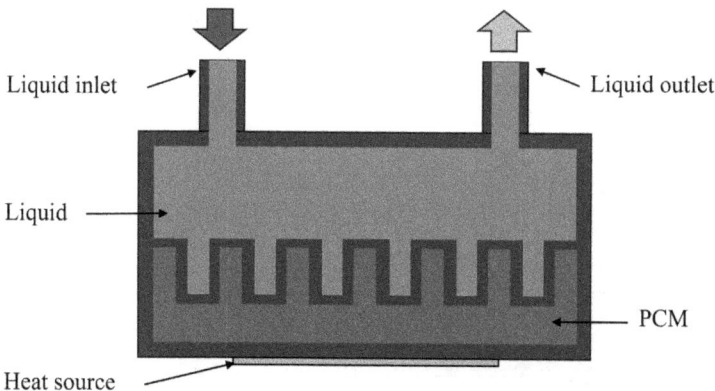

FIGURE 9.13 Illustration of a PCM–liquid cooling-based heat sink.

9.4.3 PCM Thermoelectric Cooling

PCM thermoelectric cooling refers to a cooling method that combines the use of PCMs and thermoelectric devices to remove heat from a system. Thermoelectric cooling is a method of cooling that utilizes the thermoelectric effect to transfer heat from one side of a device to the other. It relies on the properties of certain materials, known as thermoelectric materials, which can generate a temperature difference when an electric current is applied. This temperature difference allows heat to be transferred from the cold side of the device to the hot side, resulting in cooling on the cold side (Figure 9.15). Thermoelectric cooling offers advantages such as compactness, solid-state operation, and precise temperature control, making it suitable for various applications, including electronics cooling, refrigeration, and thermal management in confined spaces [87,88].

In the context of PCM thermoelectric cooling, PCM contained in a heat sink or heat pipe [90] absorbs heat generated by electronic components or other heat sources.

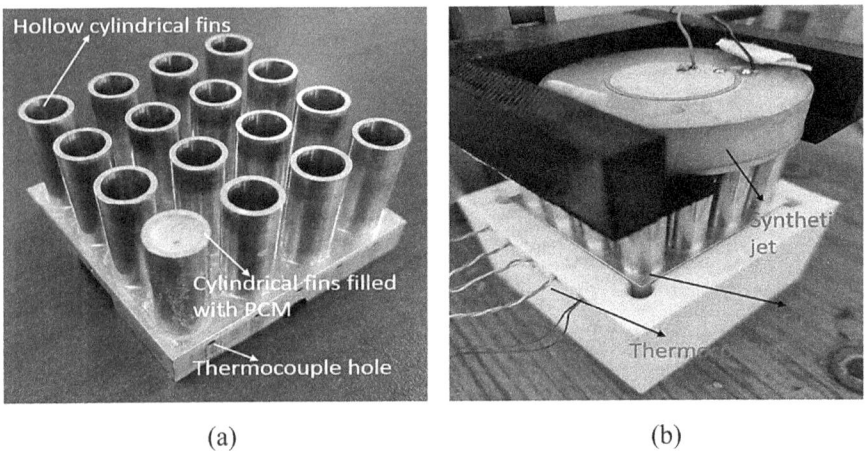

(a)   (b)

FIGURE 9.14   Jet-based PCM heat sink (a). Heat sink with PCM-filled and unfilled cylindrical fins (b) test section [86].

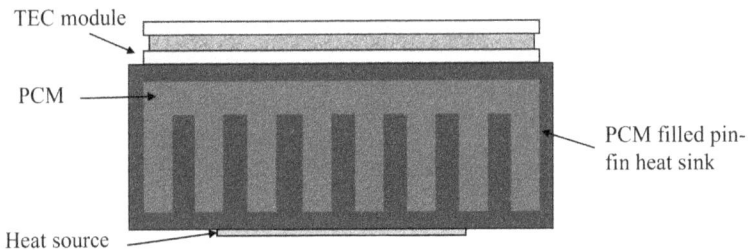

FIGURE 9.15   Illustration of thermoelectric cooling (TEC) [89].

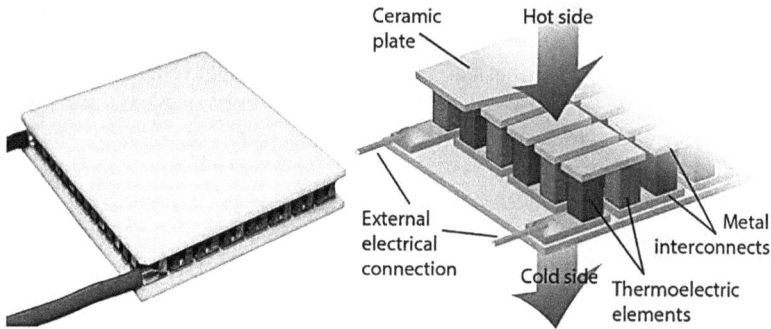

FIGURE 9.16   Illustration of PCM thermoelectric cooling system.

When the temperature of the system rises above a certain threshold, the PCM undergoes a phase change, absorbing heat and storing thermal energy. Thermoelectric modules then transfer the stored thermal energy from the PCM to an external heat sink or the surrounding environment (Figure 9.16).

PCM thermoelectric cooling can be a viable cooling approach for certain electronics applications, particularly those with moderate heat dissipation requirements and where temperature regulation and compactness are important considerations [87,88]. However, it is crucial to carefully assess the specific requirements and constraints of the system to determine the practicality and effectiveness of using PCM thermoelectric cooling in each scenario. One consideration is the integration of the thermoelectric modules and PCM into the electronic system. Proper design and engineering are required to ensure effective thermal contact between the thermoelectric modules and the PCM, as well as between the electronic components and the PCM. The overall system efficiency and power consumption should also be evaluated. Thermoelectric modules have relatively low energy conversion efficiencies. This means that a significant amount of electrical power is required to drive the thermoelectric cooling process, which can impact the overall power consumption of the electronic system.

9.5 CONCLUSION AND FUTURE RECOMMENDATIONS

In conclusion, the utilization of PCMs has emerged as a potent solution for thermal management in electronic devices. This chapter has undertaken an exhaustive exploration of the capabilities of PCMs, with a distinct emphasis on passive and hybrid cooling methodologies. The key points are delineated as follows:

- **Passive Cooling Methods**: The utilization of PCM-enhanced heat sinks, heat pipes, and thermal interface materials (TIMs) has demonstrated significant effectiveness in the absorption and dissipation of heat. These methodologies are instrumental in ensuring reliable temperature regulation in electronic components.

- **Efficacy of PCM-Based Heat Pipes**: These specialized heat pipes offer a compelling alternative to traditional heat pipes, particularly in scenarios characterized by high thermal loads or transient heat events.
- **Role of PCM-Integrated TIMs**: The integration of PCMs with TIMs serves to augment heat transfer between electronic components and heat sinks. This has a cascading positive effect on overall heat dissipation.
- **Hybrid Cooling Systems**: Incorporating PCMs with other cooling techniques, such as air cooling, liquid cooling, and thermoelectric cooling, results in systems with enhanced cooling capacity. These hybrid systems excel in precise temperature control and offer remarkable adaptability.
- **Directions for Future Research**: Future inquiries should be geared towards the exploration of solid–solid PCMs to diversify the range of available options. Moreover, the integration of PCM-based systems with Internet of Things (IoT) technology and intelligent control algorithms could lead to adaptive and smart cooling solutions.
- **Environmental and Sustainability Concerns**: It is imperative to perform comprehensive lifecycle analyses and environmental impact assessments to ascertain the sustainability of PCM-based cooling systems.
- **Application-Specific Investigations**: Research that is tailored to specific industries, such as data centers, aerospace, and consumer electronics, will furnish valuable insights into the practical implementation and efficacy of PCM-based thermal management systems.

REFERENCES

[1] Arden WM. The international technology roadmap for semiconductors—perspectives and challenges for the next 15 years. *Curr Opin Solid State Mater Sci* 2002;6:371–377.

[2] Sohel Murshed SM, Nieto de Castro CA. A critical review of traditional and emerging techniques and fluids for electronics cooling. *Renew Sustain Energy Rev* 2017;78: 821–833. https://doi.org/10.1016/j.rser.2017.04.112.

[3] Pfahl RC, McElroy J. The 2004 international electronics manufacturing initiative (iNEMI) technology roadmaps. *2005 Conference on High Density Microsystem Design and Packaging and Component Failure Analysis*; 2005–Jun 27:1–7. IEEE.

[4] Faraji H, Benkaddour A, Oudaoui K, El Alami M, Faraji M. Emerging applications of phase change materials: A concise review of recent advances. *Heat Transf* 2021;50:1443–1493. https://doi.org/10.1002/htj.21938.

[5] Lawag RA, Ali HM. Phase change materials for thermal management and energy storage: A review. *J Energy Storage* 2022;55. https://doi.org/10.1016/j.est.2022.105602.

[6] Hollis J, Sharar DJ, Bandhauer T. Effect of phase change material on dynamic thermal management performance for power electronics packages. *J Electron Packag Trans ASME* 2021;143. https://doi.org/10.1115/1.4052669.

[7] Maqbool Z, Hanief M, Parveez M. Review on performance enhancement of phase change material based heat sinks in conjugation with thermal conductivity enhancers for electronic cooling. *J Energy Storage* 2023;60. https://doi.org/10.1016/j.est.2022.106591.

[8] Afaynou I, Faraji H, Choukairy K, Arshad A, Arıcı M. Heat transfer enhancement of phase-change materials (PCMs) based thermal management systems for electronic components: A review of recent advances. *Int Commun Heat Mass Transf* 2023;143. https://doi.org/10.1016/j.icheatmasstransfer.2023.106690.

[9] Hua W, Zhang L, Zhang X. Research on passive cooling of electronic chips based on PCM: A review. *J Mol Liq* 2021;340. https://doi.org/10.1016/j.molliq.2021.117183.

[10] Mozafari M, Lee A, Mohammadpour J. Thermal management of single and multiple PCMs based heat sinks for electronics cooling. *Therm Sci Eng Prog* 2021;23. https://doi.org/10.1016/j.tsep.2021.100919.

[11] Sahoo SK, Das MK, Rath P. Application of TCE-PCM based heat sinks for cooling of electronic components: A review. *Renew Sustain Energy Rev* 2016;59:550–582. https://doi.org/10.1016/j.rser.2015.12.238.

[12] Ling Z, Zhang Z, Shi G, Fang X, Wang L, Gao X, et al. Review on thermal management systems using phase change materials for electronic components, Li-ion batteries and photovoltaic modules. *Renew Sustain Energy Rev* 2014;31:427–438. https://doi.org/10.1016/j.rser.2013.12.017.

[13] Kandasamy R, Wang XQ, Mujumdar AS. Application of phase change materials in thermal management of electronics. *Appl Therm Eng* 2007;27:2822–2832. https://doi.org/10.1016/j.applthermaleng.2006.12.013.

[14] Sohel Murshed SM, Nieto de Castro CA. A critical review of traditional and emerging techniques and fluids for electronics cooling. *Renew Sustain Energy Rev* 2017;78:821–833. https://doi.org/10.1016/j.rser.2017.04.112.

[15] Hua W, Zhang L, Zhang X. Research on passive cooling of electronic chips based on PCM: A review. *J Mol Liq* 2021;340:117183. https://doi.org/10.1016/j.molliq.2021.117183.

[16] Mohamed SA, Al-Sulaiman FA, Ibrahim NI, Zahir MH, Al-Ahmed A, Saidur R, et al. A review on current status and challenges of inorganic phase change materials for thermal energy storage systems. *Renew Sustain Energy Rev* 2017;70:1072–1089. https://doi.org/10.1016/j.rser.2016.12.012.

[17] Magendran SS, Saleem F, Khan A, Mubarak NM, Vaka M. Nano-structures & nano-objects synthesis of organic phase change materials (PCM) for energy storage applications: A review. *Nano-Struct Nano-Objects* 2019;20:100399. https://doi.org/10.1016/j.nanoso.2019.100399.

[18] Gulfam R, Zhang P, Meng Z. Advanced thermal systems driven by paraffin-based phase change materials—A review. *Appl Energy* 2019;238:582–611. https://doi.org/10.1016/j.apenergy.2019.01.114.

[19] Rathod MK, Banerjee J, Rathod MK, Banerjee J. Experimental investigations on latent heat storage unit using paraffin wax as phase change material. *Exp Heat Transf* 2014;6152. https://doi.org/10.1080/08916152.2012.719065.

[20] He B, Martin V, Setterwall F. Phase transition temperature ranges and storage density of paraffin wax phase change materials. *Energy* 2004;29:1785–1804. https://doi.org/10.1016/j.energy.2004.03.002.

[21] Afzaal M, Al-sulaiman FA. Effects of carbon-based fillers on thermal properties of fatty acids and their eutectics as phase change materials used for thermal energy storage: A review. *J Energy Storage* 2021;35:102329. https://doi.org/10.1016/j.est.2021.102329.

[22] Yuan Y, Zhang N, Tao W, Cao X, He Y. Fatty acids as phase change materials: A review. *Renew Sustain Energy Rev* 2014;29:482–498. https://doi.org/10.1016/j.rser.2013.08.107.

[23] Qi G, Liang C, Bao R, Liu Z, Yang W, Xie B, et al. Solar energy materials & solar cells polyethylene glycol based shape-stabilized phase change material for thermal energy storage with ultra-low content of graphene oxide. *Sol Energy Mater Sol Cells* 2014;123:171–177. https://doi.org/10.1016/j.solmat.2014.01.024.

[24] Karaman S, Karaipekli A, Sarı A, Bic A. Polyethylene glycol (PEG)/ diatomite composite as a novel form-stable phase change material for thermal energy storage. *Sol Energy Mater Sol Cells* 2011;95:1647–1653. https://doi.org/10.1016/j.solmat.2011.01.022.

[25] Shao X, Chen C, Yang Y, Ku X, Fan L. Rheological behaviors of sugar alcohols for low-to-medium temperature latent heat storage: Effects of temperature in both the molten and supercooled liquid states. *Sol Energy Mater Sol Cells* 2019;195:142–154. https://doi.org/10.1016/j.solmat.2019.03.006.

[26] Abouei A, Karimi-maleh H, Naddafi M, Karimi F. Application of bio-based phase change materials for effective heat management. *J Energy Storage* 2023;61:106859. https://doi.org/10.1016/j.est.2023.106859.

[27] Lin Y, Alva G, Fang G. Review on thermal performances and applications of thermal energy storage systems with inorganic phase change materials. *Energy* 2018;165:685–708. https://doi.org/10.1016/j.energy.2018.09.128.

[28] Luo J, Zou D, Wang Y, Wang S, Huang L. Battery thermal management systems (BTMs) based on phase change material (PCM): A comprehensive review. *J Chem Eng* 2022;430.

[29] Ge H, Li H, Mei S, Liu J. Low melting point liquid metal as a new class of phase change material: An emerging frontier in energy area. *Renew Sustain Energy Rev* 2013;21:331–346. https://doi.org/10.1016/j.rser.2013.01.008.

[30] Liu M, Saman W, Bruno F. Review on storage materials and thermal performance enhancement techniques for high temperature phase change thermal storage systems. *Renew Sustain Energy Rev* 2012;16:2118–2132. https://doi.org/10.1016/j.rser.2012.01.020.

[31] Kenisarin MM. High-temperature phase change materials for thermal energy storage. *Renew Sustain Energy Rev* 2010;14:955–970. https://doi.org/10.1016/j.rser.2009.11.011.

[32] Khare S, Dell'Amico M, Knight C, McGarry S. Selection of materials for high temperature latent heat energy storage. *Sol Energy Mater Sol Cells* 2012;107:20–27. https://doi.org/10.1016/j.solmat.2012.07.020.

[33] Fukahori R, Nomura T, Zhu C, Sheng N, Okinaka N, Akiyama T. Thermal analysis of Al-Si alloys as high-temperature phase-change material and their corrosion properties with ceramic materials. *Appl Energy* 2016;163:1–8. https://doi.org/10.1016/j.apenergy.2015.10.164.

[34] Kumar N, Hirschey J, LaClair TJ, Gluesenkamp KR, Graham S. Review of stability and thermal conductivity enhancements for salt hydrates. *J Energy Storage* 2019;24:100794. https://doi.org/10.1016/j.est.2019.100794.

[35] Faizan M, Ahmed R, Ali HM. A critical review on thermophysical and electrochemical properties of Ionanofluids (nanoparticles dispersed in ionic liquids) and their applications. *J Taiwan Inst Chem Eng* 2021;124:391–423. https://doi.org/10.1016/j.jtice.2021.02.004.

[36] Leong KY, Abdul Rahman MR, Gurunathan BA. Nano-enhanced phase change materials: A review of thermo-physical properties, applications and challenges. *J Energy Storage* 2019;21:18–31. https://doi.org/10.1016/j.est.2018.11.008.

[37] Salunkhe PB, Shembekar PS. A review on effect of phase change material encapsulation on the thermal performance of a system. *Renew Sustain Energy Rev* 2012;16:5603–5616. https://doi.org/10.1016/j.rser.2012.05.037.

[38] Ghasemi K, Tasnim S, Mahmud S. PCM, nano/microencapsulation and slurries: A review of fundamentals, categories, fabrication, numerical models and applications. *Sustain Energy Technol Assess* 2022;52:102084. https://doi.org/10.1016/j.seta.2022.102084.

[39] Cabeza LF, Castellón C, Nogués M, Medrano M, Leppers R, Zubillaga O. Use of microencapsulated PCM in concrete walls for energy savings. *Energy Build* 2007;39:113–119. https://doi.org/10.1016/j.enbuild.2006.03.030.

[40] Sari A, Alkan C, Karaipekli A, Uzun O. Microencapsulated n-octacosane as phase change material for thermal energy storage. *Sol Energy* 2009;83:1757–1763. https://doi.org/10.1016/j.solener.2009.05.008.

[41] Huang X, Zhu C, Lin Y, Fang G. Thermal properties and applications of microencapsulated PCM for thermal energy storage: A review. *Appl Therm Eng* 2019;147:841–855. https://doi.org/10.1016/j.applthermaleng.2018.11.007.

[42] Regin AF, Solanki SC, Saini JS. Heat transfer characteristics of thermal energy storage system using PCM capsules: A review. *Renew Sustain Energy Rev* 2008;12:2438–2458. https://doi.org/10.1016/j.rser.2007.06.009.

[43] Chai L, Wang X, Wu D. Development of bifunctional microencapsulated phase change materials with crystalline titanium dioxide shell for latent-heat storage and photocatalytic effectiveness. *Appl Energy* 2015;138:661–674. https://doi.org/10.1016/j.apenergy.2014.11.006.

[44] Ho CJ, Hsu ST, Jang JH, Hosseini SF, Yan WM. Experimental study on thermal performance of water-based nano-PCM emulsion flow in multichannel heat sinks with parallel and divergent rectangular mini-channels. *Int J Heat Mass Transf* 2020;146. https://doi.org/10.1016/j.ijheatmasstransfer.2019.118861.

[45] Ramakrishnan S, Wang X, Sanjayan J, Wilson J. Assessing the feasibility of integrating form-stable phase change material composites with cementitious composites and prevention of PCM leakage. *Mater Lett* 2017;192:88–91. https://doi.org/10.1016/j.matlet.2016.12.052.

[46] Yu S, Jeong SG, Chung O, Kim S. Bio-based PCM/carbon nanomaterials composites with enhanced thermal conductivity. *Sol Energy Mater Sol Cells* 2014;120:549–554. https://doi.org/10.1016/j.solmat.2013.09.037.

[47] Babapoor A, Azizi M, Karimi G. Thermal management of a Li-ion battery using carbon fiber-PCM composites. *Appl Therm Eng* 2015;82:281–290. https://doi.org/10.1016/j.applthermaleng.2015.02.068.

[48] Bianco V, De Rosa M, Vafai K. Phase-change materials for thermal management of electronic devices. *Appl Therm Eng* 2022;214. https://doi.org/10.1016/j.applthermaleng.2022.118839.

[49] Oya T, Nomura T, Okinaka N, Akiyama T. Phase change composite based on porous nickel and erythritol. *Appl Therm Eng* 2012;40:373–377. https://doi.org/10.1016/j.applthermaleng.2012.02.033.

[50] Atinafu DG, Chang SJ, Kim KH, Dong W, Kim S. A novel enhancement of shape/thermal stability and energy-storage capacity of phase change materials through the formation of composites with 3D porous (3,6)-connected metal–organic framework. *Chem Eng J* 2020;389:124430. https://doi.org/10.1016/j.cej.2020.124430.

[51] Cao R, Chen S, Wang Y, Han N, Liu H, Zhang X. Functionalized carbon nanotubes as phase change materials with enhanced thermal, electrical conductivity, light-to-thermal, and electro-to-thermal performances. *Carbon N Y* 2019;149:263–272. https://doi.org/10.1016/j.carbon.2019.04.005.

[52] Alkan C, Sari A. Fatty acid/poly(methyl methacrylate) (PMMA) blends as form-stable phase change materials for latent heat thermal energy storage. *Sol Energy* 2008;82:118–124. https://doi.org/10.1016/j.solener.2007.07.001.

[53] Krupa I, Miková G, Luyt AS. Phase change materials based on low-density polyethylene/paraffin wax blends. *Eur Polym J* 2007;43:4695–4705. https://doi.org/10.1016/j.eurpolymj.2007.08.022.

[54] Righetti G, Zilio C, Doretti L, Longo GA, Mancin S. On the design of Phase Change Materials based thermal management systems for electronics cooling. *Appl Therm Eng* 2021;196. https://doi.org/10.1016/j.applthermaleng.2021.117276.

[55] Krupa I, Miková G, Luyt AS. Phase change materials based on low-density polyethylene/paraffin wax blends. *Eur Polym J* 2007;43:4695–4705. https://doi.org/10.1016/j.eurpolymj.2007.08.022.

[56] Huang P, Wei G, Cui L, Wang G, Du X. A morphology optimization of enclosure shape of low melting point alloy-based PCM heat sink. *J Energy Storage* 2023;64:107153. https://doi.org/10.1016/j.est.2023.107153.

[57] Bashar M, Siddiqui K. Experimental investigation of transient melting and heat transfer behavior of nanoparticle-enriched PCM in a rectangular enclosure. *J Energy Storage* 2018;18:485–497. https://doi.org/10.1016/j.est.2018.06.006.

[58] Emam M, Ookawara S, Ahmed M. Thermal management of electronic devices and concentrator photovoltaic systems using phase change material heat sinks: Experimental investigations. *Renew Energy* 2019;141:322–339. https://doi.org/10.1016/j.renene.2019.03.151.

[59] Ali HM, Ashraf MJ, Giovannelli A, Irfan M, Irshad T Bin, Hamid HM, et al. Thermal management of electronics: An experimental analysis of triangular, rectangular and circular pin-fin heat sinks for various PCMs. *Int J Heat Mass Transf* 2018;123:272–284. https://doi.org/10.1016/j.ijheatmasstransfer.2018.02.044.

[60] Yang XH, Tan SC, Ding YJ, Wang L, Liu J, Zhou YX. Experimental and numerical investigation of low melting point metal based PCM heat sink with internal fins. *Int Commun Heat Mass Transf* 2017;87:118–124. https://doi.org/10.1016/j.icheatmasstransfer.2017.07.001.

[61] Iradukunda AC, Vargas A, Huitink D, Lohan D. Transient thermal performance using phase change material integrated topology optimized heat sinks. *Appl Therm Eng* 2020;179:115723. https://doi.org/10.1016/j.applthermaleng.2020.115723.

[62] Joseph M, Antony V, Sajith V. Characterisation of heat dissipation from PCM based heat sink using Mach–Zehnder Interferometry. *Heat Mass Transfer/Waerme- Und Stoffuebertragung* 2022;58:171–193. https://doi.org/10.1007/s00231-021-03101-1.

[63] Krishnan S, Garimella SV, Kang SS. A novel hybrid heat sink using phase change materials for transient thermal management of electronics. *IEEE Trans Compon Packag Technol* 2005;28:281–289. https://doi.org/10.1109/TCAPT.2005.848534.

[64] Duan J, Xiong Y, Yang D. Melting behavior of phase change material in honeycomb structures with different geometrical cores. *Energies (Basel)* 2019;12. https://doi.org/10.3390/en12152920.

[65] Dinesh BVS, Bhattacharya A. Comparison of energy absorption characteristics of PCM-metal foam systems with different pore size distributions. *J Energy Storage* 2020;28:101190. https://doi.org/10.1016/j.est.2019.101190.

[66] Rahmanian S, Moein-Jahromi M, Rahmanian-Koushkaki H, Sopian K. Performance investigation of inclined CPV system with composites of PCM, metal foam and nanoparticles. *Sol Energy* 2021;230:883–901. https://doi.org/10.1016/j.solener.2021.10.088.

[67] Ali HM. Heat transfer augmentation of porous media (metallic foam) and phase change material based heat sink with variable heat generations: An experimental evaluation. *Sustain Energy Technol Assess* 2022;52:102218. https://doi.org/10.1016/j.seta.2022.102218.

[68] Baby R, Balaji C. Experimental investigations on thermal performance enhancement and effect of orientation on porous matrix filled PCM based heat sink. *Int Commun Heat Mass Transf* 2013;46:27–30. https://doi.org/10.1016/j.icheatmasstransfer.2013.05.018.

[69] Rehman T ur, Ali HM. Thermal performance analysis of metallic foam-based heat sinks embedded with RT-54HC paraffin: An experimental investigation for electronic cooling. *J Therm Anal Calorim* 2020;140:979–990. https://doi.org/10.1007/s10973-019-08961-8.

[70] Ling YZ, Zhang XS, Wang F, She XH. Performance study of phase change materials coupled with three-dimensional oscillating heat pipes with different structures for electronic cooling. *Renew Energy* 2020;154:636–649. https://doi.org/10.1016/j.renene.2020.03.008.

[71] Zhao J, Rao Z, Liu C, Li Y. Experimental investigation on thermal performance of phase change material coupled with closed-loop oscillating heat pipe (PCM/CLOHP) used in thermal management. *Appl Therm Eng* 2016;93:90–100. https://doi.org/10.1016/j.applthermaleng.2015.09.018.

[72] Yu Z, Zhang J, Pan W. A review of battery thermal management systems about heat pipe and phase change materials. *J Energy Storage* 2023;62. https://doi.org/10.1016/j.est.2023.106827.

[73] Yang XH, Tan SC, He ZZ, Liu J. Finned heat pipe assisted low melting point metal PCM heat sink against extremely high power thermal shock. *Energy Convers Manag* 2018;160:467–476. https://doi.org/10.1016/j.enconman.2018.01.056.

[74] Li WQ, Li YX, Yang TH, Zhang TY, Qin F. Experimental investigation on passive cooling, thermal storage and thermoelectric harvest with heat pipe-assisted PCM-embedded metal foam. *Int J Heat Mass Transf* 2023;201. https://doi.org/10.1016/j.ijheatmasstransfer.2022.123651.

[75] Sarvar F, Whalley DC, Conway PP. Thermal interface materials-A review of the state of the art. *2006 1st Electronic Systemintegration Technology Conference 2006 Sep 5* (Vol. 2, pp. 1292–1302). IEEE.

[76] Goel N, Anoop TK, Bhattacharya A, Cervantes JA, Mongia RK, Machiroutu SV, et al. Technical review of characterization methods for Thermal Interface Materials (TIM). *2008 11th IEEE Intersociety Conference on Thermal and Thermomechanical Phenomena in Electronic Systems, I-THERM* 2008:248–258. https://doi.org/10.1109/ITHERM.2008.4544277.

[77] Shia D, Yang J. Analytical, numerical and experimental study of phase change material in TIM2 application for high-power server CPUs. *InterSociety Conference on Thermal and Thermomechanical Phenomena in Electronic Systems, ITHERM* 2020;2020-July:158–165. https://doi.org/10.1109/ITherm45881.2020.9190178.

[78] Kozak Y, Abramzon B, Ziskind G. Experimental and numerical investigation of a hybrid PCM-air heat sink. *Appl Therm Eng* 2013;59:142–152. https://doi.org/10.1016/j.applthermaleng.2013.05.021.

[79] Kalbasi R. Introducing a novel heat sink comprising PCM and air—Adapted to electronic device thermal management. *Int J Heat Mass Transf* 2021;169:120914. https://doi.org/10.1016/j.ijheatmasstransfer.2021.120914.

[80] Xie J, Choo KF, Xiang J, Lee HM. Characterization of natural convection in a PCM-based heat sink with novel conductive structures. *Int Commun Heat Mass Transf* 2019;108:104306. https://doi.org/10.1016/j.icheatmasstransfer.2019.104306.

[81] Rajabifar B. Enhancement of the performance of a double layered microchannel heat-sink using PCM slurry and nanofluid coolants. *Int J Heat Mass Transf* 2015;88:627–635. https://doi.org/10.1016/j.ijheatmasstransfer.2015.05.007.

[82] Rajabi Far B, Mohammadian SK, Khanna SK, Zhang Y. Effects of pin tip-clearance on the performance of an enhanced microchannel heat sink with oblique fins and phase change material slurry. *Int J Heat Mass Transf* 2015;83:136–145. https://doi.org/10.1016/j.ijheatmasstransfer.2014.11.082.

[83] Seyf HR, Zhou Z, Ma HB, Zhang Y. Three dimensional numerical study of heat-transfer enhancement by nano-encapsulated phase change material slurry in micro-tube heat sinks with tangential impingement. *Int J Heat Mass Transf* 2013;56:561–573. https://doi.org/10.1016/j.ijheatmasstransfer.2012.08.052.

[84] Alquaity ABS, Al-Dini SA, Wang EN, Yilbas BS. Numerical investigation of liquid flow with phase change nanoparticles in microchannels. *Int J Heat Fluid Flow* 2012;38:159–167. https://doi.org/10.1016/j.ijheatfluidflow.2012.10.001.

[85] Dammel F, Stephan P. Heat transfer to suspensions of microencapsulated phase change material flowing through minichannels. *J Heat Transf* 2012;134:1–8. https://doi.org/10.1115/1.4005062.

[86] Arumuru V, Rajput K, Nandan R, Rath P, Das M. A novel synthetic jet based heat sink with PCM filled cylindrical fins for efficient electronic cooling. *J Energy Storage* 2023;58:106376. https://doi.org/10.1016/j.est.2022.106376.

[87] Cai Y, Wang Y, Liu D, Zhao FY. Thermoelectric cooling technology applied in the field of electronic devices: Updated review on the parametric investigations and model developments. *Appl Therm Eng* 2019;148:238–255. https://doi.org/10.1016/j.applthermaleng.2018.11.014.

[88] Zhao D, Tan G. A review of thermoelectric cooling: Materials, modeling and applications. *Appl Therm Eng* 2014;66:15–24. https://doi.org/10.1016/j.applthermaleng.2014.01.074.

[89] Arora S. Selection of thermal management system for modular battery packs of electric vehicles: A review of existing and emerging technologies. *J Power Sources* 2018;400:621–640. https://doi.org/10.1016/j.jpowsour.2018.08.020.

[90] Sun X, Zhang L, Liao S. Performance of a thermoelectric cooling system integrated with a gravity-assisted heat pipe for cooling electronics. *Appl Therm Eng* 2017;116:433–444. https://doi.org/10.1016/j.applthermaleng.2016.12.094.

10 Phase Change Material Applications in Construction and Building Materials

Benjamin Duraković

HIGHLIGHTS

- A wide range of PCM applications in construction materials is explored, such as concretes, pavements, cement render, gypsum boards, cladding materials, and wood.
- A comprehensive overview of PCM integration techniques used in construction materials is given.
- Thermal and mechanical properties of PCM-based construction materials are investigated.

NOMENCLATURE

CR	Cement render
IEA	International Energy Agency
mPCM	Micro-encapsulated gypsum board
PCM	Phase change material
SSPCM	Shape-stabilized PCM
TIM	Thermal insulation material

10.1 INTRODUCTION

Due to globalization and rapid population growth, the construction sector, as one of the largest consumers of energy, is at the forefront of studies aimed at reducing heat losses, saving energy, and reducing costs. According to the International Energy Agency (IEA), the building envelope is responsible for 39% of carbon dioxide emissions and 36% of global energy consumption [1]. To address this problem, various model of sustainable building construction strategies were employed [2, 3]. One of them is integration of phase change materials (PCMs) in the building envelope is an effective solution to improve the thermal performance of buildings and reduce environmental impact [4]. PCMs integrated within building envelopes were

DOI: 10.1201/9781003331957-10

experimented with over the years. Particularly, the integration of PCM in construction materials for floors [5], walls [6], concrete [7], roofs [8], windows [9, 10], cladding [11, 12], and plastering [13] has been studied. Available results showed that PCM-based building envelopes provide indoor peak temperature reduction [14–16] and peak temperature delays [17]. This way, indoor thermal comfort and building energy efficiency are enhanced [18].

Therefore, PCM-based building envelopes are effective in improving building energy flexibility in dynamic building energy management in response to electricity price volatility [19, 20]. Current building energy management systems use conventional control methodologies that are "reactive" in nature, which prevents them from fully utilizing the potential of PCM-based construction materials in demand-responsive building control applications [21]. To reduce electricity costs and optimize demand flexibility, PCM-based construction materials may play an important role [22]. Because of their potential to store thermal energy from solar radiation, industrial waste heat, and excess heat, PCMs have attracted much attention and interest from scientists around the world.

A promising passive method [23, 24] for boosting the thermal mass of lightweight envelopes [25] is the use of PCMs. Therefore, various PCM-based construction materials and energy technologies have been investigated in the literature [26]. It is observed that PCMs have been applied in various construction materials such as cement render (CR), gypsum boards, bricks, wood etc., which significantly improves building thermal performance. Therefore, the purpose of this chapter is to present up-to-date research results on the application of PCM in construction materials, compile an overview of the properties of PCM-based construction materials, and review the main integration approaches of PCMs in construction materials. This will be a valuable source of information for researchers, practitioners, and students in this field.

10.2 PCMS IN CONSTRUCTION MATERIALS

The incorporation of PCMs into diverse construction materials is achieved through various integration methods. The building envelope is a crucial building component influencing heating/cooling load reduction, serving as a primary structure for PCM integration [27, 28]. The position of the PCM layer plays a crucial role. Typically, the layer is positioned on the interior side of the envelope or in proximity to the indoor environment. In this context, PCM layers contribute significantly to the indoor environment thermal comfort by effectively storing excess heat. This stored heat can be judiciously reused when required or expelled outdoors through adequate ventilation. Such a process not only mitigates thermal load but also diminishes discomfort hours, thereby enhancing overall energy efficiency [31, 32].

10.2.1 PCM-BASED FOAM CONCRETE

PCM-based foam concrete (FC) presents a promising avenue for elevating the thermal storage capacity and optimizing the thermal performance of buildings [29]. However, the application effectiveness is significantly influenced by diverse phase transition temperatures and prevailing climatic conditions. Foam concrete, tailored for lower densities, can be fashioned using a cement and water blend for cement paste-based FC or a mix of cement, sand, and water for mortar-based FC [30, 31]. The introduction of a foaming agent, whether protein-based or synthetic (refer to Figure 10.1), plays a pivotal role in generating intermittent air spaces within the mortar [37]. Foam concrete does not have a particular strategy for creating and acquiring the desired qualities. However, there are not many ways to accurately determine the mixing proportions needed [32]. The advantage of employing FC is that it makes the ideal substrate for PCM capsule embedding since it provides a porous and lightweight framework without adding to the structural load on buildings [33].

Paraffin PCM is the most suitable heat-storing substance because of its affordability, environmental friendliness, and stability. Due to its compatibility with building materials, foamed silica is the most suitable material to use to encapsulate PCMs [34]. The most stable structure and suitable composite for foam cement is a PCM composite with a paraffin content of 45%. The thermal conductivity and heat absorption capacity of phase change FC, according to experiments, is much higher than that of traditional FC. Additionally, when the composite PCM's composition increases, its thermal conductivity may eventually decrease and its capacity for heat storage may significantly rise [35].

10.2.2 CONCRETE

Adding PCMs directly to concrete results in lower thermal conductivity and an increase in thermal mass at specified temperatures. On the other hand, the concrete's compressive strength is an important property [36, 37] that can be affected. Some of the negative characteristics that PCM concrete has shown include lower strength, unclear long-term stability, and decreased fire resistance. Studies on PCM concrete have proven its benefits, primarily in the form of lowering interior temperatures in

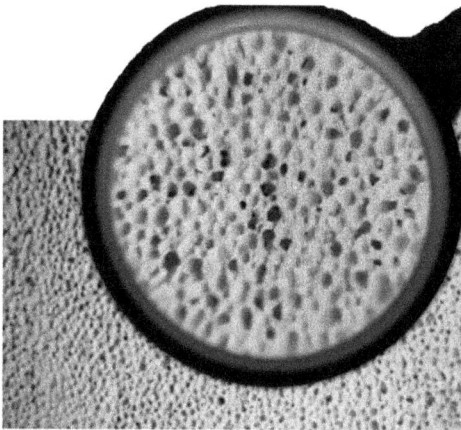

FIGURE 10.1 PCM-based foamed concrete (source: Author).

warm climates [38]. Several different approaches to fusing concrete using PCMs are shown in Figure 10.2. Drilling holes in the concrete and filling them with a PCM [39] and filling the preexisting voids in a hollow concrete floor [40] have been potential options.

In this experiment [40], paraffin with a melting point of 27.5 °C, was used to fill the concrete. The results showed that the temperature was lower on the opposite side of the hollow concrete in the summer. Because of this, during the warmer months, such flooring may be utilized as a passive thermal conditioner. However, further experiments with actual weather conditions are required to verify the results.

10.2.3 PCM-BASED CONCRETE PAVEMENT FOR ICE AND SNOW MELTING

PCM-based asphalt pavement mixtures can increase the thermal inertia of concrete and can effectively manage extreme temperatures. Extremely low pavement temperatures cause freeze–thaw damage and low-temperature cracking. The application of PCMs in pavement mixtures can reduce higher and lower temperature extremities and reduce freeze–thaw damage [41]. There are three approaches to PCM integration in concrete such as placing PCM in pipes, placing the PCM particles in a concrete mixture, and filling concrete surface voids with PCM. Figure 10.3 shows the approach to the integration of PCMs in concrete pavement.

PCMs that have temperature transition near zero degrees Celsius are used with the potential to regulate the extreme temperature of asphalt pavement and concrete [42]. Benefits are reduced speed of heating and cooling, reduced temperature peaks [43], reduced deicing needs [44], and increased asphalt surface temperature to 3.4 °C [45].

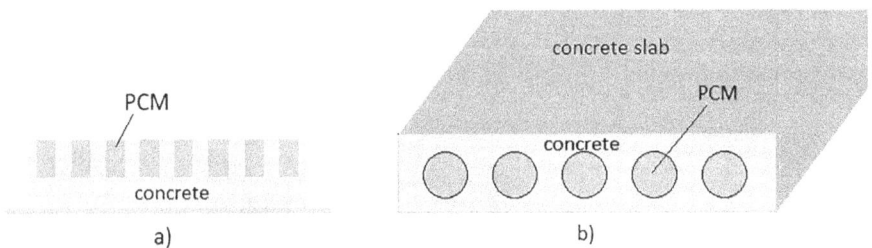

FIGURE 10.2 Methods of PCM integration in concrete: (a) PCM holes in concrete; (b) PCM pipes (source: Author).

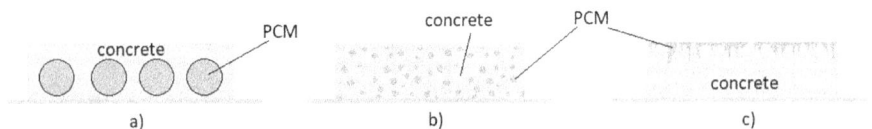

FIGURE 10.3 Methods of PCM integration in concrete pavement: (a) PCM pipes; (b) PCM particles; (c) PCM absorption by filling concrete surface voids (source: Author).

The challenges related to PCM-based concrete pavement are those associated with the heavy load and temperature. Therefore, the pavement must have a high compressive strength to sustain heavy traffic loads [46]. The encapsulation may break due to the loads, and PCM leakage may occur. The leakage may negatively affect the asphalt properties [47] and reduce the cement hydration and mechanical properties of concrete [48]. In the process of construction, the aggregates and asphalt binder are exposed to a higher temperature (140–160 °C), and the asphalt pavements are transported to the site. The installation is done using rollers; thus, the PCM and encapsulation medium must withstand the increased temperature and the pressure under rolling and heavy traffic [41].

10.2.4   PCM-Based Cement Render

A PCM-based cement mortar with a 60% PCM substitution has been found to be suitable for several construction applications [49, 50]. When selecting the CR, a cement-to-sand ratio of 1:4 by volume and a water content of 0.5 of cement were prioritized [65]. To produce PCM-based cement render, a meticulous blending process is undertaken wherein cement and sand are mixed in a mixer until a consistent shade is attained. Achieving uniformity in color involves initially mixing cement with 40% of the sand, followed by the gradual addition of PCM powder (constituting 60% by volume of the sand) at a lower speed. This careful approach is essential to prevent any damage to the PCM micro-capsules. The mixing and addition of PCM powder are iterated until a consistent color is achieved. During this process, water is gradually incorporated while maintaining constant stirring. The workability and consistency of the fresh mortar are assessed using the flow table method, as outlined in BS EN 1015-3 [51]. Finally, the necessary samples were made by pouring freshly mixed mortar into molds. After 24 hours in the molds, the samples were unmolded and stored in water until their testing ages were reached (i.e., 28 days).

In some of the previous studies, researchers replaced part of the sand in cement plaster/mortar by integrating PCM in various ratios [52]. According to Ramakrishnan et al. and Cunha et al. [53], a cement mortar with 60% PCM is suitable for numerous uses in structures to create cement render. The PCM cement render is created with 40% of the sand and is mixed until a uniform color is achieved. Following this, the PCM powder, constituting 60% of the sand by volume, is gradually incorporated while the mixer operates at a lower speed to ensure the preservation of PCM micro-capsules. The water is then gradually added and continuously mixed after that [54].

In another study, micro-encapsulated PCM was mixed with cement render to create a PCM-based composite for exterior wall finishing [55]. Following the research, a few items were found. Due to the PCM's thermal conductivity and reduced density, the PCM incorporation with CR produced PCM composites having lesser density but also thermal conductivity. There was a decrease of 41.5–59.2% in thermal conductivity, while the density decreased by 12–23%. Despite a 53% reduction in compressive strength, the CR-based composite with PCM integration nonetheless met the requirements for external finishing applications, making it appropriate for use in this industry. Perhaps CR does not produce the same outcome when combined with PCM as FC does, but it is still dependable in the application of finishing materials [56].

10.2.5 PCM GYPSUM BOARDS

The application of PCMs in gypsum boards has been examined since the 1990s [57]. Gypsum boards with PCMs reportedly decrease cooling loads of the building by 7–20% [58]. They are usually located on the inside side of the wall due to the increased thermal inertia enhanced by the PCMs [57]. Figure 10.4 represents a conceptual application of PCM in gypsum boards using micro-capsules.

PCM in gypsum boards can affect the mechanical properties of the boards by decreasing them and leading to less efficiency. However, some studies have suggested adding reinforcements that can regulate this disadvantage. The energy stored in a gypsum board with an increase of 7% in PCM percentage shows an increase of 20% in the thermal storage capacity [57]. When placed in the building envelope with hot air flow only from the interior side, the energy is also exchanged through one side. As a result, lower-thickness gypsum boards show better results in thermal exchange. One study shows that 1.5 cm boards are 66.66% affected by thermal exchange, in comparison with the 2.5 cm thick gypsum boards that gain thermal exchange of 40%. The best results occur in oblique free-standing gypsum boards in the interior, which are affected by airflow from both sides. This study proves that the position and thickness of the board, along with the percentage of the area in contact with the airflow, can affect the results of thermal exchange. Furthermore, a gypsum board with 45% PMCs can store 9.5 times more energy per unit mass when compared to brick, five times more than thermal brick, and three times more energy when compared to regular gypsum board [57].

Some studies showed that a gypsum board 1.5 cm thick with PCMs saves the same amount of thermal energy as a brick wall that is 12cm thick while maintaining a pleasant temperature range (20–30 °C) [59]. In comparison to the other building envelopes, gypsum board enhanced with PCM installed on lightweight and medium-weight envelopes is shown to deliver good energy flexibility, efficiency, and strength

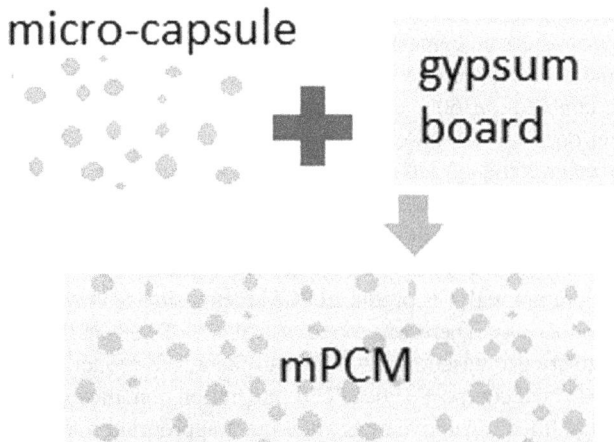

FIGURE 10.4 Micro-encapsulated gypsum board (mPCM) (source: Author).

restriction for all energy flexibility events. Additionally, it was determined that the medium-weight construction PCM envelope constituted the most effective type to take part in both morning and night energy flexibility activities [60].

In terms of heat storage and temperature lag, PCM-based gypsum board performs adequately, minimizing changes in interior temperature. In addition, compared to conventional gypsum board, the capacity to absorb moisture was significantly improved. Therefore PCM-based gypsum board is a prospective composite material used in construction that was employed to further push energy efficiency in buildings [61].

With a rise in the mass percentage of mPCM in gypsum, particularly in the range of the phase transition temperature, the thermal diffusivity of the mPCM–gypsum composite is reduced, allowing for the avoidance of summertime overheating. In exchange, the low thermal diffusivity reduces the rate of thermal energy absorption [62]. Only in the PCM subzone is that thermal efficiency (dynamic state of heat storage) of mPCM–gypsum composite boards higher than that of a gypsum board. In comparison to gypsum board, mPCM–gypsum composite boards have a lower heat absorption outside of this subzone. Thermal efficiency at the melting point is seen to rise with an increase in mass percentage in gypsum up to 30% of mPCM [63].

10.2.6 PCM Brick

The goal is to improve the storage capacity and insulation resistance of brick, a building material that is frequently employed in the construction of buildings [64]. It is important that masonry wall maintains required mechanical properties to sustain loads [65]. When selecting a PCM, latent heat and melting temperature should be considered the most important variables. While, if it is comfortable, the melting temperature is preferred to be roughly equal to the mathematical mean of the external thermal wave [66]. The bricks' thermal performance is drastically enhanced by increasing the amount of PCMs included in the brick, although this amount should be further adjusted for financial and structural strength reasons. The inner surface temperature fluctuation of the brick is only a little influenced by the external convective heat transfer coefficient; however, the internal convective heat transfer coefficient is especially critical [67].

In general, the hollow brick storage capacity and insulation power are enhanced by the incorporation of PCM. This fact is demonstrated by a sizable amplitude reduction and sizable phase shift of the thermal wave when it crosses the wall of brick containing PCMs. As a result, this technique enables building materials with optimum design (thickness). Additionally, it provides intriguing passive temperature regulation, boosts internal comfort, and thus aids in reducing building energy demand [68]. The quantity of incoming heat to a building and, subsequently, the requirement for mechanical cooling is significantly influenced by the envelope built utilizing hollow brick walls that integrate phase change materials [69].

The effect of PCM capsules on the reduction in heat transfer and the temperature variation were investigated when added to hollow bricks in a particular climate [70]. Changeable PCM amounts—namely from 6% to 20% of the brick volume—were discussed. Findings demonstrated that the bricks' thermal performance improved as

PCM quantity rose. The bricks of 16% and 20% PCM, which performed similarly and brought the indoor temperature reduction to thermal comfort standards, also produced the best results. Generally, PCMs enhanced thermal inertia, but if combined with insulation material, they enhance thermal performance and shift the peak load of the brick wall [71].

10.2.6.1 Fillings in PCM Bricks

The wall thermal performance could be greatly improved by adding thermal insulation material (TIM) and PCM to hollow bricks, although the operational mechanisms for improving the thermal characteristics of bricks differed naturally. Filling TIM contributed more to raising thermal resistance, while using PCM may clearly enhance thermal inertia [72]. Figure 10.5 shows PCM integration in hollow bricks and walls.

Filling all cavities with TIM would result in an increase from 1.50–2.20 h to 6.17–6.50 h in delay time and lower the decrement factor for inner surface temperatures from 17.01% to 11.04%. Adding PCM to the inside of cavities as well could decrease the decrement factor from 12.30%–17.00% to 1.71%–2.21% and raise the time of the delay from 1.40–2.00 h to 6.20–6.54 h [73]. When compared to outer hollow cavities, inner hollow cavities are a superior option for filling with PCM because they have a lower attenuation rate, a longer delay time, and a better ability to adjust to the climate, according to recent studies [74]. When integrating TIM and PCM simultaneously, it can lead to improving both thermal resistance and thermal inertia.

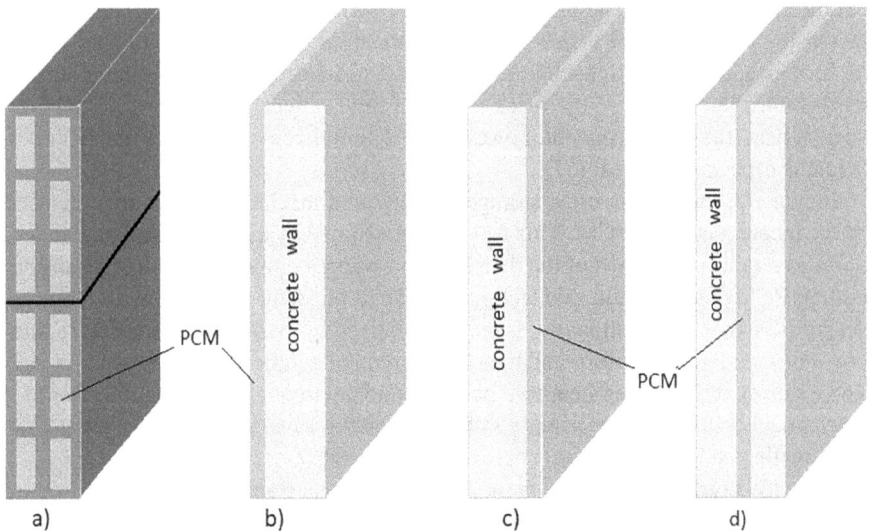

FIGURE 10.5 Methods of PCM integration in a wall: (a) brick; (b) exterior layer; (c) interior layer; (d) middle layer (source: Author).

10.2.7 PCMs in Cladding Materials

Cladding is a material that is attached to the building wall to foster thermal properties of buildings, such as insulation materials [75–77]. PCM layers are widely applied to building interiors near the internal environment to manage the indoor temperature. However, by adding PCM to the exteriors of the envelopes, their thermal insulation is improved, which can reduce heat gain from the outside environment [78]. The building envelope is where most heat input or loss occurs in buildings. Thus, the building envelope should be properly planned to limit energy use for cooling and heating. Due to their power to retain and dump huge quantities of heat at a relatively steady temperature, PCMs are primarily employed in the construction of building envelopes [79].

To incorporate PCMs into the building structure, the best position is crucial for consideration. The PCM layer should be placed nearer the heat source, nevertheless, as this is the most reasonable statement. For instance, in hot areas, the PCM should be placed close to the outer envelope layers [80]. The primary reason is that, under these circumstances, PCM acts as an insulator (heat barrier). Therefore, the stored heat needs to be as far away from the indoors as feasible to prevent any unwanted heat emittance there and to take advantage of the cooling effect at night in the evening. When it is cold outside, the PCM layer acts as a heat supply by preventing or restricting heat from escaping from indoors to the outside, storing the heat, and then releasing it back into the indoor space when the temperature decreases [81].

A study from 2022 examined the integration of PCM into foamed concrete used as building envelope PCM-based composites. The results indicate a reduction in density and thermal conductivity compared to the cement render used conventionally. Furthermore, the results show potential in using foamed concrete with PCMs to manufacture building envelopes with improved thermal characteristics in the form of lightweight cladding panels [82]. By boosting the thermal energy storage capacity of multilayer glazing facades, the applications of PCM can lower energy consumption and raise thermal comfort.

PCM thickness significantly affects the compromise between the thermal and solar transmittance performance of the facade system [83]. Thermal comfort is improved by increasing PCM thickness since the interior surface temperatures and the lag between temperature changes is increased. The heat loss from multilayered glazing facades can be efficiently reduced by increasing the PCM thickness [84]. However, the temperature decrement factor is only marginally impacted by PCM thickness. However, it was advised to incorporate the PCM thickness into a multilayered glazing facade 20 mm thick due to the significant reduction in solar transmittance. Due to their greater light transmittance and better views, multilayer glazing facades are frequently used in modern buildings. However, multilayer glazing facades have significantly lower thermal resistance, which has a significant impact on building energy demand [85]. Due to a large, glazed area that has heat loss, glazed facades are responsible for around 30% of the overall building energy consumption [86],

10.2.8 PCMs in Wood

Wood is frequently used in building construction due to its natural, renewable, and porous characteristics. Enhancing the capabilities of wood regarding energy storage helps in the reduction of carbon dioxide and control of living environment temperature considering the ongoing rise in global carbon dioxide emissions and increasingly critical environmental issues [87]. The PCM in wood composite has great mechanical strength and thermally induced flexibility because of the good mechanical properties of the support material, which is important for the application of thermal management [88]. To achieve thermal energy storage in buildings, interior panels that are wood based have a latent heat storage and phase transition temperature. They have a strong potential in thermal energy storage since they could encapsulate a significant quantity of energy, up to 56.1 J/g at a melting temperature of 22.3 °C, that could be comparable with conventional wallboards integrated with PCMs [89]. An inexpensive and simple way to embed PCM into the wood is through direct impregnation. The fundamental drawback of this approach is that when PCM depletes over time, the thermal mass and heat storage capabilities of the structure deteriorate [90].

Prolonged exposure to 50 °C storage temperature and room temperature showed that UV treatment on the wood surface significantly improved PCM retention in the wood. The time it takes for a fire to start, the peak heat release rate, the rate of smoke production, and the overall amount of smoke produced are all increased when PCM is present in the wood. The time to ignite the PCM-filled woods and peak heat release rate are both improved by blending propyl ester with ultraviolet coating, but smoke generation is also increased. The amount of PCM in wood is reduced over time due to evaporation or leaking, which is considered the main disadvantage [91].

10.3 PCM INTEGRATION IN CONSTRUCTION MATERIALS

The integration of PCM into building materials can be done using approaches such as macro-encapsulation, micro-encapsulation, and shape stabilization [92–94]. Figure 10.6 illustrates the conceptual integration approaches of PCMs in construction materials.

Macro-encapsulation and micro-encapsulation are traditional and common ways of integrating PCMs with construction materials. Shape stabilization is a newer method, which is cheaper and simpler. Shape stabilization through direct incorporation of PCM or through immersion into a solid or liquid phase is the most cost-effective integration approach [95, 96].

Challenges associated with the incorporation of PCMs in construction materials are liquid phase leakage, a recurring issue associated with aesthetic discoloration, reactivity with building materials, and diminished performance through PCM volume loss [97, 98]. These challenges result in a multitude of issues. Conversely, the utilization of micro-encapsulated PCMs addresses these concerns by enabling direct incorporation into building materials. This method not only accelerates heat transfer due to increased surface area but also mitigates the risk of liquid PCM leakage [99]. Moreover, micro-encapsulated PCM offers versatility in construction applications, serving as additives, complete replacements, or substitutes for sand [100].

The decrease in binder quantity leads to a reduction in the material's compressive strength. Moreover, given that PCM often possesses a lower density than sand,

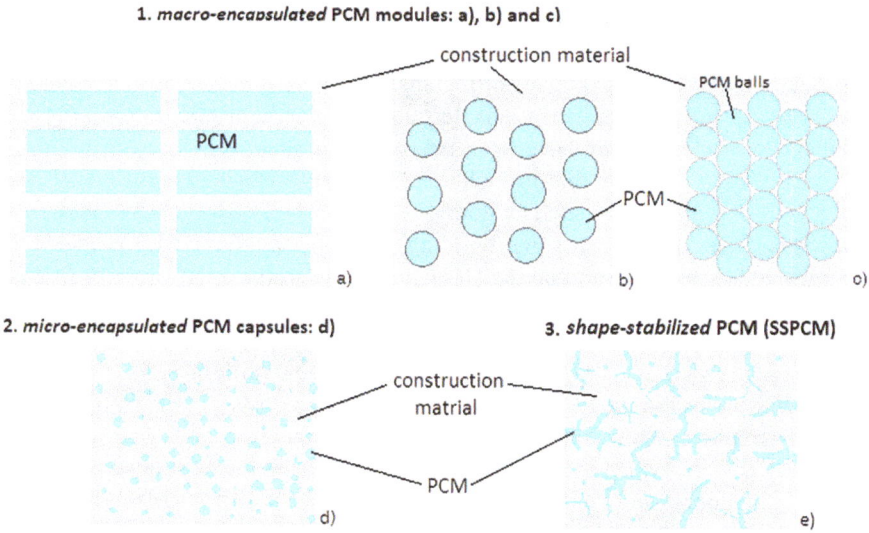

FIGURE 10.6 PCM integration approaches in construction materials: (a), (b), and (c) macro-encapsulation; (d) micro-encapsulation; (e) shape-stabilized PCM (source: Author).

opting for a mass-based replacement approach introduces substantial volume variations [101]. In practical scenarios, the technique most relevant and widely employed is the sand replacement by volume method, as validated in numerous studies. Although this approach commonly results in lower strength, its application can be strategically directed towards the production of non-structural items [102], thereby mitigating the challenge associated with diminished strength.

10.3.1 MACRO-ENCAPSULATION

Major advantages of macro-encapsulation include high durability and high thermal stability, which made encapsulation the most used integration method. The technique of enclosing the PCM in a shell material larger than 5 mm [103] (or 10 mm [104] in size) is known as macro-encapsulation. The enclosing shell's shape can take any form (tubes, pouches, cylinders, cubes, etc.). This allows macro-encapsulated PCMs to be easily included in building envelopes of any form, size, or dimension [103]. Once incorporated, macro-encapsulation enhances interior thermal characteristics of the building [105].

To encapsulate PCM using macro-capsules, no technique is required before the process itself begins. The most cost-effective containers on the market can be found as high-density polyethylene bottles, mild steel cans, and tin-plated metal cans [106]. Furthermore, macro-encapsulated PCM is rarely damaged during integration, has less leakage of PCMs, has a larger quantity for integration, and has a larger ratio of incorporation [107]. Many different PCM encasing types have been manufactured such as various panels [108], bricks, slabs, slats, blades, pouches, spheres, and tubes.

Based on their underlying geometries, these macro-capsules may be further classified as either rectangular, spherical, or cylindrical. Panels, bricks, slabs, slats, and blades are all examples of rectangular shells, although they are not yet distinguished by a clear description. Because of the ease of integration afforded by their flat surfaces, they have been the most common shape used in building envelopes [109]. It is possible to include a rectangular shell in a variety of different building envelopes, such as floors, walls, roofs, ceilings, and windows [9, 110]. It has been found that some building components having cavities, such as bricks, window blinds, and glazing [111], are useful in the application of PCMs. Pouch shells are typically rectangular and fashioned from pliable materials like aluminum foil or nylon. Heat bonding devices are used to split them into many compartments so that PCM does not pool in one area while it is a liquid [112]. Their key benefits are their inexpensive manufacturing costs, simple construction, and easy removal. However, their limited pressure-bearing ability causes them to quickly distort, which may lead to leaks. The outside of buildings, including walls [112], floors, and roofs, have all been covered with pouch shells. Shells may be either spherical (referring to a sphere) or cylindrical (to a tube). There have been a few efforts to use PCM spheres as flooring, where the PCM tubes are fastened to wooden frames [113] or placed into horizontal polystyrene foam slots. The fact that they are used for a smaller percentage of building envelopes than rectangular shells has received surprisingly little attention. They are difficult to repair, particularly when mounted on walls [113]. Table 10.1 describes typical macro-encapsulation containers and the conditions under which they may be used in building envelopes.

Low thermal conductivity and bonding with the matrix material are considered disadvantages of macro-encapsulated PCMs that lead to performance reduction. Common improvements to these shortcomings include suitable container design that increases the thermal conductivity and stability of the PCM [105, 126].

10.3.2 MICRO-ENCAPSULATION

Since micro-encapsulation containers are made less than 1 mm or 1 cm in diameter, it is difficult to produce them; therefore, they require a different manufacturing approach. This approach includes a spray-drying process [127] or interfacial polymerization [128] and requires organic PCM, such as paraffin wax [129]. Micro-capsules are usually spherical, while macro-capsules can take rectangular, cylindrical, or spherical shapes. The micro-capsule size range is 1–1000 μm. The synthesis of the micro-capsules includes physical, chemical, or physico-chemical processes [130, 127].

Physical processes such as drying and dehydration are all involved in making micro-capsule shells. There are two salient physical techniques, spray drying and solvent evaporation, that encapsulate PCM [127]. Chemical methods include different types of polymerizations such as in situ and interfacial [131], suspension [127], and emulsion [132]. In in situ polymerization, the average micro-capsule diameter fell rapidly, and encapsulation effectiveness improved with higher stirring speeds to 97.4% [133]. The physico-chemical technique combines physical and chemical

TABLE 10.1

Typical Macro-Encapsulation Containers

Shape	Categories	Application	Requirement for use	Reference
Panel	Rectangular shells	Wall	A variety of clamps and metal frames are used to secure the panels.	[114]
Panel	Rectangular shells	Wall	Thickness must be determined by the thickness of commercially available panels.	[115]
Panel	Rectangular shells	Ceiling	Plasterboards are supported by metal canisters, which are then placed on the bearing.	[116]
Brick	Rectangular shells	Solar Wall	The system's prefabricated status is preserved by a wooden frame.	[117]
Slab	Rectangular shells	Wall	In order to reduce phase separation, PCM is divided into many levels and then bracketed to an insulating wall.	[118]
Slab	Rectangular shells	Window	Two PCMs next to one another create an air passageway.	[119]
Slat	Rectangular shells	Window	To maximize reflectivity, a white, highly reflective cloth covers the PCM-filled slats.	[120]
Blade	Rectangular shells	Window	Before using a blade, make sure it passes a water and airtightness test.	[121]
Pouch	Rectangular shells	Wall	The pouch has been laminated on both sides with aluminum foil and is separated into many compartments.	[122]
Pouch	Rectangular shells	Wall	To keep PCM from shifting to one side, the pouch is split into three sections.	[123]
Sphere	Spherical shells	Floor	As a PCM packed bed, the spherical PCM was implanted right below the floor boards.	[124]
Tube	Cylindrical shells	Wall	Both ends of the tubes are sealed with caps to prevent leaks, and they are mounted to the wall using brackets.	[113]
Pipe	Cylindrical shells	Wall	The polystyrene foam is slotted horizontally to accommodate the pipes.	[125]

(*source:* Author)

processes. Hydrolysis, cross-linking, condensation, and phase separation are used to microencapsulate. Coacervation and the sol–gel method are the main physico-chemical approaches [134].

Due to their chemical and thermic stability, PCM micro-capsules are well integrated into construction materials such as wallboard, cement mortar, gypsum plaster, sandwich panels, and slabs [135, 136]. PCM-based concrete greatly increases its overall mechanical resistance and rigidity [137] and improves the thermal and acoustic insulation of walls [138]. Micro-capsules of PCM may also be included.

10.3.3 Shape-Stabilized PCMs

Composites where the liquid PCMs are held by another porous support material by capillary forces are called shape-stabilized PCMs (SSPCMs).[1] Therefore, the shape stabilization mainly depends on the porosity support material. Support materials can be natural or synthetic. Natural materials that appear porous include some types of clay, while some materials produced by the template method can be considered synthetic. This containment method is considered less expensive than the micro and macro encapsulations [139]. A general manufacturing concept of the SSPCM is shown in Figure 10.7.

Over the last decade, SSPCMs have garnered a lot of attention owing to their remarkable shape-retaining properties after several heat cycles without encapsulation [140–142]. The direct impregnation in porous building materials is the simplest way to integrate PCMs. During manufacturing, the PCM is added as liquid or powder to construction materials and fills small pores in materials such as concrete, plaster, gypsum, wood, and cement. The heat transfers through the material structure to the pores and a phase transition takes place. With this method, leakage is observed [143, 144].

Shape stabilization manufacturing techniques reported in the literature are much simpler compared to the macro- and micro-encapsulation techniques. Most of these techniques have a physical nature such as absorption/adsorption that uses direct impregnation or vacuum impregnation [145]. Figure 10.8 shows the other techniques that include templating to produce hierarchical porosity, centrifugal spinning, electrospinning, and intercalation.

The *templating* technique is used to create hierarchical porous matrices in which the structure of these materials is made of interconnected pores that have different lengths and scales. *Intercalation* techniques are used for basal spacing of layered materials such as clay and graphite supports [146]. *Electrospinning* and *centrifugal spinning* are novel techniques used for fiber fabrication. Electrospinning has lower

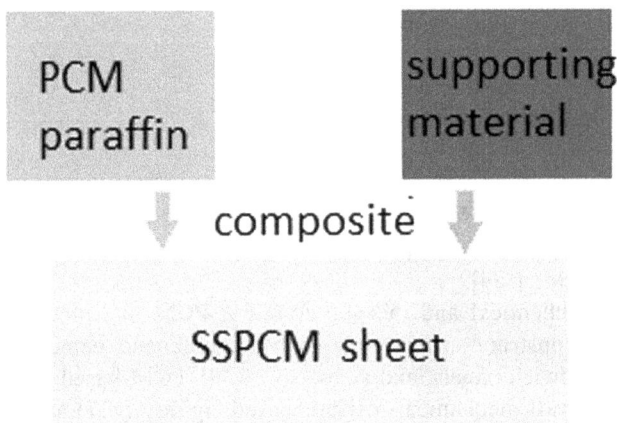

FIGURE 10.7　Shape-stabilized PCMs (source: Author).

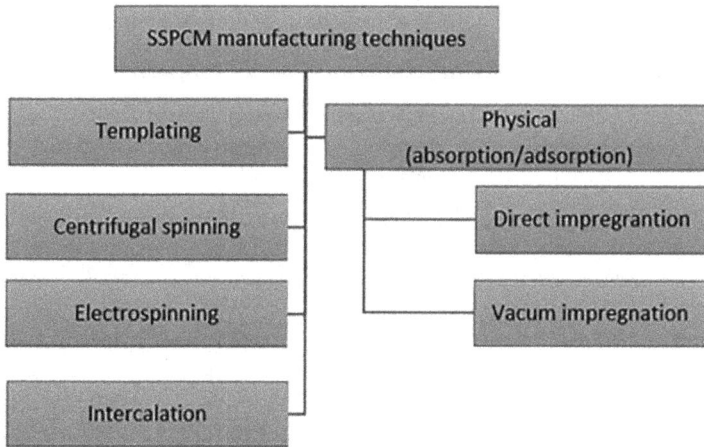

FIGURE 10.8 Manufacturing methods for SSPCM (source: Author).

productivity compared to centrifugal spinning; thus, centrifugal spinning looks less expensive and has higher productivity, but both techniques are effective, economic, and versatile. *Vacuum impregnation* is a simple and cheap technique to integrate the liquid phase of PCM in the porous support material. Vacuum method provides a higher rate of incorporated PCM in the support material than direct impregnation [147]. The amount of PCM in the construction materials may be reduced over time due to evaporation or leaking, which is considered the main disadvantage of this method.

10.4 PROPERTIES OF PCM-BASED CONSTRUCTION MATERIALS

Recent research highlights the successful integration of PCMs into a spectrum of elements, spanning both structural and non-structural components, encompassing walls, ceilings, roofs, and windows [148]. Introducing a PCM layer to walls not only enhances the energy efficiency of architectural elements [149] but may also introduce potential trade-offs, diminishing the mechanical properties and durability of the underlying building materials [150]. Refer to Table 10.2 for a concise overview of average values pertaining to key properties of PCM-based construction materials, including density, compressive strength, thermal conductivity, and latent/sensible heat.

The literature reveals a thorough exploration of methods aimed at enhancing temperature regulation, heat generation during cement hydration, fracture resistance, and thermal shrinkage in concrete elements, particularly mass concrete, through the incorporation of PCMs [166, 167]. Researchers have also delved into strategies to mitigate thermal stress and damage caused by freezing–thawing cycles or temperature gradients, such as curling stresses, in concrete pavements integrated with PCMs [168]. Notably, an inverse relationship between the PCM/concrete ratio and compressive strength has been observed. Figure 10.9 illustrates the consistent pattern of compressive strength variation in PCM-based concrete across different PCM/concrete ratios.

TABLE 10.2

Compilation of PCM-Based Construction Material Properties

Material	Average density (kg/m³)	Average compressive strength (MPa)	Average thermal conductivity (W/m K)	Latent heat (J/kg)	Reference
Foam concrete	896	32.38 (over 7 days)	0.53		[151–153]
Foam concrete	896 to 1808	5.8	0.54	41.78	[154]
Foam concrete	910	6.8	0.54	42	[154]
Cement render	1392	14.5 (over 28 days)	0.54	180	[154]
Cement render	896	4.9	0.18	180	[155]
Brick	1600	21.2	0.32		[156]
Brick	930	8.73	0.21	190	[157]
Brick	2049	23.75	0.358	7360	[158]
Gypsum board	1120	4.8	0.15		[159, 160]
Gypsum board	956	2.5	0.417	1.13 (sensible)	[161]
Glass	750	15	0.280	15.3	[162]
Glass	880	1000+	0.2	18.4	[163]
Wood	315	30–60	0.08	21.12	[164]
Wood	400–900	30–60	0.32	25.12	[165]

(*source:* Author)

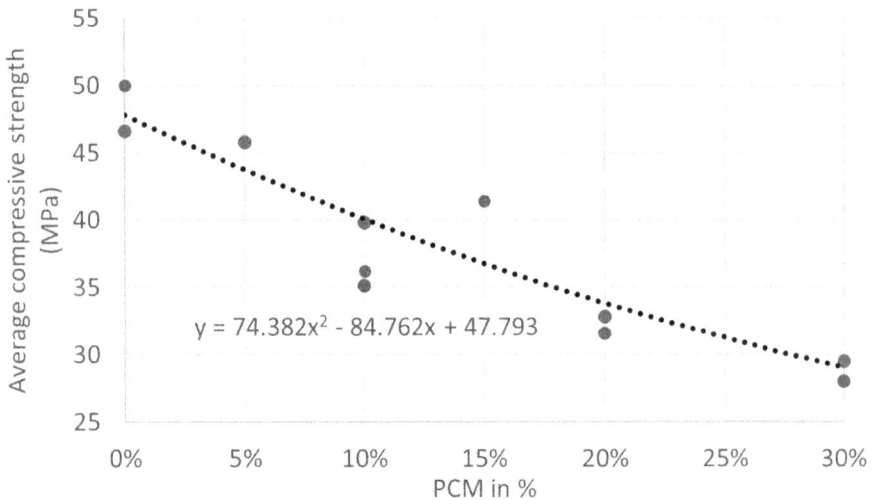

$$y = 74.382x^2 - 84.762x + 47.793$$

FIGURE 10.9　Compressive strength changes with respect to PCM (source: Author).

Since the addition of PCMs decreases composite strength, various studies have investigated how to improve the strength. To increase the strength, the addition of firm quartz with an increased volume proportion was investigated. However, the level of strength loss varies with the water-to-cement ratio (w/c), where lower w/c results in smaller strength reductions. A portion of the strength loss may also be attributable to PCM release into the paste and chemical reactions within [169]. The inclusion of a firm mineral aggregate that is found in concrete is predicted to minimize the magnitude of the strength losses observed with the PCM application. Also, it is known that PCMs can reduce crack formation in concrete, which refers to its thermal properties [170]. The already indicated durability, water absorption, and drying shrinkage have no negative impact on the durability of mortars using PCM micro-capsules [171]. In the context of cracking, one specific advantage of including compliant inclusions in quasi-brittle materials is an increase in fracture toughness. The insertion of compliant inclusions in quasi-brittle materials improves fracture toughness when such soft elements are integrated into a brittle matrix, which is a significant advantage in the context of cracking. Such positive behaviors are principally caused by the elastic mismatch between the host matrix and the inclusion phase, which leads to fracture blunting and twisting [172]. It has also been reported that incorporating PCM micro-capsules into concrete reduces its elasticity modulus [173].

Based on the economic analyses available in the literature, the payback period ranges from 9 to 30 years if used as energy storage in construction materials in buildings [174, 175]. Different studies evaluated that the payback period of 12.94 years is optimal for RT21–RT25–RT28 triple PCM [176]. In addition, studies have shown that this will be more economical in the future if energy rates rise and PCM installation cost falls.

10.5   CONCLUSION

In this chapter, PCM-based construction materials have been discussed, including material types, blending methods, and material properties. The main findings are as follows:

- The integration of PCM in construction materials for building applications is done mostly in widely used construction materials such as concrete, asphalt pavement, gypsum board, bricks, cement render, and wood. Studies indicate that PCM-based materials have significantly improved thermal performance through the improved thermal mass of the materials.
- Although research has indicated quite a large energy savings potential, the grades of phase change materials now on the market do not seem ideal for wide applications in construction. The heat of the fusion of several known materials with a transition around the comfort temperature is relatively low.
- Current integration of PCMs in construction materials is focused on macro-encapsulation and micro-encapsulation, which have common challenges such as the prevention of leakage and the preservation of PCM. PCM macro-flexible encapsulation's shell design and PCM selection possibilities make it

appealing. Shape-stabilized PCM is an intriguing new approach that promises to reduce the risk of leakage. The main disadvantage of SSPCM is a reduction in PCM over time due to leakage or evaporation.

- Adding PCM to construction material will improve the energy savings but it will gradually reduce the mechanical properties (compressive strength) and durability of the materials. Thus, it is required to find innovative chemical or physical integration techniques that would lessen the negative effects in the aspect of mechanical properties. However, revolutionary techniques are required to effectively implement the PCM's impact on the storage of energy and general control of buildings' thermal capabilities.

NOTE

1 Known also as a form-stabilized composite phase change material (FSCPCM).

REFERENCES

[1] Iea, "2019 Global Status Report for Buildings and Constructi on Towards a zero-emissions, efficient and resilient buildings and constructi on sector", Accessed: Jan. 19, 2023 [Online]. Available: www.iea.org

[2] J. A. Nebrida and F. E. Gomba, "Sustainable construction strategies for building construction projects in the Kingdom of Bahrain: A model," *Sustainable Engineering and Innovation*, vol. 5, no. 1, pp. 31–47, May 2023, doi:10.37868/SEI.V5I1.ID193.

[3] M. M. Adrian, E. P. Purnomo, A. Enrici, and T. Khairunnisa, "Energy transition towards renewable energy in Indonesia," *Heritage and Sustainable Development*, vol. 5, no. 1, pp. 107–118, May 2023, doi:10.37868/HSD.V5I1.108.

[4] M. Halilovic and M. Alibegovic, "Potential of air quality improvements in Sarajevo using innovative architecture approach," *Periodicals of Engineering and Natural Sciences*, vol. 5, no. 2, pp. 128–135, Mar. 2017, doi:10.21533/pen.v5i2.89.

[5] S. Lu, B. Xu, and X. Tang, "Experimental study on double pipe PCM floor heating system under different operation strategies," *Renew Energy*, vol. 145, pp. 1280–1291, Jan. 2020, doi:10.1016/j.renene.2019.06.086.

[6] M. N. Islam and D. H. Ahmed, "Delaying the temperature fluctuations through PCM integrated building walls—Room conditions, PCM placement, and temperature of the heat sources," *Energy Storage*, vol. 3, no. 5, p. e245, Oct. 2021, doi:10.1002/EST2.245.

[7] A. Figueiredo, J. Lapa, R. Vicente, and C. Cardoso, "Mechanical and thermal characterization of concrete with incorporation of microencapsulated PCM for applications in thermally activated slabs," *Construction and Building Materials*, vol. 112, pp. 639–647, Jun. 2016, doi:10.1016/j.conbuildmat.2016.02.225.

[8] I. Baskar and M. Chellapandian, "Experimental and finite element analysis on the developed real-time form stable PCM based roof system for thermal energy storage applications," *Energy Build*, vol. 276, p. 112514, Dec. 2022, doi:10.1016/J.ENBUILD.2022.112514.

[9] B. Duraković and S. Mešetović, "Thermal performances of glazed energy storage systems with various storage materials: An experimental study," *Sustainable Cities and Society*, vol. 45, pp. 422–430, 2019, doi:10.1016/j.scs.2018.12.003.

[10] B. Duraković, "PCM-based glazing systems and components," in *Green Energy and Technology*, Springer, 2020, pp. 89–119. doi:10.1007/978-3-030-38335-0_5.

[11] C. Baldassarri, S. Sala, A. Caverzan, and M. Lamperti Tornaghi, "Environmental and spatial assessment for the ecodesign of a cladding system with embedded Phase Change Materials," *Energy Build*, vol. 156, pp. 374–389, Dec. 2017, doi:10.1016/J. ENBUILD.2017.09.011.

[12] Z. A. Al-Absi, M. I. M. Hafizal, and M. Ismail, "Experimental study on the thermal performance of PCM-based panels developed for exterior finishes of building walls," *Journal of Building Engineering*, vol. 52, p. 104379, Jul. 2022, doi:10.1016/J. JOBE.2022.104379.

[13] M. Frigione, M. Lettieri, and A. Sarcinella, "Phase change materials for energy efficiency in buildings and their use in mortars," *Materials*, vol. 12, no. 8, Apr. 2019, doi:10.3390/MA12081260.

[14] Z. A. Al-Absi, M. I. Mohd Hafizal, M. Ismail, A. Mardiana, and A. Ghazali, "Peak indoor air temperature reduction for buildings in hot-humid climate using phase change materials," *Case Studies in Thermal Engineering*, vol. 22, p. 100762, Dec. 2020, doi:10.1016/J.CSITE.2020.100762.

[15] E. A. Mccullough and H. Shim, "The use of phase change materials in outdoor clothing," *Intelligent Textiles and Clothing*, pp. 63–81, Jan. 2006, doi:10.1533/9781845691622.1.63.

[16] B. Pause, "Phase change materials and their application in coatings and laminates for textiles," *Smart Textile Coatings and Laminates: A Volume in Woodhead Publishing Series in Textiles*, pp. 236–250, Jan. 2010, doi:10.1533/9781845697785.2.236.

[17] B. Duraković, "Passive solar heating/cooling strategies," in *PCM-Based Building Envelope Systems: Innovative Energy Solutions for Passive Design*, Springer Nature, 2020, pp. 39–62. doi:10.1007/978-3-030-38335-0_3.

[18] B. Duraković, M. Hadziabdić, and O. Buyukdagli, "Building energy demand management strategies and methods," in *Building Energy Flexibility and Demand Management*, Academic Press, 2023, pp. 63–85. doi:10.1016/B978-0-323-99588-7.00007-9.

[19] S. Yang, H. Oliver Gao, and F. You, "Model predictive control in phase-change-material-wallboard-enhanced building energy management considering electricity price dynamics," *Applied Energy*, vol. 326, p. 120023, Nov. 2022, doi:10.1016/J. APENERGY.2022.120023.

[20] B. Duraković, "Phase change materials for building envelope," in *Green Energy and Technology*, Springer, 2020, pp. 17–37. doi:10.1007/978-3-030-38335-0_2.

[21] B. Duraković, "PCMs in building structure," in *PCM-Based Building Envelope Systems: Innovative Energy Solutions for Passive Design*, Springer Nature, 2020, pp. 63–87. doi:10.1007/978-3-030-38335-0_4.

[22] C. Arumugam and S. Shaik, "Air-conditioning cost saving and CO2 emission reduction prospective of buildings designed with PCM integrated blocks and roofs," *Sustainable Energy Technologies and Assessments*, vol. 48, p. 101657, Dec. 2021, doi:10.1016/J. SETA.2021.101657.

[23] M. Halilovic, "Vernacular architecture sustainability principles: A case study of Bosnian stone houses in Idbar village," *Periodicals of Engineering and Natural Sciences*, vol. 8, no. 4, pp. 2564–2574, Dec. 2020, Accessed: Feb. 10, 2021 [Online]. Available: http://pen.ius.edu.ba/index.php/pen/article/view/1760

[24] H. I. Altintas, "Investigation of zero energy house design: Principles concepts opportunities and challenges," *Heritage and Sustainable Development, ISSN 2712-0554*, vol. 1, no. 1, pp. 21–32, Jun. 2019.

[25] H. M. Ali et al., "Advances in thermal energy storage: Fundamentals and applications," *Progress in Energy and Combustion Science*, vol. 100, p. 101109, Jan. 2024, doi:10.1016/J.PECS.2023.101109.

[26] B. Duraković, M. Halilović, and H. M. Ali, "Phase change materials applications in buildings," *Phase Change Materials for Heat Transfer*, pp. 225–248, Jan. 2023, doi:10.1016/B978-0-323-91905-0.00005-8.

[27] Q. Al-Yasiri and M. Szabó, "Effect of encapsulation area on the thermal performance of PCM incorporated concrete bricks: A case study under Iraq summer conditions," *Case Studies in Construction Materials*, vol. 15, p. e00686, Dec. 2021, doi:10.1016/J. CSCM.2021.E00686.

[28] M. Mäkinen, "Introduction to phase change materials," *Intelligent Textiles and Clothing*, pp. 21–33, Jan. 2006, doi:10.1533/9781845691622.1.21.

[29] Q. Li, et al., "Thermal performance and economy of PCM foamed cement walls for buildings in different climate zones," *Energy Build*, vol. 277, p. 112470, Dec. 2022, doi:10.1016/J.ENBUILD.2022.112470.

[30] F. Findik, "Green concrete for structural buildings," *Heritage and Sustainable Development*, vol. 4, no. 1, pp. 67–76, Jun. 2022, doi:10.37868/HSD.V4I1.84.

[31] Y. H. M. Amran, N. Farzadnia, and A. A. A. Ali, "Properties and applications of foamed concrete: A review," *Construction and Building Materials*, vol. 101, pp. 990–1005, Dec. 2015, doi:10.1016/J.CONBUILDMAT.2015.10.112.

[32] B. Xu and Z. Li, "Paraffin/diatomite composite phase change material incorporated cement-based composite for thermal energy storage," *Applied Energy*, vol. 105, pp. 229–237, May 2013, doi:10.1016/J.APENERGY.2013.01.005.

[33] T. Huo, et al., "China's energy consumption in the building sector: A Statistical Yearbook-Energy Balance Sheet based splitting method," *Journal of Cleaner Production*, vol. 185, pp. 665–679, 2018, doi:10.1016/j.jclepro.2018.02.283.

[34] J. Dallaire, H. M. Adeel Hassan, J. H. Bjernemose, M. P. Rudolph Hansen, I. Lund, and C. T. Veje, "Performance analysis of a dual-stack Air-PCM heat exchanger with novel air flow configuration for cooling applications in buildings," *Build Environ*, vol. 223, p. 109450, Sep. 2022, doi:10.1016/J.BUILDENV.2022.109450.

[35] A. Marani and M. L. Nehdi, "Integrating phase change materials in construction materials: Critical review," *Construction and Building Materials*, vol. 217, pp. 36–49, Aug. 2019, doi:10.1016/j.conbuildmat.2019.05.064.

[36] T. Adagba, A. Abubakar, and A. S. Baba, "The effect of cement replacement with metakaolin and sugarcane bagasse ash as supplementary cementitious materials on the properties of concrete," *Sustainable Engineering and Innovation*, vol. 5, no. 2, pp. 117–126, Oct. 2023, doi:10.37868/SEI.V5I2.ID197.

[37] A. A. Adday and A. S. Ali, "Flexural behavior of steel fiber reinforced concrete beams comprising coarse and fine rubber and strengthened by CFRP sheets," *Heritage and Sustainable Development*, vol. 5, no. 2, pp. 280–308, Oct. 2023, doi:10.37868/HSD. V5I2.257.

[38] L. F. Cabeza, C. Castellón, M. Nogués, M. Medrano, R. Leppers, and O. Zubillaga, "Use of microencapsulated PCM in concrete walls for energy savings," *Energy Build*, vol. 39, no. 2, pp. 113–119, Feb. 2007, doi:10.1016/J.ENBUILD.2006.03.030.

[39] H. J. Alqallaf and E. M. Alawadhi, "Concrete roof with cylindrical holes containing PCM to reduce the heat gain," *Energy Build*, vol. 61, pp. 73–80, Jun. 2013, doi:10.1016/J. ENBUILD.2013.01.041.

[40] L. Royon, L. Karim, and A. Bontemps, "Thermal energy storage and release of a new component with PCM for integration in floors for thermal management of buildings," *Energy Build*, vol. 63, pp. 29–35, Aug. 2013, doi:10.1016/J.ENBUILD.2013.03.042.

[41] B. R. Anupam, U. C. Sahoo, and P. Rath, "Phase change materials for pavement applications: A review," *Construction and Building Materials*, vol. 247, p. 118553, Jun. 2020, doi:10.1016/J.CONBUILDMAT.2020.118553.

[42] Y. Farnam, H. S. Esmaeeli, P. D. Zavattieri, J. Haddock, and J. Weiss, "Incorporating phase change materials in concrete pavement to melt snow and ice," *Cement and Concrete Composites*, vol. 84, pp. 134–145, Nov. 2017, doi:10.1016/J.CEMCONCOMP.2017.09.002.

[43] M. Yang, X. Zhang, X. Zhou, B. Liu, X. Wang, and X. Lin, "Research and exploration of phase change materials on solar pavement and asphalt pavement: A review," *Journal of Energy Storage*, vol. 35, p. 102246, Mar. 2021, doi:10.1016/J.EST.2021.102246.

[44] R. Methode Kalombe, S. Sobhansarbandi, and J. Kevern, "Low-cost phase change materials based concrete for reducing deicing needs," *Construction and Building Materials*, vol. 363, p. 129129, Jan. 2023, doi:10.1016/J.CONBUILDMAT.2022.129129.

[45] M. Mahedi, B. Cetin, and K. S. Cetin, "Freeze-thaw performance of phase change material (PCM) incorporated pavement subgrade soil," *Construction and Building Materials*, vol. 202, pp. 449–464, Mar. 2019, doi:10.1016/J.CONBUILDMAT.2018.12.210.

[46] N. K. Al-Bayati and M. Q. Ismael, "Effect of differently treated recycled concrete aggregates on Marshall properties and cost-benefit of asphalt mixtures," *Sustainable Engineering and Innovation*, vol. 5, no. 2, pp. 127–140, Oct. 2023, doi:10.37868/SEI.V5I2.ID201.

[47] Y. Du, et al., "Laboratory investigation of phase change effect of polyethylene glycolon on asphalt binder and mixture performance," *Construction and Building Materials*, vol. 212, pp. 1–9, Jul. 2019, doi:10.1016/j.conbuildmat.2019.03.308.

[48] L. Haurie, S. Serrano, M. Bosch, A. I. Fernandez, and L. F. Cabeza, "Single layer mortars with microencapsulated PCM: Study of physical and thermal properties, and fire behaviour," *Energy Build*, vol. 111, pp. 393–400, Jan. 2016, doi:10.1016/J.ENBUILD.2015.11.028.

[49] S. Ramakrishnan, X. Wang, J. Sanjayan, and J. Wilson, "Thermal energy storage enhancement of lightweight cement mortars with the application of phase change materials," *Procedia Engineering*, vol. 180, pp. 1170–1177, Jan. 2017, doi:10.1016/J.PROENG.2017.04.277.

[50] S. Cunha, I. Aguiar, and J. B. Aguiar, "Phase change materials composite boards and mortars: Mixture design, physical, mechanical and thermal behavior," *Journal of Energy Storage*, vol. 53, p. 105135, Sep. 2022, doi:10.1016/J.EST.2022.105135.

[51] 标准分享网, www.bzfxw.com, 1998.

[52] P. H. Shaikh, N. B. M. Nor, A. A. Sahito, P. Nallagownden, I. Elamvazuthi, and M. S. Shaikh, "Building energy for sustainable development in Malaysia: A review," *Renewable and Sustainable Energy Reviews*, vol. 75, pp. 1392–1403, Aug. 2017, doi:10.1016/J.RSER.2016.11.128.

[53] Z. A. Al-Absi, M. I. M. Hafizal, M. Ismail, H. Awang, and A. Al-Shwaiter, "Properties of PCM-based composites developed for the exterior finishes of building walls," *Case Studies in Construction Materials*, vol. 16, p. e00960, Jun. 2022, doi:10.1016/J.CSCM.2022.E00960.

[54] D. Wu, M. Rahim, M. el Ganaoui, R. Djedjig, R. Bennacer, and B. Liu, "Experimental investigation on the hygrothermal behavior of a new multilayer building envelope integrating PCM with bio-based material," *Build Environ*, vol. 201, p. 107995, Aug. 2021, doi:10.1016/J.BUILDENV.2021.107995.

[55] P. H. Shaikh, N. B. M. Nor, A. A. Sahito, P. Nallagownden, I. Elamvazuthi, and M. S. Shaikh, "Building energy for sustainable development in Malaysia: A review," *Renewable and Sustainable Energy Reviews*, vol. 75, pp. 1392–1403, Aug. 2017, doi:10.1016/J.RSER.2016.11.128.

[56] B. Németh, A. Ujhidy, J. Tóth, J. Gyenis, and T. Feczkó, "Testing of microencapsulated phase-change heat storage in experimental model houses under winter weather conditions," *Build Environ*, vol. 204, p. 108119, Oct. 2021, doi:10.1016/J.BUILDENV.2021.108119.

[57] A. Oliver, "Thermal characterization of gypsum boards with PCM included: Thermal energy storage in buildings through latent heat," *Energy Build*, vol. 48, pp. 1–7, May 2012, doi:10.1016/J.ENBUILD.2012.01.026.

[58] N. Shukla, A. Fallahi, and J. Kosny, "Performance characterization of PCM impregnated gypsum board for building applications," *Energy Procedia*, vol. 30, pp. 370–379, Jan. 2012, doi:10.1016/J.EGYPRO.2012.11.044.

[59] Y. Yang, Z. Shen, W. Wu, H. Zhang, Y. Ren, and Q. Yang, "Preparation of a novel diatomite-based PCM gypsum board for temperature-humidity control of buildings," *Build Environ*, vol. 226, p. 109732, Dec. 2022, doi:10.1016/J.BUILDENV.2022.109732.

[60] A. Oliver, "Thermal characterization of gypsum boards with PCM included: Thermal energy storage in buildings through latent heat," *Energy Build*, vol. 48, pp. 1–7, May 2012, doi:10.1016/J.ENBUILD.2012.01.026.

[61] M. Saffari, C. Roe, and D. P. Finn, "Improving the building energy flexibility using PCM-enhanced envelopes," *Applied Thermal Engineering*, vol. 217, p. 119092, Nov. 2022, doi:10.1016/J.APPLTHERMALENG.2022.119092.

[62] L. Derradji, A. Hamid, B. Zeghmati, M. Amara, A. Bouttout, and F. B. Errebai, "Experimental study on the use of microencapsulated phase change material in walls and roofs for energy savings," *Journal of Energy Engineering*, vol. 141, no. 4, p. 04014046, Oct. 2014, doi:10.1061/(ASCE)EY.1943-7897.0000238.

[63] P. Schossig, H. M. Henning, S. Gschwander, and T. Haussmann, "Micro-encapsulated phase-change materials integrated into construction materials," *Solar Energy Materials and Solar Cells*, vol. 89, no. 2–3, pp. 297–306, Nov. 2005, doi:10.1016/J.SOLMAT.2005.01.017.

[64] N. A. Azmi, M. Arıcı, and A. Baharun, "A review on the factors influencing energy efficiency of mosque buildings," *Journal of Cleaner Production*, vol. 292, p. 126010, Apr. 2021, doi:10.1016/J.JCLEPRO.2021.126010.

[65] F. M. Wani, et al., "Finite element analysis of unreinforced masonry walls with different bond patterns," *Sustainable Engineering and Innovation*, vol. 5, no. 1, pp. 58–72, Jun. 2023, doi:10.37868/SEI.V5I1.ID194.

[66] Q. Al-Yasiri and M. Szabó, "Incorporation of phase change materials into building envelope for thermal comfort and energy saving: A comprehensive analysis," *Journal of Building Engineering*, vol. 36, p. 102122, Apr. 2021, doi:10.1016/J.JOBE.2020.102122.

[67] E. Tunçbilek, M. Arıcı, M. Krajčík, S. Nižetić, and H. Karabay, "Thermal performance based optimization of an office wall containing PCM under intermittent cooling operation," *Applied Thermal Engineering*, vol. 179, p. 115750, Oct. 2020, doi:10.1016/J.APPLTHERMALENG.2020.115750.

[68] Q. Al-Yasiri and M. Szabó, "Effect of encapsulation area on the thermal performance of PCM incorporated concrete bricks: A case study under Iraq summer conditions," *Case Studies in Construction Materials*, vol. 15, p. e00686, Dec. 2021, doi:10.1016/J.CSCM.2021.E00686.

[69] Q. Al-Yasiri and M. Szabó, "Thermal performance of concrete bricks based phase change material encapsulated by various aluminium containers: An experimental study under Iraqi hot climate conditions," *Journal of Energy Storage*, vol. 40, p. 102710, Aug. 2021, doi:10.1016/J.EST.2021.102710.

[70] M. Mahdaoui, et al., "Building bricks with phase change material (PCM): Thermal performances," *Construction and Building Materials*, vol. 269, p. 121315, Feb. 2021, doi:10.1016/J.CONBUILDMAT.2020.121315.

[71] C. Jia, X. Geng, F. Liu, and Y. Gao, "Thermal behavior improvement of hollow sintered bricks integrated with both thermal insulation material (TIM) and Phase-Change Material (PCM)," *Case Studies in Thermal Engineering*, vol. 25, p. 100938, Jun. 2021, doi:10.1016/J.CSITE.2021.100938.

[72] X. Wang, H. Yu, L. Li, and M. Zhao, "Experimental assessment on a kind of composite wall incorporated with shape-stabilized phase change materials (SSPCMs)," *Energy Build*, vol. 128, pp. 567–574, Sep. 2016, doi:10.1016/J.ENBUILD.2016.07.031.

[73] C. Jia, X. Geng, F. Liu, and Y. Gao, "Thermal behavior improvement of hollow sintered bricks integrated with both thermal insulation material (TIM) and Phase-Change Material (PCM)," *Case Studies in Thermal Engineering*, vol. 25, p. 100938, Jun. 2021, doi:10.1016/J.CSITE.2021.100938.

[74] P. Principi and R. Fioretti, "Thermal analysis of the application of pcm and low emissivity coating in hollow bricks," *Energy Build*, vol. 51, pp. 131–142, Aug. 2012, doi:10.1016/J.ENBUILD.2012.04.022.

[75] B. Rudalija and B. Duraković, "Thermal characterization of straw-based panels made out of straw and natural binders," in *Advanced Technologies, Systems, and Applications VI. IAT 2021. Lecture Notes in Networks and Systems*, Springer, Jun. 2022, pp. 297–304. doi:10.1007/978-3-030-90055-7_22.

[76] B. Durakovic, G. Yildiz, and M. E. Yahia, "Comparative performance evaluation of conventional and renewable thermal insulation materials used in building envelops," *Tehnicki Vjesnik*, vol. 27, no. 1, pp. 283–289, Feb. 2020, doi:10.17559/TV-20171228212943.

[77] G. Yildiz, B. Duraković, and A. Abd Almisreb, "Performances study of natural and conventional building insulation materials," *Int J Adv Sci Eng Inf Technol*, vol. 11, no. 4, pp. 1395–1404, Aug. 2021, doi:10.18517/IJASEIT.11.4.11139.

[78] M. A. Wahid, S. E. Hosseini, H. M. Hussen, H. J. Akeiber, S. N. Saud, and A. T. Mohammad, "An overview of phase change materials for construction architecture thermal management in hot and dry climate region," *Applied Thermal Engineering*, vol. 112, pp. 1240–1259, Feb. 2017, doi:10.1016/J.APPLTHERMALENG.2016.07.032.

[79] A. H. N. Al-mudhafar, M. T. Hamzah, and A. L. Tarish, "Potential of integrating PCMs in residential building envelope to reduce cooling energy consumption," *Case Studies in Thermal Engineering*, vol. 27, p. 101360, Oct. 2021, doi:10.1016/J.CSITE.2021.101360.

[80] P. Pirdavari and S. Hossainpour, "Numerical study of a Phase Change Material (PCM) embedded solar thermal energy operated cool store: A feasibility study," *International Journal of Refrigeration*, vol. 117, pp. 114–123, Sep. 2020, doi:10.1016/J.IJREFRIG.2020.04.028.

[81] S. Jaber and S. Ajib, "Novel cooling unit using PCM for residential application," *International Journal of Refrigeration*, vol. 35, no. 5, pp. 1292–1303, Aug. 2012, doi:10.1016/J.IJREFRIG.2012.03.023.

[82] Z. A. Al-Absi, M. I. M. Hafizal, M. Ismail, H. Awang, and A. Al-Shwaiter, "Properties of PCM-based composites developed for the exterior finishes of building walls," *Case Studies in Construction Materials*, vol. 16, p. e00960, Jun. 2022, doi:10.1016/J.CSCM.2022.E00960.

[83] X. Chen, H. Yang, and L. Lu, "A comprehensive review on passive design approaches in green building rating tools," *Renewable and Sustainable Energy Reviews*, vol. 50, pp. 1425–1436, Oct. 2015, doi:10.1016/J.RSER.2015.06.003.

[84] E. Cuce, S. B. Riffat, and C. H. Young, "Thermal insulation, power generation, lighting and energy saving performance of heat insulation solar glass as a curtain wall application in Taiwan: A comparative experimental study," *Energy Convers Manag*, vol. 96, pp. 31–38, May 2015, doi:10.1016/J.ENCONMAN.2015.02.062.

[85] D. Li, T. Ma, C. Liu, Y. Zheng, Z. Wang, and X. Liu, "Thermal performance of a PCM-filled double glazing unit with different optical properties of phase change material," *Energy Build*, vol. 119, pp. 143–152, May 2016, doi:10.1016/J.ENBUILD.2016.03.036.

[86] M. Ozel, "Determination of optimum insulation thickness based on cooling transmission load for building walls in a hot climate," *Energy Convers Manag*, vol. 66, pp. 106–114, Feb. 2013, doi:10.1016/J.ENCONMAN.2012.10.002.

[87] Y. Li, et al., "Processing wood into a phase change material with high solar-thermal conversion efficiency by introducing stable polyethylene glycol-based energy storage polymer," *Energy*, vol. 254, p. 124206, Sep. 2022, doi:10.1016/J.ENERGY.2022.124206.

[88] R. Hemmati and H. Saboori, "Short-term bulk energy storage system scheduling for load leveling in unit commitment: modeling, optimization, and sensitivity analysis," *Journal of Advanced Research*, vol. 7, no. 3, pp. 360–372, May 2016, doi:10.1016/J.JARE.2016.02.002.

[89] E. Oró, A. de Gracia, A. Castell, M. M. Farid, and L. F. Cabeza, "Review on phase change materials (PCMs) for cold thermal energy storage applications," *Applied Energy*, vol. 99, pp. 513–533, Nov. 2012, doi:10.1016/J.APENERGY.2012.03.058.

[90] M. M. Farid and R. M. Husian, "An electrical storage heater using the phase-change method of heat storage," *Energy Conversion and Management*, vol. 30, no. 3, pp. 219–230, Jan. 1990, doi:10.1016/0196-8904(90)90003-H.

[91] "Sci-Hub | The effect of ultraviolet coating on containment and fire hazards of phase change materials impregnated wood structure. *Journal of Energy Storage*, vol. 32, p. 101727. Accessed: Dec. 09, 2022 [Online]. Available: https://sci-hub.se/10.1016/j.est.2020.101727.

[92] F. Kuznik, D. David, K. Johannes, and J. J. Roux, "A review on phase change materials integrated in building walls," *Renewable and Sustainable Energy Reviews*, vol. 15, no. 1, pp. 379–391, Jan. 2011, doi:10.1016/J.RSER.2010.08.019.

[93] R. Aridi and A. Yehya, "Review on the sustainability of phase-change materials used in buildings," *Energy Conversion and Management: X*, vol. 15, p. 100237, Aug. 2022, doi:10.1016/J.ECMX.2022.100237.

[94] Z. A. Al-Absi, M. I. M. Hafizal, M. Ismail, H. Awang, and A. Al-Shwaiter, "Properties of PCM-based composites developed for the exterior finishes of building walls," *Case Studies in Construction Materials*, vol. 16, p. e00960, Jun. 2022, doi:10.1016/J.CSCM.2022.E00960.

[95] D. Vérez, E. Borri, G. Zsembinszki, and L. F. Cabeza, "Thermal energy storage co-benefits in building applications transferred from a renewable energy perspective," *Journal of Energy Storage*, vol. 58, p. 106344, Feb. 2023, doi:10.1016/J.EST.2022.106344.

[96] Y. Cui, J. Xie, J. Liu, and S. Pan, "Review of phase change materials integrated in building walls for energy saving," *Procedia Engineering*, vol. 121, pp. 763–770, Jan. 2015, doi:10.1016/J.PROENG.2015.09.027.

[97] Q. Al-Yasiri and M. Szabó, "Selection of phase change material suitable for building heating applications based on qualitative decision matrix," *Energy Conversion and Management: X*, vol. 12, p. 100150, Dec. 2021, doi:10.1016/J.ECMX.2021.100150.

[98] Z. A. Al-Absi, M. I. M. Hafizal, M. Ismail, H. Awang, and A. Al-Shwaiter, "Properties of PCM-based composites developed for the exterior finishes of building walls," *Case Studies in Construction Materials*, vol. 16, p. e00960, Jun. 2022, doi:10.1016/J.CSCM.2022.E00960.

[99] "Editorial Board," *Journal of Energy Storage*, vol. 24, p. 100857, Aug. 2019, doi:10.1016/s2352-152x(19)30712-1.

[100] M. Theodoridou, L. Kyriakou, and I. Ioannou, "PCM-enhanced lime plasters for vernacular and contemporary architecture," *Energy Procedia*, vol. 97, pp. 539–545, Nov. 2016, doi:10.1016/J.EGYPRO.2016.10.070.

[101] M. M. Alsaadawi, M. Amin, and A. M. Tahwia, "Thermal, mechanical and microstructural properties of sustainable concrete incorporating Phase change materials,"

Construction and Building Materials, vol. 356, p. 129300, Nov. 2022, doi:10.1016/J.CONBUILDMAT.2022.129300.

[102] A. A. Aliabdo, A. E. M. Abd-Elmoaty, and H. H. Hassan, "Utilization of crushed clay brick in concrete industry," *Alexandria Engineering Journal*, vol. 53, no. 1, pp. 151–168, Mar. 2014, doi:10.1016/J.AEJ.2013.12.003.

[103] P. K. S. Rathore and S. K. Shukla, "Potential of macroencapsulated PCM for thermal energy storage in buildings: A comprehensive review," *Construction and Building Materials*, vol. 225, pp. 723–744, Nov. 2019, doi:10.1016/J.CONBUILDMAT.2019.07.221.

[104] O. Pons, A. Aguado, A. I. Fernández, L. F. Cabeza, and J. M. Chimenos, "Review of the use of phase change materials (PCMs) in buildings with reinforced concrete structures," *Materiales de Construcción*, vol. 64, no. 315, pp. e031–e031, Sep. 2014, doi:10.3989/MC.2014.05613.

[105] P. K. S. Rathore and S. K. Shukla, "Potential of macroencapsulated PCM for thermal energy storage in buildings: A comprehensive review," *Construction and Building Materials*, vol. 225, pp. 723–744, Nov. 2019, doi:10.1016/J.CONBUILDMAT.2019.07.221.

[106] A. F. Regin, S. C. Solanki, and J. S. Saini, "Heat transfer characteristics of thermal energy storage system using PCM capsules: A review," *Renewable and Sustainable Energy Reviews*, vol. 12, no. 9, pp. 2438–2458, Dec. 2008, doi:10.1016/J.RSER.2007.06.009.

[107] S. A. Memon, "Phase change materials integrated in building walls: A state of the art review," *Renewable and Sustainable Energy Reviews*, vol. 31, pp. 870–906, Mar. 2014, doi:10.1016/J.RSER.2013.12.042.

[108] B. Durakovic and M. Halilovic, "Thermal performance analysis of PCM solar wall under variable natural conditions: An experimental study," *Energy for Sustainable Development*, vol. 76, p. 101274, Oct. 2023, doi:10.1016/J.ESD.2023.101274.

[109] R. Jacob and F. Bruno, "Review on shell materials used in the encapsulation of phase change materials for high temperature thermal energy storage," *Renewable and Sustainable Energy Reviews*, vol. 48, pp. 79–87, Aug. 2015, doi:10.1016/J.RSER.2015.03.038.

[110] B. Durakovic and M. Torlak, "Experimental and numerical study of a PCM window model as a thermal energy storage unit," *International Journal of Low-Carbon Technologies*, vol. 12, no. 3, pp. 272–280, 2017, doi:10.1093/ijlct/ctw024.

[111] B. Durakovic and M. Torlak, "Simulation and experimental validation of phase change material and water used as heat storage medium in window applications," *Journal of Materials and Environmental Sciences*, vol. 8, no. 5, pp. 1837–1846. *ISSN*, 2017.

[112] S. J. Chang, S. Wi, S. G. Jeong, and S. Kim, "Thermal performance evaluation of macro-packed phase change materials (PCMs) using heat transfer analysis device," *Energy Build*, vol. 117, pp. 120–127, Apr. 2016, doi:10.1016/J.ENBUILD.2016.02.014.

[113] M. Zhang, M. A. Medina, and J. B. King, "Development of a thermally enhanced frame wall with phase-change materials for on-peak air conditioning demand reduction and energy savings in residential buildings," *International Journal of Energy Research*, vol. 29, no. 9, pp. 795–809, Jul. 2005, doi:10.1002/ER.1082.

[114] M. Ahmad, A. Bontemps, H. Sallée, and D. Quenard, "Thermal testing and numerical simulation of a prototype cell using light wallboards coupling vacuum isolation panels and phase change material," *Energy Build*, vol. 38, no. 6, pp. 673–681, Jun. 2006, doi:10.1016/J.ENBUILD.2005.11.002.

[115] K. O. Lee, M. A. Medina, and X. Sun, "On the use of plug-and-play walls (PPW) for evaluating thermal enhancement technologies for building enclosures: Evaluation of a thin phase change material (PCM) layer," *Energy Build*, vol. 86, pp. 86–92, Jan. 2015, doi:10.1016/J.ENBUILD.2014.10.020.

[116] G. Zhou and M. Pang, "Experimental investigations on the performance of a collector–storage wall system using phase change materials," *Energy Conversion and Management*, vol. 105, pp. 178–188, Aug. 2015, doi:10.1016/J.ENCONMAN.2015.07.070.

[117] Y. Berthou, P. H. Biwole, P. Achard, H. Sallée, M. Tantot-Neirac, and F. Jay, "Full scale experimentation on a new translucent passive solar wall combining silica aerogels and phase change materials," *Solar Energy*, vol. 115, pp. 733–742, May 2015, doi:10.1016/J.SOLENER.2015.03.038.

[118] G. Zhou and M. Pang, "Experimental investigations on the performance of a collector–storage wall system using phase change materials," *Energy Convers Manag*, vol. 105, pp. 178–188, Nov. 2015, doi:10.1016/J.ENCONMAN.2015.07.070.

[119] Y. Xiang and G. Zhou, "Thermal performance of a window-based cooling unit using phase change materials combined with night ventilation," *Energy Build*, vol. 108, pp. 267–278, Dec. 2015, doi:10.1016/J.ENBUILD.2015.09.030.

[120] H. Weinlaeder, W. Koerner, and M. Heidenfelder, "Monitoring results of an interior sun protection system with integrated latent heat storage," *Energy Build*, vol. 43, no. 9, pp. 2468–2475, Sep. 2011, doi:10.1016/J.ENBUILD.2011.06.007.

[121] T. Silva, R. Vicente, F. Rodrigues, A. Samagaio, and C. Cardoso, "Performance of a window shutter with phase change material under summer Mediterranean climate conditions," *Applied Thermal Engineering*, vol. 84, pp. 246–256, Jun. 2015, doi:10.1016/J.APPLTHERMALENG.2015.03.059.

[122] K. O. Lee, M. A. Medina, E. Raith, and X. Sun, "Assessing the integration of a thin phase change material (PCM) layer in a residential building wall for heat transfer reduction and management," *Applied Energy*, vol. 137, pp. 699–706, Jan. 2015, doi:10.1016/J.APENERGY.2014.09.003.

[123] S. J. Chang, S. Wi, S. G. Jeong, and S. Kim, "Thermal performance evaluation of macro-packed phase change materials (PCMs) using heat transfer analysis device," *Energy Build*, vol. 117, pp. 120–127, Apr. 2016, doi:10.1016/J.ENBUILD.2016.02.014.

[124] K. Nagano, S. Takeda, T. Mochida, K. Shimakura, and T. Nakamura, "Study of a floor supply air conditioning system using granular phase change material to augment building mass thermal storage—Heat response in small scale experiments," *Energy Build*, vol. 38, no. 5, pp. 436–446, May 2006, doi:10.1016/J.ENBUILD.2005.07.010.

[125] M. A. Medina, J. B. King, and M. Zhang, "On the heat transfer rate reduction of structural insulated panels (SIPs) outfitted with phase change materials (PCMs)," *Energy*, vol. 33, no. 4, pp. 667–678, Apr. 2008, doi:10.1016/J.ENERGY.2007.11.003.

[126] B. Duraković, "PCMs in separate heat storage modules," *Green Energy and Technology*, pp. 121–146, 2020, doi:10.1007/978-3-030-38335-0_6/COVER.

[127] G. Alva, Y. Lin, L. Liu, and G. Fang, "Synthesis, characterization and applications of microencapsulated phase change materials in thermal energy storage: A review," *Energy Build*, vol. 144, pp. 276–294, Jun. 2017, doi:10.1016/J.ENBUILD.2017.03.063.

[128] Y. Fang, "A comprehensive study of phase change materials (PCMs) for building walls applications," Mar. 2009, Accessed: Dec. 18, 2022. [Online]. Available: https://kuscholarworks.ku.edu/handle/1808/5537

[129] A. R. Vakhshouri and A. R. Vakhshouri, "Paraffin as phase change material," *Paraffin—an Overview*, Dec. 2019, doi:10.5772/INTECHOPEN.90487.

[130] J. Giro-Paloma, M. Martínez, L. F. Cabeza, and A. I. Fernández, "Types, methods, techniques, and applications for microencapsulated phase change materials (MPCM): A review," *Renewable and Sustainable Energy Reviews*, vol. 53. Elsevier Ltd, pp. 1059–1075, Jan. 01, 2016. doi:10.1016/j.rser.2015.09.040.

[131] Y. Ma, X. Chu, G. Tang, and Y. Yao, "The effect of different soft segments on the formation and properties of binary core microencapsulated phase change materials with polyurea/polyurethane double shell," *Journal of Colloid and Interface Science*, vol. 392, no. 1, pp. 407–414, Feb. 2013, doi:10.1016/J.JCIS.2012.10.052.

[132] N. Şahan, D. Nigon, S. C. Mantell, J. H. Davidson, and H. Paksoy, "Encapsulation of stearic acid with different PMMA-hybrid shell materials for thermotropic materials," *Solar Energy*, vol. 184, pp. 466–476, May 2019, doi:10.1016/J.SOLENER.2019.04.026.

[133] J. F. Su, X. Y. Wang, S. B. Wang, Y. H. Zhao, and Z. Huang, "Fabrication and properties of microencapsulated-paraffin/gypsum-matrix building materials for thermal energy storage," *Energy Conversion and Management*, vol. 55, pp. 101–107, Mar. 2012, doi:10.1016/J.ENCONMAN.2011.10.015.

[134] A. Arshad, M. Jabbal, Y. Yan, and J. Darkwa, "The micro-/nano-PCMs for thermal energy storage systems: A state of art review," *International Journal of Energy Research*, vol. 43, no. 11, pp. 5572–5620, Sep. 2019, doi:10.1002/ER.4550.

[135] S. Y. Kong, X. Yang, S. C. Paul, L. S. Wong, and B. Šavija, "Thermal response of mortar panels with different forms of macro-encapsulated phase change materials: A finite element study," *Energies (Basel)*, vol. 12, no. 13, 2019, doi:10.3390/EN12132636.

[136] Y. Konuklu, M. Ostry, H. O. Paksoy, and P. Charvat, "Review on using microencapsulated phase change materials (PCM) in building applications," *Energy Build*, vol. 106, pp. 134–155, Nov. 2015, doi:10.1016/J.ENBUILD.2015.07.019.

[137] J. Giro-Paloma, R. Al-Shannaq, A. I. Fernández, and M. M. Farid, "Preparation and characterization of microencapsulated phase change materials for use in building applications," *Materials*, vol. 9, no. 1, 2016, doi:10.3390/MA9010011.

[138] L. F. Cabeza, C. Castellón, M. Nogués, M. Medrano, R. Leppers, and O. Zubillaga, "Use of microencapsulated PCM in concrete walls for energy savings," *Energy Build*, vol. 39, no. 2, pp. 113–119, Feb. 2007, doi:10.1016/J.ENBUILD.2006.03.030.

[139] C. Cárdenas-Ramírez, F. Jaramillo, and M. Gómez, "Systematic review of encapsulation and shape-stabilization of phase change materials," *Journal of Energy Storage*, vol. 30, p. 101495, Aug. 2020, doi:10.1016/J.EST.2020.101495.

[140] N. Zhu, S. Wang, X. Xu, and Z. Ma, "A simplified dynamic model of building structures integrated with shaped-stabilized phase change materials," *International Journal of Thermal Sciences*, vol. 49, no. 9, pp. 1722–1731, Sep. 2010, doi:10.1016/J.IJTHERMALSCI.2010.03.020.

[141] N. Zhang, Y. Yuan, Y. Yuan, T. Li, and X. Cao, "Lauric–palmitic–stearic acid/expanded perlite composite as form-stable phase change material: Preparation and thermal properties," *Energy Build*, vol. 82, pp. 505–511, Oct. 2014, doi:10.1016/J.ENBUILD.2014.07.049.

[142] G. Zhou and J. He, "Thermal performance of a radiant floor heating system with different heat storage materials and heating pipes," *Applied Energy*, vol. 138, pp. 648–660, Jan. 2015, doi:10.1016/J.APENERGY.2014.10.058.

[143] M. Xiao, B. Feng, and K. Gong, "Preparation and performance of shape stabilized phase change thermal storage materials with high thermal conductivity," *Energy Conversion and Management*, vol. 43, no. 1, pp. 103–108, Jan. 2002, doi:10.1016/S0196-8904(01)00010-3.

[144] L. F. Cabeza, C. Castellón, M. Nogués, M. Medrano, R. Leppers, and O. Zubillaga, "Use of microencapsulated PCM in concrete walls for energy savings," *Energy Build*, vol. 39, no. 2, pp. 113–119, Feb. 2007, doi:10.1016/J.ENBUILD.2006.03.030.

[145] X. Zhang, et al., "Shape-stabilized composite phase change materials with high thermal conductivity based on stearic acid and modified expanded vermiculite," *Renew Energy*, vol. 112, pp. 113–123, Nov. 2017, doi:10.1016/J.RENENE.2017.05.026.

[146] M. Li and Z. Wu, "A review of intercalation composite phase change material: Preparation, structure and properties," *Renewable and Sustainable Energy Reviews*, vol. 16, no. 4, pp. 2094–2101, May 2012, doi:10.1016/J.RSER.2012.01.016.

[147] A. Karaipekli and A. Sari, "Development and thermal performance of pumice/organic PCM/gypsum composite plasters for thermal energy storage in buildings," *Solar Energy Materials and Solar Cells*, vol. 149, pp. 19–28, May 2016, doi:10.1016/j.solmat.2015.12.034.

[148] L. F. Cabeza, et al., "Behaviour of a concrete wall containing micro-encapsulated PCM after a decade of its construction," *Solar Energy*, vol. 200, pp. 108–113, Apr. 2020, doi:10.1016/J.SOLENER.2019.12.003.

[149] X. Shi, S. A. Memon, W. Tang, H. Cui, and F. Xing, "Experimental assessment of position of macro encapsulated phase change material in concrete walls on indoor temperatures and humidity levels," *Energy Build*, vol. 71, pp. 80–87, Mar. 2014, doi:10.1016/J.ENBUILD.2013.12.001.

[150] S. A. Memon, H. Z. Cui, H. Zhang, and F. Xing, "Utilization of macro encapsulated phase change materials for the development of thermal energy storage and structural lightweight aggregate concrete," *Applied Energy*, vol. 139, pp. 43–55, Feb. 2015, doi:10.1016/J.APENERGY.2014.11.022.

[151] O. Gencel, et al., "Properties of eco-friendly foam concrete containing PCM impregnated rice husk ash for thermal management of buildings," *Journal of Building Engineering*, vol. 58, p. 104961, Oct. 2022, doi:10.1016/J.JOBE.2022.104961.

[152] A. S. Tarasov, E. P. Kearsley, A. S. Kolomatskiy, and H. F. Mostert, "Heat evolution due to cement hydration in foamed concrete," *Magazine of Concrete Research*, vol. 62, no. 12, pp. 895–906, May 2015, doi:10.1680/MACR.2010.62.12.895.

[153] H. Umar, et al., "Mechanical properties of concrete containing beeswax/dammar gum as phase change material for thermal energy storage," *AIMS Energy*, vol. 6, no. 3, pp. 521–529, 2018, doi:10.3934/ENERGY.2018.3.521.

[154] Z. A. Al-Absi, M. I. M. Hafizal, M. Ismail, H. Awang, and A. Al-Shwaiter, "Properties of PCM-based composites developed for the exterior finishes of building walls," *Case Studies in Construction Materials*, vol. 16, p. e00960, Jun. 2022, doi:10.1016/J.CSCM.2022.E00960.

[155] Z. A. Al-Absi, M. I. M. Hafizal, and M. Ismail, "Experimental study on the thermal performance of PCM-based panels developed for exterior finishes of building walls," *Journal of Building Engineering*, vol. 52, p. 104379, Jul. 2022, doi:10.1016/J.JOBE.2022.104379.

[156] X. Sun, M. A. Medina, K. O. Lee, and X. Jin, "Laboratory assessment of residential building walls containing pipe-encapsulated phase change materials for thermal management," *Energy*, vol. 163, pp. 383–391, Nov. 2018, doi:10.1016/J.ENERGY.2018.08.159.

[157] Q. Al-Yasiri and M. Szabó, "Effect of encapsulation area on the thermal performance of PCM incorporated concrete bricks: A case study under Iraq summer conditions," *Case Studies in Construction Materials*, vol. 15, p. e00686, Dec. 2021, doi:10.1016/J.CSCM.2021.E00686.

[158] S. Shaik, C. Arumugam, S. V. Shaik, M. Arıcı, A. Afzal, and Z. Ma, "Strategic design of PCM integrated burnt clay bricks: Potential for cost-cutting measures for air conditioning and carbon dioxide extenuation," *Journal of Cleaner Production*, vol. 375, p. 134077, Nov. 2022, doi:10.1016/J.JCLEPRO.2022.134077.

[159] X. Kong, L. Jiang, Y. Yuan, and X. Qiao, "Experimental study on the performance of an active novel vertical partition thermal storage wallboard based on composite phase change material with porous silica and microencapsulation," *Energy*, vol. 239, p. 122451, Jan. 2022, doi:10.1016/J.ENERGY.2021.122451.

[160] M. Bake, A. Shukla, and S. Liu, "Development of gypsum plasterboard embodied with microencapsulated phase change material for energy efficient buildings," *Materials Science for Energy Technologies*, vol. 4, pp. 166–176, Jan. 2021, doi:10.1016/J.MSET.2021.05.001.

[161] S. G. Jeong, S. Wi, S. J. Chang, J. Lee, and S. Kim, "An experimental study on applying organic PCMs to gypsum-cement board for improving thermal performance of buildings in different climates," *Energy Build*, vol. 190, pp. 183–194, May 2019, doi:10.1016/J.ENBUILD.2019.02.037.

[162] C. Molinari, et al., "Effect of scale-up on the properties of PCM-impregnated tiles containing glass scraps," *Case Studies in Construction Materials*, vol. 14, p. e00526, Jun. 2021, doi:10.1016/J.CSCM.2021.E00526.

[163] R. I. Hatamleh, N. H. Abu-Hamdeh, A. Khoshaim, and M. A. Alzahrani, "Using phase change material (PCM) to improve the solar energy capacity of glass in solar collectors by enhancing their thermal performance via developed MD approach," *Engineering Analysis with Boundary Elements*, vol. 143, pp. 163–169, Oct. 2022, doi:10.1016/J.ENGANABOUND.2022.06.010.

[164] A. M. Mohammed, A. Elnokaly, and A. M. M. Aly, "Empirical investigation to explore potential gains from the amalgamation of phase changing materials (PCMs) and wood shavings," *Energy and Built Environment*, vol. 2, no. 3, pp. 315–326, Jul. 2021, doi:10.1016/J.ENBENV.2020.07.001.

[165] H. Chen, J. Xuan, Q. Deng, and Y. Gao, "WOOD/PCM composite with enhanced energy storage density and anisotropic thermal conductivity," *Progress in Natural Science: Materials International*, vol. 32, no. 2, pp. 190–195, Apr. 2022, doi:10.1016/J.PNSC.2022.01.002.

[166] B. Šavija and E. Schlangen, "Use of phase change materials (PCMs) to mitigate early age thermal cracking in concrete: Theoretical considerations," *Construction and Building Materials*, vol. 126, pp. 332–344, Nov. 2016, doi:10.1016/J.CONBUILDMAT.2016.09.046.

[167] F. Fernandes, et al., "On the feasibility of using phase change materials (PCMs) to mitigate thermal cracking in cementitious materials," *Cement and Concrete Composites*, vol. 51, pp. 14–26, Aug. 2014, doi:10.1016/J.CEMCONCOMP.2014.03.003.

[168] S. Pilehvar, et al., "Effect of freeze-thaw cycles on the mechanical behavior of geopolymer concrete and Portland cement concrete containing micro-encapsulated phase change materials," *Construction and Building Materials*, vol. 200, pp. 94–103, Mar. 2019, doi:10.1016/J.CONBUILDMAT.2018.12.057.

[169] A. Barde, G. Mazzotta, and J. Weiss, "Early-age flexural strength: The role of aggregates and their influence on maturity predictions | Request PDF," in *Pecial Volume of the Material Science of Concrete VII*, Wiley Publishers, 2005. Accessed: Jan. 09, 2023 [Online]. Available: www.researchgate.net/publication/290796140_Early-age_flexural_strength_The_role_of_aggregates_and_their_influence_on_maturity_predictions

[170] B. A. Young, et al., "A general method for retrieving thermal deformation properties of microencapsulated phase change materials or other particulate inclusions in cementitious composites," *Materials & Design*, vol. 126, pp. 259–267, Jul. 2017, doi:10.1016/J.MATDES.2017.04.023.

[171] Z. Wei, et al., "The durability of cementitious composites containing microencapsulated phase change materials," *Cement and Concrete Composites*, vol. 81, pp. 66–76, Aug. 2017, doi:10.1016/J.CEMCONCOMP.2017.04.010.

[172] K.-J. Shin, B. Bucher, and J. Weiss, "Role of lightweight synthetic particles on the restrained shrinkage cracking behavior of mortar," *Journal of Materials in Civil Engineering*, vol. 23, no. 5, pp. 597–605, May 2011, doi:10.1061/(ASCE)MT.1943-5533.0000213.

[173] L. Haurie, S. Serrano, M. Bosch, A. I. Fernandez, and L. F. Cabeza, "Single layer mortars with microencapsulated PCM: Study of physical and thermal properties, and fire behaviour," *Energy Build*, vol. 111, pp. 393–400, Jan. 2016, doi:10.1016/J. ENBUILD.2015.11.028.

[174] J. Bohórquez-Órdenes, A. Tapia-Calderón, D. A. Vasco, O. Estuardo-Flores, and A. N. Haddad, "Methodology to reduce cooling energy consumption by incorporating PCM envelopes: A case study of a dwelling in Chile," *Build Environ*, vol. 206, p. 108373, Dec. 2021, doi:10.1016/J.BUILDENV.2021.108373.

[175] S. Kenzhekhanov, S. A. Memon, and I. Adilkhanova, "Quantitative evaluation of thermal performance and energy saving potential of the building integrated with PCM in a subarctic climate," *Energy*, vol. 192, p. 116607, Feb. 2020, doi:10.1016/J. ENERGY.2019.116607.

[176] M. Salihi, et al., "Evaluation of global energy performance of building walls integrating PCM: Numerical study in semi-arid climate in Morocco," *Case Studies in Construction Materials*, vol. 16, p. e00979, Jun. 2022, doi:10.1016/J.CSCM.2022.E00979.

11 Passive Thermal Regulation of Batteries

Bilal Lamrani, Badr Eddine Lebrouhi, and Tarik Kousksou

HIGHLIGHTS

- PCMs enhance battery cooling efficiency.
- A simulation model was successfully validated.
- A 3°C temperature reduction was achieved.
- Combining a PCM with air cooling further lowers temperature.
- PCMs prove highly effective in battery thermal control.

NOMENCLATURE

A: Heat transfer area
C: Heat capacity (J/kgK)
h: Coefficient of heat transfer by convection (W/m²K)
L_s: Heat of fusion (J/kg)
m: Mass (kg)
Q: Battery heat generation (W)
T: Temperature (°C or K)

SUBSCRIPTS

b: Battery pack
base: Base
ext: Exterior
irr: Irreversible
m: melting
pcm: Phase change material
rev: Reversible
t: Total

11.1 INTRODUCTION

To attain the ambitious target of achieving zero carbon emissions by 2050, there is an urgent need to accelerate the adoption of clean energy technologies. This entails a multifaceted approach encompassing energy efficiency,

DOI: 10.1201/9781003331957-11

the harnessing of renewable energy sources like solar and wind power, and a fundamental shift toward electric mobility [1–3]. Electric vehicles (EVs) assume a pivotal role in the pursuit of this objective, as an increasing number of countries are phasing out conventional fossil fuel-powered vehicles in favor of EVs and plug-in hybrid electric vehicles (PHEVs) [3, 4]. Despite the numerous advantages that EVs offer, their reliance on advanced lithium-ion batteries demands consistent and efficient thermal management to avert potential issues such as battery failure and thermal runaway, which can lead to fires and associated safety hazards. These incidents not only pose risks to users, passengers, and communities but also raise concerns about environmental damage. Furthermore, the operating temperature of a lithium-ion battery has a profound impact on its overall performance and longevity. Maintaining the battery within the temperature range of 15–35°C is considered optimal for ensuring performance and durability. Elevated temperatures can lead to reduced capacity and shortened battery lifespan, while lower temperatures can increase internal resistance and diminish capacity [5]. Moreover, variations in temperature across the battery, known as temperature gradients, can also adversely affect performance. To mitigate these challenges and maximize the benefits of lithium-ion batteries in EVs, it is imperative to implement continuous and efficient thermal control mechanisms. These systems are essential for maintaining the desired temperature range during the battery's charging and discharging cycles, ultimately ensuring the safety, performance, and longevity of electric vehicle batteries [6–8]. This comprehensive approach not only advances environmental and sustainability goals but also fosters the widespread adoption of electric mobility as a means of achieving a carbon-neutral future.

The quest for efficient battery thermal management (BTM) strategies has given rise to a diverse array of innovative approaches designed to tackle the intricate challenges associated with temperature control and the overall performance of lithium-ion batteries [9–14]. These strategies encompass a wide spectrum of techniques, which can be broadly categorized based on several critical factors: the transfer medium employed (whether it is air, liquid, or phase change material (PCM), the placement of the thermal management system (internal or external to the battery pack), the primary purpose of the strategy (cooling or heating), and the energy utilization method (active or passive) [5]. In recent years, there has been a significant surge in research efforts, with extensive investigations and experimentation conducted to explore these various BTM strategies from a multitude of angles [15, 16]. These studies have contributed significantly to our understanding of the inherent strengths and limitations of different BTM methods. By evaluating these strategies from multiple perspectives, researchers have provided valuable insights into the complexities of thermal management in lithium-ion batteries. This collective body of research and experimentation serves as a crucial resource, offering insights into the advantages and constraints of distinct BTM approaches. It has significantly advanced our ability to enhance the performance, safety, and durability of lithium-ion batteries. By bridging the gap between theoretical understanding and practical implementation, these studies have paved the way for improved battery technologies, ultimately facilitating the sustainable growth of electric mobility and energy storage applications.

Within the extensive landscape of techniques for managing battery thermal conditions, PCM-based systems have emerged as a particularly efficient and compelling choice for battery cooling, all while avoiding the need for active power consumption. These systems have garnered substantial attention due to their remarkable advantages, making them a promising solution in the field of battery thermal management [17, 18]. One of the key benefits of PCM-based systems is their straightforward design. They are relatively easy to integrate into battery packs and require minimal structural modifications. This simplicity not only streamlines the manufacturing process but also makes maintenance and repairs more manageable. Precision in temperature control is another notable advantage. PCM-based systems enable fine-tuned and consistent regulation of battery temperature. This level of control is pivotal for ensuring optimal battery performance, longevity, and safety. Affordability is a crucial factor in the adoption of any technology. PCM-based systems are often cost effective, which makes them an attractive choice for a broad range of applications. Moreover, these systems exhibit a remarkable degree of chemical stability and reliability. They maintain their properties over time, ensuring long-term effectiveness and safety. This stability is particularly vital in applications where battery integrity and safety are paramount. The versatility of PCM-based systems is another compelling feature. They find applicability in a wide array of contexts, from electric vehicles and consumer electronics to renewable energy storage systems and grid applications. Their adaptability underscores their importance in the broader landscape of thermal management solutions. However, the selection of an appropriate PCM is a critical decision that involves considering various factors. These include thermal properties, physical characteristics, chemical kinetics, and economic viability [19]. To be effective, the PCM must have a melting temperature within the operational range of the specific battery application, which typically falls between 15–35°C. The choice of PCM should align with the specific requirements and constraints of the system to achieve optimal thermal management outcomes. For those seeking a selection of PCMs suitable for BTM applications, an extensive list is available [19]. This resource serves as a valuable guide for researchers, engineers, and manufacturers to make informed decisions when implementing PCM-based solutions for battery thermal management.

Research on using PCM for battery thermal management typically focuses on three areas:

- Using PCMs as a passive system by studying the impact of different factors (such as the type of PCM and the amount used) on battery performance. [13, 20–23].
- Improving the poor thermal conductivity of PCMs by combining them with other materials such as metal foams (like aluminum, copper, and nickel foams) [19, 24] or using metal fins [25, 26], composite PCMs [20, 27], or other materials such as metal mesh and carbon-based materials like expanded graphite, carbon nanotubes, and carbon fiber with conventional PCM [28, 29].
- Combining PCM with other transfer mediums, such as air cooling [30, 31], liquid cooling [32, 33], or heat pipe cooling [34, 35]. These research investigations have demonstrated that these arrangements are successful in reducing the battery's maximum temperature, thus ensuring optimal efficiency, upholding consistent cell temperature distribution, and averting the occurrence of thermal runaway.

The insights gleaned from the extensive literature review shed light on a notable gap in the existing body of research. A significant portion of studies that have explored the integration of PCMs with liquid cooling for battery thermal management have typically not subjected their systems to rigorous, real-world, continuous charging and discharging cycles for comprehensive evaluation. Instead, many of these investigations have heavily relied on computational fluid dynamics (CFD) tools, which can be time consuming and computationally intensive. To address these shortcomings and enhance the practicality and relevance of research in this field, there is a compelling need for a more efficient and versatile approach. A fast-lumped model emerges as an ideal solution, as it can provide accurate predictions regarding the behavior of coupled PCM-battery pack systems during actual battery operation. Such a model can offer valuable insights into how these systems perform under dynamic conditions and provide a practical basis for decision making in the design and implementation of battery thermal management solutions. The core focus of the research presented in this chapter revolves around the comprehensive examination of BTM systems that leverage PCMs as a means of passive thermal control. By bridging the gap between theoretical understanding and practical application, this research aims to contribute significantly to the field of thermal management in the context of energy storage and electric mobility. Through the use of fast-lumped modeling, it seeks to provide a pragmatic and efficient means of improving the performance, safety, and sustainability of battery systems, ultimately advancing the transition to cleaner and more energy-efficient technologies.

11.2   DESCRIPTION OF THE INVESTIGATED SYSTEM

The tested battery bank is composed of a substantial number of individual lithium-ion (Li-ion) batteries, specifically $LiNiMnCoAlO_2$ (NMC) batteries, totaling 21,700 units. These batteries are organized into 24 series-connected banks, each configured as 6S4P (six batteries in series and four parallel banks). The choice of this battery configuration is highly relevant due to its current popularity in the market and the impressive energy density it offers. Scenario 1 involves operating the battery while allowing air to circulate between the individual cells. In this arrangement, the goal is to maintain a consistent temperature within the battery pack. In Scenario 2, a PCM is introduced around the battery cells. This addition aims to regulate the thermal response of the battery and absorb the heat generated during its operational phases. Table 11.1 presents essential characteristics of both the PCM material and the battery cells, providing a detailed insight into their properties. Scenario 3 incorporates a hybrid battery thermal management system (BTMS), combining different techniques for effective thermal control. The complexity of this system is visualized in Figure 11.1, where the PCM is an integral part of the overall thermal management strategy. This chapter aims to comprehensively explore and evaluate these scenarios, shedding light on their individual merits and potential contributions to efficient battery thermal management.

TABLE 11.1

Characteristics of the Analyzed Li-Ion Battery Cells and Phase Change Materials (PCMs)

Parameter	Value
Li-ion cell mass	69 ± 2 g
Li-ion cell diameter	21.7 ± 0.2 mm
Li-Ion cell Height	70.9 ± 0.2 mm
Li-ion cell nominal capacity	4 Ah
Li-ion cell voltage	3.65 V
Density of the phase change material	870 kg m^{-3}
Heat capacity of the phase change material	2412 J kg^{-1} K^{-1}
Thermal conductivity of the phase change material	5.023 W m^{-1} K^{-1}
Heat of fusion of the phase change material	119.24 kJ kg^{-1}
Melting interval of the phase change material	$31–36°C$

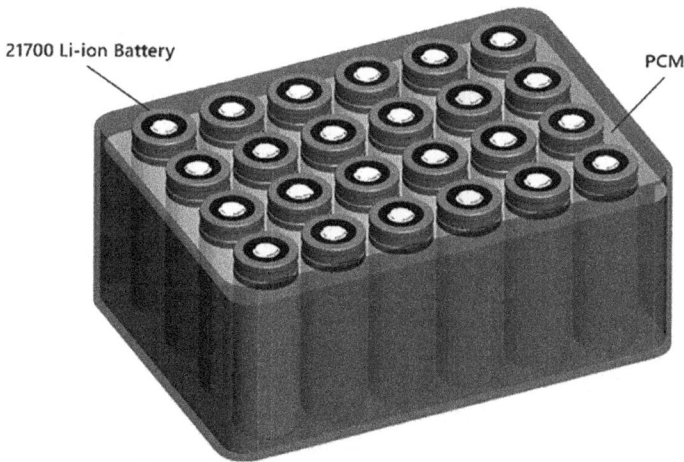

FIGURE 11.1 The investigated battery pack with PCM.

11.3 MODEL DEVELOPMENT AND VERIFICATION

The model developed in this research uniquely combines a heat generation model for energy storage devices with a lumped thermal model. This innovative approach offers several significant advantages. It empowers us to make highly reliable predictions regarding the actual performance of batteries, providing invaluable insights into their behavior under various conditions. Moreover, it accomplishes this with notable efficiency, significantly reducing the time required to obtain these crucial predictions. The evolved lumped thermal method applied to the investigated Li-ion battery system, integrated with PCM, is a focal point of this work. This method has been meticulously tailored to encapsulate the intricate interplay between heat generation within the batteries and the ensuing thermal dynamics. To elucidate this further, we will delve into the specifics of this evolved lumped method, which forms the backbone of our ability to simulate and assess the battery system's performance effectively.

The battery cells' governing energy equation is as follows:

$$\left(mc_p \right)_b \frac{\partial T_b}{\partial t} = Q_t + h_{pcm} A_b \left(T_{pcm} - T_b \right)$$

(11.1)

In this expression, Q_t represents the total heat produced by the battery packs, and it is expressed as:

$$Q_t = Q_{irr} + Q_{rev}$$

(11.2)

In this critical section of our work, we delve into the fundamental principles governing our thermal modeling approach. The distinction between Q_{irr} and Q_{rev} is pivotal. Q_{irr} embodies the irreversible heat generation within the system, representing the heat that cannot be undone or reclaimed. Conversely, Q_{rev} pertains to the reversible heat generation, signifying the heat that can be reversed, particularly when managing the battery's thermal behavior.

During the battery's charging or discharging phases, the PCM plays a crucial role in absorbing the heat generated. This absorption process is at the core of our approach to replicate and comprehend the phenomenon. To accurately model the PCM's phase change process, encompassing both melting and solidification, we adopt the effective heat capacity method. In this method, the PCM specific heat is expressed as a variable function contingent upon the PCM's temperature and the heat of fusion associated with it. This dynamic approach allows us to represent the PCM's behavior in a more detailed and responsive manner during phase changes. It takes into account the fluctuations in specific heat capacity as the PCM transitions between solid and liquid phases in response to temperature variations.

The expression we employ for the PCM heat capacity is a central element of our modeling, and it is given as:

$$C_{p,pcm}(T) = C_{base}(T) + L_s \frac{\dfrac{2\gamma}{\Delta T}}{\pi \left[\left((T - T_m) \left(\dfrac{2\gamma}{\Delta T} \right) \right)^2 + 1 \right]}$$

(11.3)

This comprehensive section forms the foundation of our modeling framework, ensuring a more accurate and nuanced representation of the battery and PCM's thermal dynamics during various operational scenarios. It is worth noting that the previous expression is integrated into the following PCM energy balance:

$$\left(mc_p\right)_{pcm}\frac{\partial T_{pcm}}{\partial t} = h_{pcm}A_b\left(T_b - T_{pcm}\right) + h_{ext}A_{ext}\left(T_{ext} - T_{pcm}\right)$$

(11.4)

The purpose of this study is to evaluate the robustness of the lumped concept for the PCM battery pack that was constructed. Obtained simulation results were compared with measured results from the literature, and obtained results are presented in Figure 11.2. These findings indicate the viability of the suggested modeling approach for accurately replicating the actual performance of the battery pack employing PCM as a passive thermal regulation mode during its operational phases. The model's maximum relative error, approximately 6%, attests to its accuracy in representing real-world conditions. This thoroughly developed and validated model will be subsequently employed in the upcoming section to delve into the thermal characteristics and conduct an in-depth parameter analysis of the pack with a PCM-based passive thermal regulation system.

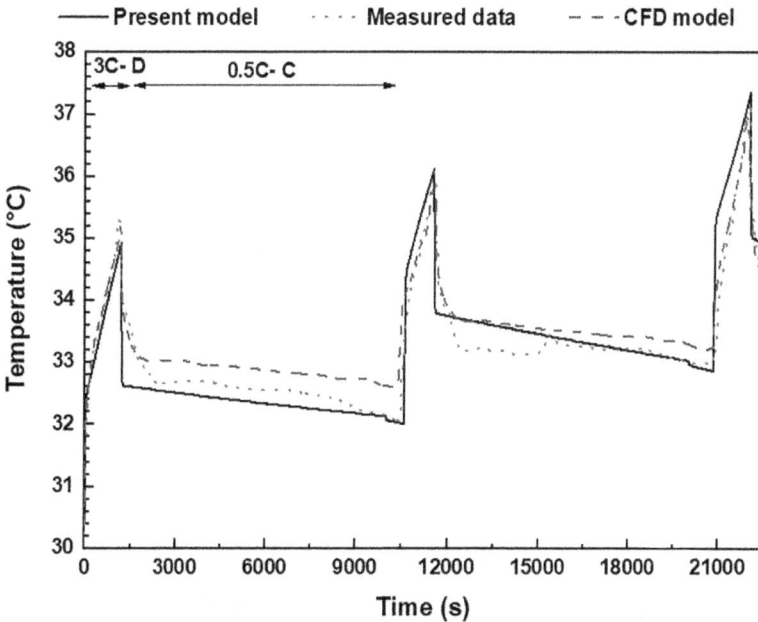

FIGURE 11.2   Temporal progression of average battery temperature: A comparison between simulation outcomes and CFD analysis, alongside experimental data [36].

11.4 RESULTS AND DISCUSSION

Our research involved the rigorous evaluation of a lithium battery integrated with a PCM as a thermal management system. To comprehensively assess the system's performance and gain insights into the impact of various design and usage factors on battery temperature, we executed a series of three cycles. These cycles were conducted under controlled conditions, employing a discharge mode at a high 3C rate for a duration of 1200 seconds, followed by a charging mode at a 0.5C rate for 7200 seconds. The choice of these specific charge and discharge rates was essential to simulate and evaluate the battery's behavior under conditions that resemble real-world scenarios. To maintain consistency and ensure accurate observations, we imposed a 10-minute interval between each charging and discharging cycle. This pause between cycles allowed the system to stabilize and return to its baseline state, ensuring that the subsequent cycle started from a well-defined condition. In Figure 11.3, we have graphically represented the relationship between the state of charge (SOC) of the battery and the magnitude of the current drawn from it. This figure provides a visual depiction of how the battery's state of charge changes in response to its operational status, offering a critical insight into the battery's performance over the course of these cycles. These data serve as key components of our research, facilitating a deeper understanding of how the PCM-equipped battery system responds to the challenges of real-world operation and how different factors influence its thermal behavior and performance.

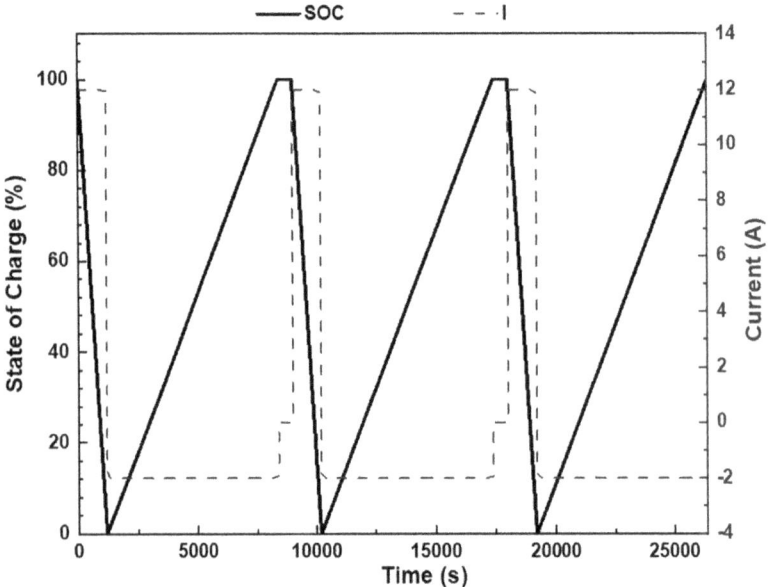

FIGURE 11.3 Evolution of battery pack state of charge and current [13].

11.4.1 EFFECT OF AMBIENT TEMPERATURE

By using a PCM, Figure 11.4 depicts how the average battery temperature in the investigated pack changed over time for various environmental temperatures. It was noticed that the temperature of a battery pack during charging and discharging cycles is considerably impacted when the battery pack is operated in a hot environment, i.e., a high outside temperature. When the ambient temperature is highest, the batteries may achieve temperatures of up to 44 degrees Celsius, even when PCM is used. When the ambient temperature rises, less heat is lost from the batteries and the PCM to the environment.

11.4.2 EFFECT OF AMBIENT CONVECTIVE COEFFICIENT

Figure 11.5 depicts the effect of boosting convective heat flow on the storage temperature when the battery storage system is running. It becomes apparent that the heat dissipation approach employed is highly appropriate for the considered system, as it significantly bolsters the efficiency of the battery system's thermal management. This improvement is achieved by effectively reducing the battery temperature, which is a pivotal factor in optimizing the system's overall performance. In reality, the capacity of the PCM to retain heat in charging/discharging periods is enhanced by increasing the amount of transferred heat, which in turn boosts the amount of dissipated thermal energy to the environment. When increasing h_{ext} from 5 W/m^2K to 15 W/m^2K, the battery's maximum temperature may be lowered by around 2.5°C when PCM thermal management is used in conjunction with this passive air cooling technique.

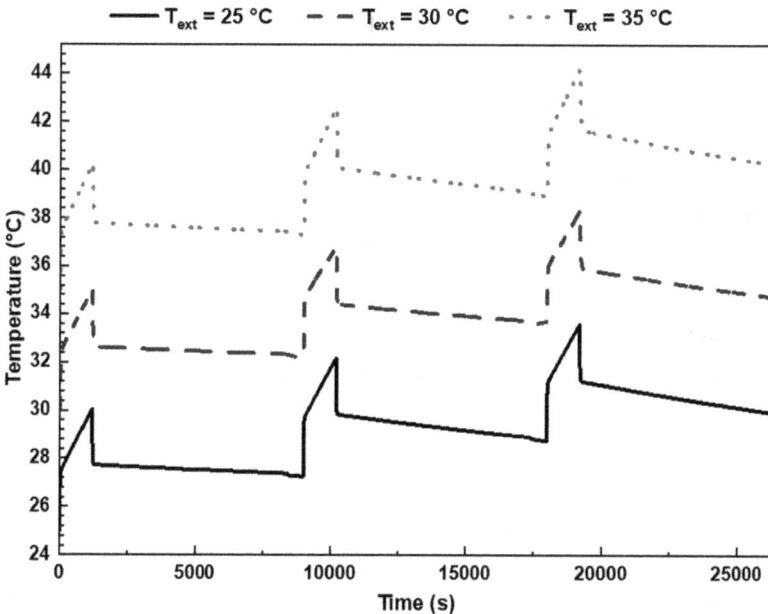

FIGURE 11.4   Influence of ambient temperature on average pack temperature.

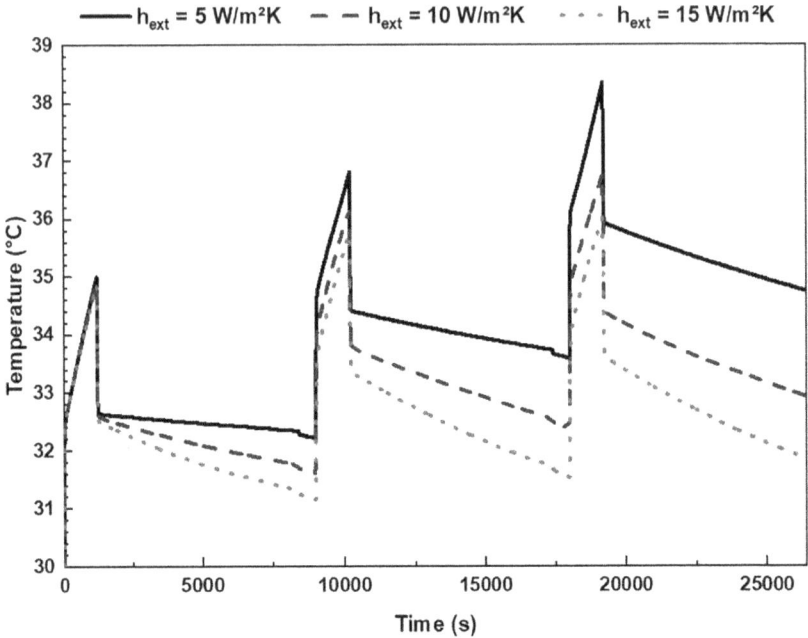

FIGURE 11.5   Influence of exterior heat convection coefficient ($T_{ext} = 30°C$).

11.4.3   EFFECT OF PCM

As illustrated in Figure 11.6, the PCM emerges as an effective thermal management solution for battery packs, adept at absorbing the heat generated by the batteries and maintaining them within a more controllable temperature range. In practical operation, the battery pack's peak temperature is notably reduced by up to 3 degrees Celsius thanks to the utilization of a PCM. The data suggest that PCM indeed proves to be a highly suitable thermal management technology for battery packs, given its capacity to dissipate the heat produced by the batteries.

Nonetheless, it is worth noting that in order to uphold the battery pack's temperature within an acceptable range, continuous heat dissipation from the PCM to the surroundings is essential throughout the system's operation. Furthermore, minimizing the liquid phase percentage of the PCM after each battery discharge procedure becomes pivotal to ensure consistent and effective heat management.

11.5   CONCLUSION

In this chapter, we have delved into a comprehensive study of a battery pack featuring the innovative use of PCMs as a thermal management system. Our research approach involved the development of a dynamic simulation model, which incorporated transient energy equations to account for both the PCM and the Li-ion cell domains. Crucially, the phase transition process of the PCM was modeled using the effective heat capacity method.

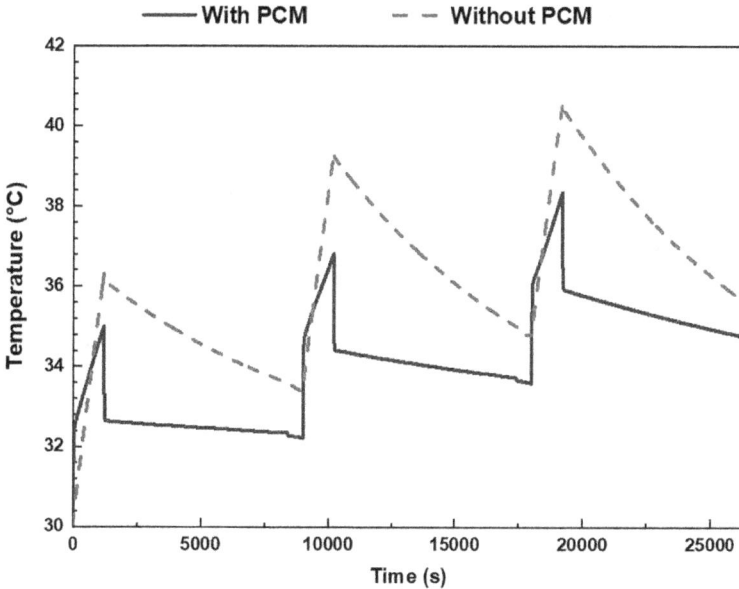

FIGURE 11.6 Impact of PCM usage on the average temperature of the Li-ion storage system.

This model was not merely a theoretical construct; it underwent a rigorous valida-tion process, during which it was compared against experimental data to ensure its accuracy and reliability. With a validated model at our disposal, we set out to assess the practical efficacy of thermal control in the battery pack.

Our findings have unveiled the remarkable potential of PCMs as an optimal thermal regulation technology for battery packs. PCMs are a standout solution thanks to their exceptional ability to efficiently dissipate the generated heat by the Li-ion cells. Most notably, the implementation of PCM resulted in a significant reduction in the maximum temperature of the storage system, lowering it by up to 3°C. This temperature reduction is of paramount importance for enhancing both battery performance and safety.

Moreover, our research demonstrated that coupling PCM thermal management with a passive air cooling technique, featuring an increased exterior convective coef-ficient from 5 W/m^{2}K to 15 W/m^{2}K, further contributed to lowering the highest bat-tery temperature by approximately 2.5°C. This highlights the synergistic benefits of combining PCM with passive air cooling, making it an effective strategy for enhanc-ing battery thermal control.

In summary, the key takeaways from this study are that:

- PCM proves to be a highly effective thermal management solution for bat-tery packs, reducing maximum temperatures by up to 3°C.
- Coupling PCM with passive air cooling techniques can further decrease the highest battery temperatures by approximately 2.5°C, enhancing over-all thermal control.

- The developed simulation model, validated against experimental data, offers a valuable tool for understanding and optimizing battery thermal management, contributing to both performance and safety goals.

REFERENCES

[1] "Net Zero by 2050—Analysis—IEA," www.iea.org/reports/net-zero-by-2050 (accessed Nov. 08, 2022).

[2] B. E. Lebrouhi, E. Schall, B. Lamrani, Y. Chaibi, and T. Kousksou, "Energy transition in France," *Sustainability (Switzerland)*, vol. 14, no. 10, 2022, doi:10.3390/su14105818.

[3] B. E. Lebrouhi, Y. Khattari, B. Lamrani, M. Maaroufi, Y. Zeraouli, and T. Kousksou, "Key challenges for a large-scale development of battery electric vehicles: A comprehensive review," *J Energy Storage*, vol. 44, no. PB, p. 103273, Dec. 2021, doi:10.1016/j. est.2021.103273.

[4] M. Boulakhbar, et al., "Towards a large-scale integration of renewable energies in Morocco," *J Energy Storage*, vol. 32, p. 101806, Dec. 2020, doi:10.1016/j. est.2020.101806.

[5] Q. L. Yue, C. X. He, M. C. Wu, and T. S. Zhao, "Advances in thermal management systems for next-generation power batteries," *Int J Heat Mass Transf*, vol. 181, p. 121853, Dec. 2021, doi:10.1016/J.IJHEATMASSTRANSFER.2021.121853.

[6] X. Shu, W. Yang, Y. Guo, K. Wei, B. Qin, and G. Zhu, "A reliability study of electric vehicle battery from the perspective of power supply system," *J Power Sources*, vol. 451, no. Jan., p. 227805, 2020, doi:10.1016/j.jpowsour.2020.227805.

[7] G. P. Beauregard and A. Z. Phoenix, "Report of investigation: Hybrids plus plug in hybrid electric vehicle," *National Rural Electric Cooperative Association, Inc. and US Department of Energy, Idaho National Laboratory by ETEC*, Online im Internet: URL: www. evworld. com/library/prius_fire_forensics. pdf [Stand 11.04. 2012], 2008.

[8] J. Kim, J. Oh, and H. Lee, "Review on battery thermal management system for electric vehicles," *Appl Therm Eng*, vol. 149, Nov. 2018, pp. 192–212, 2019, doi:10.1016/j. applthermaleng.2018.12.020.

[9] M. K. Tran, S. Panchal, T. D. Khang, K. Panchal, R. Fraser, and M. Fowler, "Concept review of a cloud-based smart battery management system for lithium-ion batteries: Feasibility, logistics, and functionality," *Batteries*, vol. 8, no. 2, p. 19, Feb. 2022, doi:10.3390/BATTERIES8020019.

[10] X. Li, et al., "Simulation of cooling plate effect on a battery module with different channel arrangement," *J Energy Storage*, vol. 49, p. 104113, May 2022, doi:10.1016/J. EST.2022.104113.

[11] M. K. Tran, C. Cunanan, S. Panchal, R. Fraser, and M. Fowler, "Investigation of individual cells replacement concept in lithium-ion battery packs with analysis on economic feasibility and pack design requirements," *Processes 2021*, vol. 9, no. 12, p. 2263, Dec. 2021, doi:10.3390/PR9122263.

[12] K. Benabdelaziz, B. Lebrouhi, A. Maftah, and M. Maaroufi, "Novel external cooling solution for electric vehicle battery pack," *Energy Reports*, vol. 6, no. Sept., pp. 262–272, Feb. 2019, doi: 10.1016/j.egyr.2019.10.043.

[13] B. Lamrani, B. E. Lebrouhi, Y. Khattari, and T. Kousksou, "A simplified thermal model for a lithium-ion battery pack with phase change material thermal management system," *J Energy Storage*, vol. 44, p. 103377, Dec. 2021, doi:10.1016/J.EST.2021.103377.

[14] B. E. Lebrouhi, B. Lamrani, M. Ouassaid, M. Abd-Lefdil, M. Maaroufi, and T. Kousksou, "Low-cost numerical lumped modelling of lithium-ion battery pack with phase change material and liquid cooling thermal management system," *J Energy Storage*, vol. 54, no. Mar., p. 105293, 2022, doi:10.1016/j.est.2022.105293.

[15] J. Lin, X. Liu, S. Li, C. Zhang, and S. Yang, "A review on recent progress, challenges and perspective of battery thermal management system," *International Journal of Heat and Mass Transfer*, vol. 167. Elsevier Ltd, p. 120834, Mar. 01, 2021. doi:10.1016/j.ijheatmasstransfer.2020.120834.

[16] P. R. Tete, M. M. Gupta, and S. S. Joshi, "Developments in battery thermal management systems for electric vehicles: A technical review," *Journal of Energy Storage*, vol. 35. Elsevier Ltd, p. 102255, Mar. 01, 2021. doi: 10.1016/j.est.2021.102255.

[17] Z. Ling, J. Cao, W. Zhang, Z. Zhang, X. Fang, and X. Gao, "Compact liquid cooling strategy with phase change materials for Li-ion batteries optimized using response surface methodology," *Appl Energy*, vol. 228, pp. 777–788, Oct. 2018, doi:10.1016/j.apenergy.2018.06.143.

[18] M. Subramanian, et al., "A technical review on composite phase change material based secondary assisted battery thermal management system for electric vehicles," *J Clean Prod*, p. 129079, Sep. 2021, doi:10.1016/J.JCLEPRO.2021.129079.

[19] S. Babu Sanker and R. Baby, "Phase change material based thermal management of lithium ion batteries: A review on thermal performance of various thermal conductivity enhancers," *J Energy Storage*, vol. 50, no. Sept. 2021, p. 104606, 2022, doi:10.1016/j.est.2022.104606.

[20] W. Wu, et al., "An innovative battery thermal management with thermally induced flexible phase change material," *Energy Convers Manag*, vol. 221, p. 113145, Oct. 2020, doi:10.1016/j.enconman.2020.113145.

[21] Z. Sun, R. Fan, and N. Zheng, "Thermal management of a simulated battery with the compound use of phase change material and fins: Experimental and numerical investigations," *Int J Therm Sci*, vol. 165, p. 106945, Jul. 2021, doi:10.1016/j.ijthermalsci.2021.106945.

[22] Y. Wang, Z. Wang, H. Min, H. Li, and Q. Li, "Performance investigation of a passive battery thermal management system applied with phase change material," *J Energy Storage*, vol. 35, p. 102279, Mar. 2021, doi:10.1016/j.est.2021.102279.

[23] K. Jiang, G. Liao, J. E, F. Zhang, J. Chen, and E. Leng, "Thermal management technology of power lithium-ion batteries based on the phase transition of materials: A review," *J Energy Storage*, vol. 32. Elsevier Ltd, p. 101816, Dec. 01, 2020. doi:10.1016/j.est.2020.101816.

[24] W. Situ, et al., "A thermal management system for rectangular LiFePO4 battery module using novel double copper mesh-enhanced phase change material plates," *Energy*, vol. 141, pp. 613–623, Dec. 2017, doi:10.1016/J.ENERGY.2017.09.083.

[25] P. Ping, R. Peng, D. Kong, G. Chen, and J. Wen, "Investigation on thermal management performance of PCM-fin structure for Li-ion battery module in high-temperature environment," *Energy Convers Manag*, vol. 176, pp. 131–146, Nov. 2018, doi:10.1016/j.enconman.2018.09.025.

[26] M. H. Shojaeefard, G. R. Molaeimanesh, and Y. S. Ranjbaran, "Improving the performance of a passive battery thermal management system based on PCM using lateral fins," *Heat and Mass Transfer*, vol. 55, no. 6, pp. 1753–1767, Jan. 2019, doi:10.1007/S00231-018-02555-0.

[27] W. Wang, X. Zhang, C. Xin, and Z. Rao, "An experimental study on thermal management of lithium ion battery packs using an improved passive method," *Appl Therm Eng*, vol. 134, pp. 163–170, Apr. 2018, doi:10.1016/J.APPLTHERMALENG.2018.02.011.

[28] F. Samimi, A. Babapoor, M. Azizi, and G. Karimi, "Thermal management analysis of a Li-ion battery cell using phase change material loaded with carbon fibers," *Energy*, vol. 96, pp. 355–371, Feb. 2016, doi:10.1016/J.ENERGY.2015.12.064.

[29] Y. Li, et al., "Optimization of thermal management system for Li-ion batteries using phase change material," *Appl Therm Eng*, vol. 131, pp. 766–778, Feb. 2018, doi:10.1016/J.APPLTHERMALENG.2017.12.055.

[30] M. Safdari, R. Ahmadi, and S. Sadeghzadeh, "Numerical investigation on PCM encapsulation shape used in the passive-active battery thermal management," *Energy*, vol. 193, p. 116840, Feb. 2020, doi:10.1016/j.energy.2019.116840.

[31] Y. Yang, L. Chen, L. Yang, and X. Du, "Numerical study of combined air and phase change cooling for lithium-ion battery during dynamic cycles," *Int J Therm Sci*, vol. 165, p. 106968, Jul. 2021, doi:10.1016/j.ijthermalsci.2021.106968.

[32] D. Kong, R. Peng, P. Ping, J. Du, G. Chen, and J. Wen, "A novel battery thermal management system coupling with PCM and optimized controllable liquid cooling for different ambient temperatures," *Energy Convers Manag*, vol. 204, no. Sept., p. 112280, 2020, doi:10.1016/j.enconman.2019.112280.

[33] P. Ping, Y. Zhang, D. Kong, and J. Du, "Investigation on battery thermal management system combining phase changed material and liquid cooling considering non-uniform heat generation of battery," *J Energy Storage*, vol. 36, p. 102448, Apr. 2021, doi:10.1016/j.est.2021.102448.

[34] K. Chen, J. Hou, M. Song, S. Wang, W. Wu, and Y. Zhang, "Design of battery thermal management system based on phase change material and heat pipe," *Appl Therm Eng*, vol. 188, p. 116665, Apr. 2021, doi:10.1016/j.applthermaleng.2021.116665.

[35] S. Abbas, Z. Ramadan, and C. W. Park, "Thermal performance analysis of compact-type simulative battery module with paraffin as phase-change material and flat plate heat pipe," *Int J Heat Mass Transf*, vol. 173, p. 121269, Jul. 2021, doi:10.1016/j.ijheatmasstransfer.2021.121269.

[36] Kong D, Peng R, Ping P, Du J, Chen G, Wen J. A novel battery thermal management system coupling with PCM and optimized controllable liquid cooling for different ambient temperatures. *Energy Convers Manag*, vol. 204, p. 112280, 2020. doi:10.1016/j.enconman.2019.112280.

12 Application of a Solid–Solid Nanocomposite PCM for Thermal Management of a Solar PV Panel

Praveen Bhaskaran Pillai, K. P. Venkitaraj, S. Suresh, Hafiz Muhammad Ali, Aswin G., Abhishek R., Aravind S., Arjun P. Suresh, and Arun Kumar C. S.

12.1 INTRODUCTION

Over the past few decades, there has been a significant increase in global energy consumption due to the combination of the world's population growth and impressive technical advancements. But the increasing energy demand, which is mostly satisfied by burning finite fossil fuels, has had serious negative effects on the environment, most notably the alarming acceleration of climate change and global warming [1,2]. The search for sustainable and regenerative energy sources is made more urgent by the knowledge that non-renewable energy reserves, which have supported civilization for generations, are now in danger of running out. In this urgently needed shift to a sustainable energy source, renewable resources have become a ray of hope [3]. Particularly in recent times, solar energy has received a lot of praise and attention [4]. This is because of its natural qualities of affordability, accessibility, and ready availability, which make it a desirable answer to the energy problems of our day [5]. This chapter explores the creative approaches and techniques that maximize the potential of solar energy while reducing the drawbacks of its erratic and temperature-sensitive nature as we set out to investigate the relationship between solar energy and phase change materials.

12.1.1 SOLID–SOLID PCMs

Solid–liquid phase change materials (PCMs) have long been a mainstay of innovation in the field of thermal energy storage (TES) and heat transfer applications [5]. These substances have the potential to effectively absorb and release thermal

DOI: 10.1201/9781003331957-12

energy [6]. But using them entails a number of serious drawbacks, such as concerns with volume change and the possibility of liquid leakage as they shift from the solid phase to the liquid phase. Researchers have turned to solid–solid PCMs, a class of materials that undergo phase transitions between solid states while absorbing and releasing considerable amounts of heat, in pursuit of a simpler approach that gets around these problems [6]. Solid–solid PCMs change between crystalline forms at certain, well-defined temperatures, in contrast to their liquid-phase counterparts [7]. Latent temperatures accompanying these transitions are comparable to those of the most efficient solid–liquid PCMs. Among the solid–solid PCMs, polyalcohols such as pentaerythritol (PE), pentaglycerine (PG), and neopentyl glycol (NPG) are notable candidates. The phase transition enthalpies of these materials are comparable to those of many paraffin-based PCMs [8]. These polyalcohols and their amine derivatives take on a body-centered tetrahedral molecular structure (α-phase) at lower temperatures. Nevertheless, they smoothly transform into a uniform face-centered cubic crystalline structure (γ-phase) as they approach the specified solid–solid phase transition temperature. Their exceptional thermal properties are a result of both this structural modification and the absorption of hydrogen bond energy [9].

Many researchers have examined the suitability and effectiveness of polyalcohols and amine derivatives as solid–solid PCMs for TES systems [9–13]. Research has yielded important insights into the underlying phase transition mechanisms in polyalcohols, such as those studied out by these researchers [8–13]. These understandings also include the fascinating phenomenon of subcooling, in which the energy required for the reformation of hydrogen bonds during cooling is less than that required for their initial breaking during the heating transition [13]. This characteristic of polyalcohols shows great potential for improving the efficacy and efficiency of solid–solid PCMs in TES applications [6,9,13,14].

12.1.2 Nanocomposite PCMs

One innovative approach to the development of enhanced TES systems is the application of PCMs. Because of their exceptional ability to both absorb and release large amounts of thermal energy, PCMs are useful for controlling temperature changes in a variety of applications. Low thermal conductivity, however, is an intrinsic constraint that has prevented PCM-based TES systems from reaching their full potential [15,16]. Because of this restriction, heat transfer has not been as efficient, and greater surface areas are now needed to compensate for these drawbacks [17,18].

Acknowledging the necessity of tackling this underlying issue, scientists have started investigating new avenues for improving PCM heat conductivity. Sharma and colleagues [19] conducted a noteworthy study on the advancements in organic solid–liquid PCMs, highlighting a common feature that is their intrinsic low thermal conductivity, which typically falls between 0.15 and 0.35 W/m·K [20]. In order to maximize heat transfer rates and maintain system compactness, novel approaches to increasing PCM thermal conductivity are crucial.

Adding thermally conductive nanoparticles to PCMs is one interesting approach that has attracted a lot of interest [16]. These nanocomposite PCMs increase the overall efficiency of TES systems by making use of the exceptional heat-conducting capabilities of nanoparticles [21]. The TES capabilities of both organic and inorganic PCMs enhanced with various nanoparticle additions have been the subject of extensive research. Although adding nanoparticles does improve heat conductivity, research shows that this improvement may also result in changed phase transition qualities. [21]

Carbon nanotubes (CNTs) and carbon nanofibers (CNFs), Al_2O_3, CuO, TiO_2, are the best candidates for nanoparticle-enhanced PCMs [19]. They have excellent thermo-physical properties that make them very suitable for use in PCM-based latent heat thermal energy storage (LHTES) systems [9–17]. The research on nanocomposite PCMs is explored in depth in this chapter, revealing the possibility of enhanced heat conductivity without sacrificing the essential phase transition features. Through examining the distinct interaction between nanomaterials and PCMs, our aim is to unleash the potential of the upcoming generation of high-efficiency TES devices.

12.1.3 PCMs for Thermal Management of PV Cells

The capacity of solar electricity has increased dramatically over the last 10 years, which is a significant turning point in the development of sustainable energy solutions. But even with the significant advancements, solar energy output still makes up only a small portion of the world's energy supply [4]. Currently, solar electricity only makes up a small portion of the global energy consumption, which highlights the enormous unrealized potential that needs to be realized [4]. Photovoltaic (PV) cells function in an intriguing way since they are part of a special system in which part of the desired solar energy is not converted into electricity [22]. The fact that a sizable amount of incident solar energy is absorbed as heat highlights the complex interplay between photovoltaics, wavelength selectivity, and efficiency. In order to maximize solar energy conversion and strengthen the role of solar power in the global energy paradigm, this chapter explores the complex interactions between solar cell technologies and their operating environments.

Our switch to sustainable energy sources has been made possible in a substantial way by the installation of solar panels that are readily available for purchase [4]. These panels are mostly made of two different types of silicon, monocrystalline and polycrystalline, each of which has advantages and characteristics of its own [4]. Due to their exceptional efficiency, monocrystalline solar panels have raised the benchmark for solar technology [4]. But it is important to recognize that, even in this sophisticated system, only a small portion of the abundant solar energy that reaches the panel is actually converted to electricity—usually between 15 and 20% [22].

The optimization of PV cell output is contingent upon the efficient regulation of operating temperatures [23,24]. A key factor in this pursuit of increased efficiency is adequate cooling. There are now two main types of cooling techniques: passive cooling and active cooling, each with their own benefits and drawbacks [24].

Active cooling applies the circulation of fluids, usually liquids or gases, to dissipate heat from PV cells [25]. Fans or water pumps are used to transfer coolant to the PV cell's surface. Although active cooling works well to lower temperatures, it can be cost prohibitive and energy intensive, which works against the primary goal of sustainability. When it comes to PV cell cooling, passive cooling techniques are the more effective alternative [21]. These techniques cover an extensive variety of approaches, all of which are intended to control temperature passively. In accordance with the concepts of sustainability and economic viability, passive cooling lowers the consumption of energy as well as running expenses.

Regardless of the various passive cooling methods readily accessible, this work explores the captivating field of PCMs. PV cell temperature regulation is made practical by PCMs, which are renowned for their ability to absorb and store latent heat during phase transitions. This method is both novel and efficient. Investigating the addition of PCMs to PV cell cooling arsenals reveals new approaches to increase energy efficiency while lowering operating costs and adverse environmental impacts [26]. The research conducted by Stritih, Uroš, and associates [27] illuminated the possibility of PCMs to usher in a new phase of PV cell performance enhancement. Their work was significant since it focused on simulation techniques and experimental configurations for effective heat extraction from PV cells. Most importantly, they conducted their research in actual outdoor climate scenarios rather than only in laboratory settings. Their findings clearly demonstrated the exceptional cooling power of PCMs, showing a convincing 35.6°C drop in temperature for PV-PCM panels compared to conventional panels. Moreover, the results highlighted a 9.2% average rise in power generation, highlighting the possibility of significant performance gains with PCM incorporation.

Nevertheless, there are certain difficulties in using PCMs for TES. One significant drawback is that many PCMs have naturally inadequate thermal conductivities, which can reduce total heat transfer rates and TES efficiency. This was thoroughly reviewed by Sharma et al. [22], who concentrated on organic solid–liquid PCMs and their various uses in TES systems. Their understanding of the distinct thermal characteristics of organic PCMs—which are characterized by low thermal conductivity values between 0.15 and 0.35 W/m·K—emphasized the need for surface area optimization for efficient heat transfer in PCM-based TES systems.

We are on the verge of a ground-breaking investigation as we set out on this chapter's expedition into the amazing field of using nanocomposite PCMs to maximize photovoltaic (PV) cell efficiency [28]. Novel approaches are required to address the basic problems with conventional photovoltaic cells, such as temperature control and energy conversion efficiency. We have proposed the idea of incorporating a high-performance composite solid–solid composite PCM, NGP with Al_2O_3 added, into the framework of photovoltaic cells. This creative approach makes use of PCMs' exceptional thermal characteristics to improve PV cells' electrical performance in addition to addressing the thermal limitations they face. The upcoming chapters will examine the procedure of producing PCM, characterize this composite, and analyze its effects on PV cell thermal and electrical performance.

Our goal as we move through this chapter is to reveal the revolutionary potential of solid–solid PCMs that are nanocomposites. Our research is motivated by the goal of increasing PV cell efficiency while addressing the difficulties related to heat control. By the end of the chapter, we hope to have shed light on the enormous potential of these solid–solid composite materials, bringing us one step closer to a time when PV cell technology will be distinguished by previously unheard of levels of efficiency, sustainability, and dependability.

12.2 MATERIALS AND METHODS

12.2.1 MATERIALS

Neopentyl glycol (NPG): NPG is a remarkable organic chemical compound denoted by the molecular formula $C_5H_{12}O_2$. What makes NPG particularly intriguing is its distinctive thermal behavior. NPG exhibits a solid–solid phase transition, transitioning from a face-centered lattice to a body-centered lattice, at a temperature of 41.9°C [15]. Additionally, it undergoes a solid-to-liquid phase transition at 129.1°C. At lower temperatures, NPG exists in a heterogeneous state, but as it approaches its phase transition temperature, it transforms into homogeneous face-centered cubic crystals characterized by high symmetry. This transition is accompanied by the absorption of a substantial amount of hydrogen bond energy. The NPG used in our study was in powdered form, with a high purity level of 99%, and was sourced from Alfa Aesar, USA. We selected NPG as our phase change material due to its solid–solid phase transition temperature (41.9°C), which aligns perfectly with the operational temperature range of our experimental setup. Additionally, its solid–liquid phase transition temperature (129.1°C) is well above the working temperature, making it an ideal choice for our purposes.

Al₂O₃ Nanoparticles: The second essential component of our research involves the incorporation of alumina nanoparticles into the NPG. These nanoparticles serve as a powerful additive to enhance the thermal properties of the composite material. The alumina nanoparticles we utilized in our study are characterized by their small particle size, falling within the range of 20–30 nm, and a remarkable purity level of 99.9% [7,14]. Furthermore, their molecular weight is 101.96 g/mol, as specified by the manufacturer. Al_2O_3 boasts several other noteworthy properties, including a high density of 3880 kg/m³, a specific heat capacity of 730 J/kgK, and an exceptional thermal conductivity of 40 W/mK. Additionally, alumina exhibits a low thermal expansion and a high melting temperature, making it an ideal candidate for integration with NPG. These properties collectively position alumina as a potent thermal enhancer in our composite material, allowing us to harness its unique attributes for our study's objectives.

Photovoltaic Cell: In this study, we employed a specific photovoltaic cell (PV cell), which played a central role in our investigations. The PV cell is crafted from monocrystalline silicon and is manufactured by 'Vikocell.' This high-performance PV cell boasts a maximum power rating of 4.7 W and generates an output voltage of 0.5 volts. Notably, it possesses a slender profile, measuring at just 200 micrometers in thickness, and covers a sizable area of 15.6 cm by 15.6 cm. The manufacturer of this PV cell has rigorously tested and specified its efficiency, which stands at an impressive 19.4%. This metric is indicative of the cell's remarkable capacity to convert incident solar energy into electricity, making it an ideal candidate for our study's objectives. As we delve deeper into our research, this PV cell serves as a cornerstone in our exploration of the integration of phase change materials and nanocomposites, with the ultimate goal of optimizing energy conversion efficiency and overall performance.

12.2.2 PREPARATION OF SOLID–SOLID NANOCOMPOSITE MATERIAL

The first step involved the precise addition of alumina nanoparticles to NPG, where weight fractions of 0.1%, 0.5%, and 1% were carefully measured using a high-precision weighing machine with an accuracy of 0.001 grams. The subsequent task revolved around achieving a homogeneous dispersion of these nanoparticles within the NPG matrix.

To achieve this, we employed a low-energy ball mill, which is shown in Figure 12.1, with specific parameters: 0.5 HP, 230 V, 50 Hz, and an operating speed of 300 rpm. The ball mill featured an internal diameter of 12 inches and a length of 13.5 inches, providing a total volume of 1 cubic foot. Notably, an opening of 4 inches in width extended along the full length of the drum. This ball mill operated at a controlled rate of 28–30 rpm for an approximate duration of 180 minutes.

It is important to emphasize that the dispersion process of alumina nanoparticles within NPG was entirely physical, facilitated by the ball mill. Consequently, the weight fractions of NPG combined with alumina remained heterogeneous. This heterogeneity stemming from mechanical mixing was instrumental in unravelling the intriguing variations in the solid–solid transition properties of NPG, especially in response to thermal cycling. This session forms a crucial foundation for our subsequent investigations, offering valuable insights into the behavior of our nanocomposite PCM.

FIGURE 12.1 Photograph of the low-energy ball mill.

12.3 CHARACTERIZATION OF SOLID–SOLID NANOCOMPOSITE PCMS

In the pursuit of comprehensive insights into our nanocomposite PCMs, required characterization techniques were diligently applied. The objective was to assess the thermal and chemical stability of the nano-enhanced PCMs over the course of 200 thermal cycles.

To achieve this, an array of tests was executed, including thermal cycling, differential scanning calorimetry (DSC), thermogravimetric analysis (TGA), and Fourier-transform infrared spectroscopy (FTIR). These tests provided invaluable data at both the outset and the conclusion of the 200 thermal cycles, facilitating an in-depth examination of the PCMs' performance under varying conditions. Furthermore, the thermos-physical properties of the PCM samples, such as specific heat and thermal conductivity, were methodically determined through the T-history method. This comprehensive suite of characterization techniques served as a pivotal foundation for our research, enabling us to unravel the intricate dynamics of our nanocomposite PCMs and to assess their suitability for the intended applications.

12.3.1 THERMAL CYCLING TEST

This study embarked on a meticulous examination of the cycling stability of various PCMs, including pure NPG and NPG augmented with Al_2O_3 nanoparticles at weight fractions of 0.1%, 0.5%, and 1%. This evaluation was facilitated through a rigorous thermal cycling test designed to assess the materials' resilience to repeated charge–discharge cycles.

The thermal cycling unit featured a laboratory hot plate equipped with a substantial power rating of 1.5 kW. The hot plate was capable of reaching temperatures up to 300°C and was meticulously monitored through a digital temperature controller. Precise temperature measurements were ensured by K-type thermocouples, which were rigorously calibrated using a handheld dry-well calibrator boasting a temperature accuracy of ± 0.25°C.

During the thermal cycling test, the PCMs underwent a repetitive sequence of rapid heating and cooling cycles within the specified temperature range. Each thermal cycle encompassed distinct phases. The process of thermal cycling is shown in Figure 12.2. The cycle was initiated at 30°C, where the heating process commenced at a controlled rate of 3°C per minute until a maximum temperature of 60°C was reached. Subsequently, the samples were maintained at 60°C for 5 minutes before being efficiently cooled back to 30°C, following the same controlled cooling rate of 3°C per minute. The samples were held at 30°C for an additional 5 minutes, and this entire thermal cycle was then repeated. Each complete cycle, marked by its detailed phases, required an average of 30 minutes to conclude. In pursuit of a comprehensive understanding of thermal stability, this procedure was iterated 200 times, allowing us to gauge the cumulative effects of cycling on the materials' thermal performance.

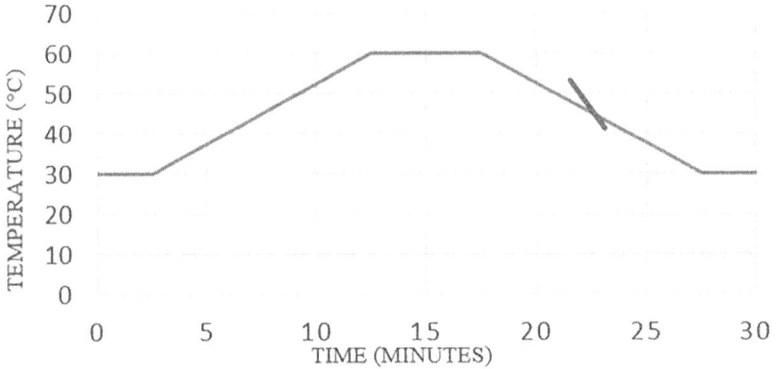

FIGURE 12.2   Thermal cycling process.

12.3.2 THERMOGRAVIMETRIC ANALYSIS (TGA)

In thermal storage and heatsink applications with PCMs, ensuring thermal stability is paramount. Particularly in the integration of PCMs with photovoltaic cells, it is imperative to assess the material's resilience to temperature variations, including exposure to a maximum temperature of 90°C [28]. The significance of this assessment lies in the operational dynamics of our application, where the PCM heat exchanger unit encounters cycles of heating and cooling. To rigorously evaluate the thermal stability of the PCM under conditions that replicate real-world usage (including repeated heating and cooling cycles), the PCM that has undergone thermal cycling testing is subjected to TGA.

TGA is a robust method of thermal analysis that systematically measures variations in the physical and chemical properties of materials as a function of temperature, with a constant heating rate [14]. This dynamic analysis subjects the sample to continuously increasing temperature conditions, typically employing a linear time-based heating rate. Throughout this analysis, the changes in sample mass resulting from various thermal events—such as phase transitions, decomposition, or oxidation—are meticulously tracked [30]. The typical outcome of TGA analysis is a plotted representation of mass change as a percentage versus temperature or time, as exemplified in Figure 12.3. These data provide invaluable insights into the material's behavior under varying thermal conditions, crucial for assessing its suitability for our specific application.

12.3.3 FOURIER-TRANSFORM INFRARED SPECTROSCOPY (FTIR)

Investigating the chemical stability of the PCM after nanoparticle integration and thermal cycling is imperative to ascertain its suitability for thermal energy applications. The potential occurrence of a chemical reaction between the PCM and nanoparticles could render them unsuitable for our intended use, as the samples will endure continuous heating and cooling processes. Hence, it is essential to evaluate the chemical stability of these samples following thermal cycling to detect any potential chemical alterations.

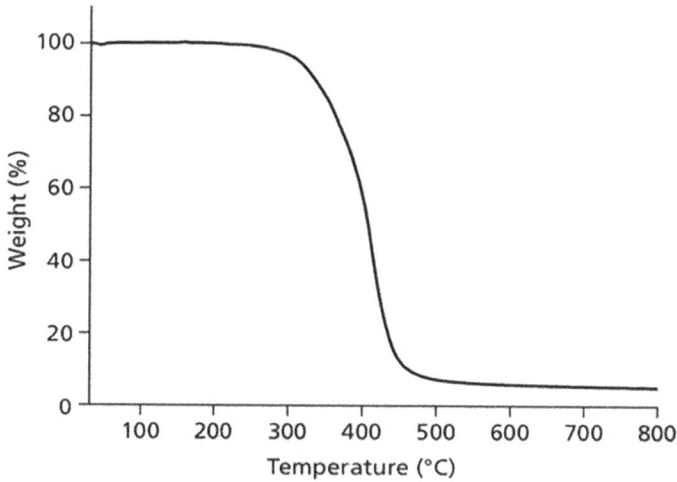

FIGURE 12.3 TGA curve.

FTIR spectroscopy, a valuable molecular spectroscopy technique, is employed for this purpose. FTIR spectroscopy exposes the samples to infrared radiation. When the frequency of the infrared radiation aligns with the natural frequency of the chemical bonds within the sample, it leads to an increase in the amplitude of molecular vibration, with the infrared radiation being absorbed. The output of FTIR analysis is a graph characterized by numerous peaks, each of which corresponds to specific chemical bonds present in the sample.

FTIR spectroscopy distinguishes between two distinct regions: the fingerprint region and the functional group region. The fingerprint region is unique to each compound, allowing for compound identification. Meanwhile, the functional group region is indicative of bond addition or dissociation. By comparing the FTIR spectrum of the NPG–alumina mixture with that of pure NPG, any changes in chemical composition can be identified. In particular, alterations in the wavenumbers associated with the peaks in the functional group region signify chemical reactivity. If these wavenumbers remain unaltered, it is indicative of chemical stability, confirming the PCM's suitability for our thermal energy applications.

12.3.4 Differential Scanning Calorimetry (DSC)

DSC is a vital method for characterizing the thermal properties of PCMs. DSC operates by measuring the heat difference required to raise the temperature of a sample of a reference material as a function of temperature [14]. The resulting DSC curve, which relates heat flow to temperature, offers crucial insights into various thermal parameters. These include transition temperatures, enthalpy associated with phase changes, onset, endset, and peak temperatures. The DSC curve presents distinctive peaks that correspond to phase transitions within the samples. In the case of solid–solid phase change materials like NPG, two peaks emerge: the solid–solid phase transition peak (or glass transition peak) and the solid–liquid phase transition peak (or fusion peak) [31]. The area beneath these peaks signifies the enthalpy of phase transition, while the onset temperature marks the initiation of phase transition.

DSC measurements were conducted using the NETZSCH DSC 204 instrument, featuring a temperature range extending from room temperature (RT in °C) to a maximum of 150°C. The PCM samples were subjected to heating at a rate of 10°C per minute, enabling the precise determination of their thermal properties.

12.3.5 T-History Method

We employed the T-history method, as introduced by Zhang et al. [29], to determine temperature-dependent properties, specifically heat capacity and thermal conductivity, across a range of samples. To ensure the application of the lumped heat capacity method, we aimed for a Biot number of less than 0.1 within our experimental setup. This involved comparing the temperature evolution of the samples with that of a reference material during the cooling process relative to ambient temperature.

For a reference material, it is preferable to use a substance with well-established thermal properties, such as water. In our study, we utilized glycerin as the reference material. The key advantage of the T-history method lies in its ability to accommodate PCMs intended for various applications, offering flexibility in terms of heating and cooling rates and temperature ranges. The instrumental setup is straightforward, and each T-history installation, like our own, is established individually.

We used 15×150 mm test tubes to contain both the samples and the reference material, applying the lumped system approach. In comparison to other techniques like conventional calorimetric methods, differential thermal analysis, and differential scanning calorimetry, the T-history method stands out due to its ability to simultaneously measure the heat of fusion, specific heat, and thermal conductivity of multiple PCM samples. Additionally, it allows for the observation of the phase change process for each PCM sample.

$$C_{p,p} = \frac{m_g C_{p,g} + m_t C_{p,t}}{m_p} + \frac{A_p}{A_g} - \frac{m_t}{m_p} C_{p,t} \tag{12.1}$$

$$H_m = \frac{m_g C_{p,w} + m_t C_{p,t}}{m_t} + \frac{A_p}{A_g}(T_o - T_s) \tag{12.2}$$

$$k = \left(1 + \frac{C_p(T_m - T_{\infty,w})}{H_m}\right) / 4\left(\frac{t_f(T_m - T_{\infty,w})}{\rho_p R^2 H_m}\right) \tag{12.3}$$

The equations used to determine specific heat and thermal conductivity values are shown as Equations 12.1–12.3. These equations consider factors such as the specific heat of glycerin and the test tube, area under the cooling curve, time of full solidification, heat of fusion, density of the sample, and mass of the glycerin sample, and test tube, as well as the thermal conductivity and various temperature parameters.

12.4 EXPERIMENTATION

12.4.1 DESIGN AND FABRICATION OF SETUP

The experimental setup and measurement unit were designed for the integration of the PV cell with PCM. The PV-PCM module was constructed using a 30 mm thick acrylic sheet with dimensions of 160×160×30 mm. At the top of the module, a 0.5 mm thick metal plate measuring 160×160 mm was affixed, serving as the platform for the PV cell placement. Within the acrylic enclosure, provisions were included for accommodating a copper coil, and the base of the enclosure was designed to be slidable, facilitating the loading of the PCM material. Figure 12.4 shows the details of fabricated experimental setup.

To ensure efficient and uniform heat transfer, a thin layer of highly conductive thermal paste was applied between the PV cell and the metal sheet. Additionally, the design incorporated holes for inserting thermocouples, which were connected to a data logger for monitoring and recording temperature data. Table 12.1 shows the detailed specifications of the PV cell used in this setup.

The experimental setup comprises three distinct PV cells, illustrated in Figures 12.5 and 12.6. These include the PV cell with a heatsink composed of NPG + 1% alumina (positioned on the left), the PV cell with a pure NPG heatsink (positioned in the middle), and the reference PV cell (positioned on the right side of the diagram). Each

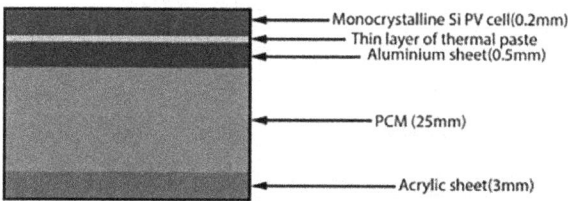

Monocrystalline Si PV cell(0.2mm)
Thin layer of thermal paste
Aluminium sheet(0.5mm)

PCM (25mm)

Acrylic sheet(3mm)

Integration of setup

Copper tube inserted in the container to perform discharge test of PCM

Photograph of assembled PCM container with heat extraction coil

FIGURE 12.4 Experimental setup.

TABLE 12.1
Specifications of the PV Cell Used

Parameter	Details/Value
Size	156 × 156 mm
Brand name	Vikocell
Nominal capacity	0.5 V
Material	Monocrystalline silicon
Max. power	4.7 W
Model number	NS6ML
Grade	Grade A
Color	Blue
Thickness	200 μm
Efficiency	19.4%

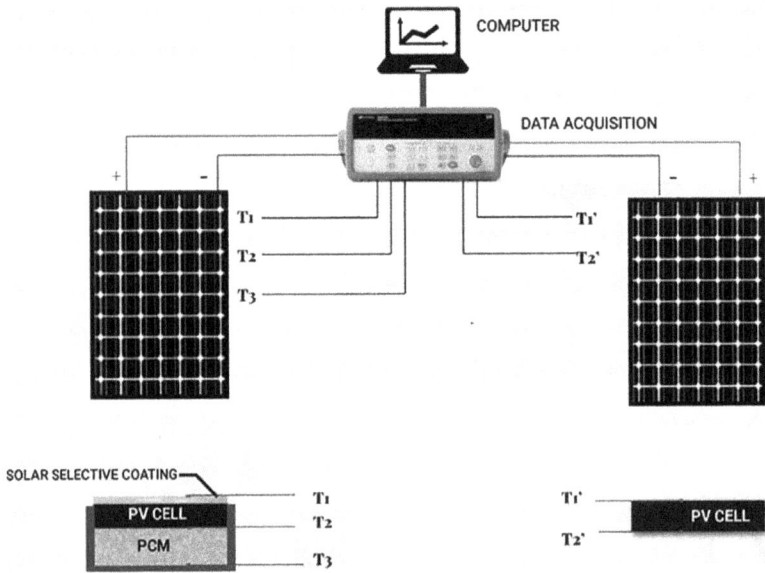

FIGURE 12.5 Schematic diagram of the experimental setup for data acquisition.

of the PV-PCM modules and PV cells was equipped with thermocouples to capture temperature data at various locations, including the surfaces, inlet, and outlet of the heat exchanger. Ambient temperature measurements were also recorded, along with voltage and current readings for each cell. Data collection occurred at 1-second intervals. The PCM-based heatsink was housed within an acrylic box with an embedded copper tube for effective heat transfer. The top section of the box was secured with a

FIGURE 12.6 Actual photograph of the setup.

thin aluminum sheet, onto which the PV cell was affixed using thermally conductive paste. This setup ensured efficient heat transfer to the PCM. Before commencing the experiment, simultaneous measurements of power output were taken from all PV cells to confirm their performance consistency. It was determined that the power output from each PV cell was nearly identical, with only a negligible margin of error, affirming the reliability of the experiment without the need for further calibration.

12.4.2 Thermal Performance Analysis

The evaluation of the thermal system's performance entails the comprehensive assessment of heat absorption and dissipation throughout the system. This assessment is carried out by considering various parameters derived from numerous tests and experiments. To quantify the heat absorbed by the PCM, temperature measurements collected from both the upper and lower boundaries of the PCM layer are utilized. The total heat absorbed by the PCM encompasses both sensible heat and latent heat contributions, providing a comprehensive understanding of its thermal behavior.

12.4.3 Electrical Performance Analysis

The assessment of the PV cells' electrical performance is conducted to gauge how the incorporation of different heatsinks, namely pure NPG and NPG with alumina by weight, affects the PV cells' efficiency. During experimentation, we closely monitor the temporal evolution of each PV cell's surface temperature. At regular intervals, the data logger records both current and voltage parameters. By integrating the data on current, voltage, and surface temperature, we can accurately calculate and evaluate the electrical performance of the PV cells.

12.5 RESULTS AND DISCUSSION

12.5.1 CHARACTERIZATION RESULTS

12.5.1.1 Thermogravimetric Analysis (TGA)

To gain insight into the impact of alumina nanoparticles on the thermal degradation behavior of NPG, we conducted TGA under normal atmospheric conditions. The TGA experiments were executed at the Sophisticated Test & Instrumentation Centre, Cochin University of Science and Technology, Kerala, India, employing a Perkin Elmer Diamond TG/DTA instrument. The samples, both before and after undergoing 200 thermal cycles, were subjected to TGA. During the TGA procedure, the samples were heated with a uniform heating rate of 20°C per minute, ranging from 40°C to 700°C. The continuous recording of sample weight concerning temperature or time yielded valuable insights into their thermal degradation characteristics. Mass loss graphs that depict the outcomes of these TGA analyses are visually presented in Figures 12.7–12.10.

Figure 12.7 illustrates the TGA curves obtained for pure NPG both before and after undergoing 200 thermal cycles. In the TGA experiments, sample weights of 8.754 mg and 6.448 mg were employed. Notably, the weight losses observed were minimal, with a decrease of less than 0.7028% and 0.4738% recorded for samples before and after thermal cycling, respectively. Importantly, these weight losses remained well below the working temperature range of 75°C. The thermal degradation behavior was further characterized by identifying specific temperature points. For the samples before thermal cycling, the 1%, 10%, 50%, and 90% degradation temperatures were determined as 82.3°C, 130.4°C, 166.4°C, and 182.5°C, respectively. Following the thermal cycling process, these degradation temperatures exhibited slight variations, with values of 88.6°C, 127.8°C, 155°C, and 172.3°C, demonstrating the resilience of the PCM to thermal cycling. Ultimately, both sample sets experienced complete degradation at elevated temperatures, reaching 190°C before thermal cycling and 176.5°C after thermal cycling. This quantitative analysis of TGA results provides essential insights into the thermal stability and degradation behavior of pure NPG, critical for evaluating its suitability as a phase change material in practical applications, such as photovoltaic cell thermal management.

Figure 12.8 presents the TGA curves depicting the thermal degradation characteristics of NPG + 0.1% Al_2O_3 both before and after the imposition of 200 thermal cycles. To investigate this, samples with initial weights of 9.161 mg and 13.765 mg were employed in the TGA analysis. It is noteworthy that the observed weight losses were minimal, registering at values below 0.5574% and 0.5437% for samples before and after the thermal cycling procedure, respectively. Importantly, these weight losses remained well below the designated working temperature range of 75°C. The thermal degradation behavior was further quantified by identifying specific temperature thresholds. For the samples before thermal cycling, the 1%, 10%, 50%, and 90%

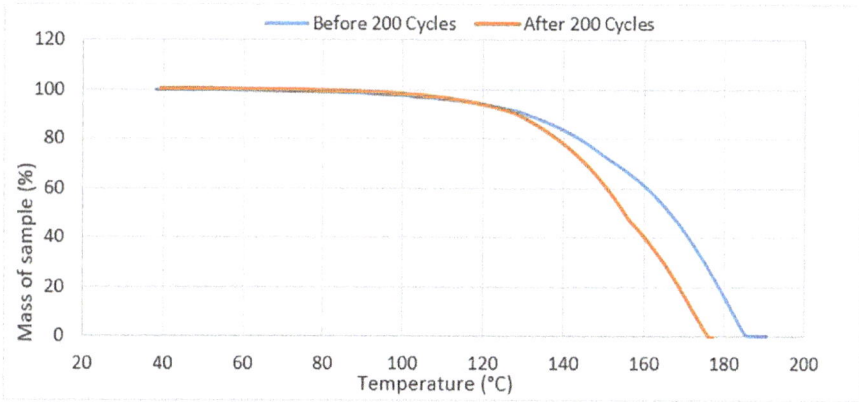

FIGURE 12.7   TGA curve for pure NPG before and after 200 thermal cycles.

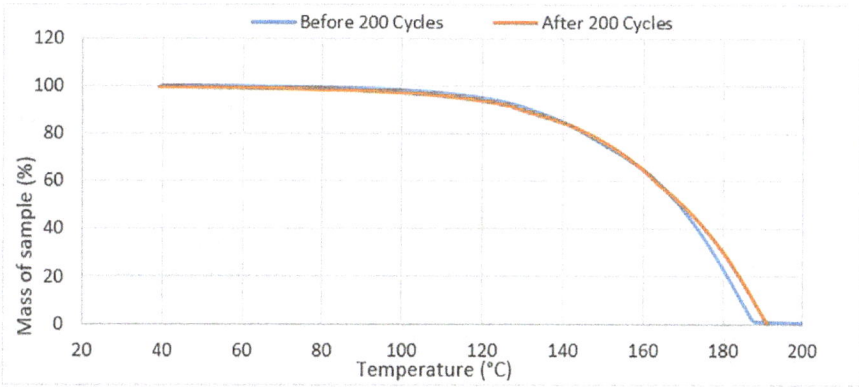

FIGURE 12.8   TGA curve for NPG + 0.1% Al_2O_3 before and after 200 thermal cycles.

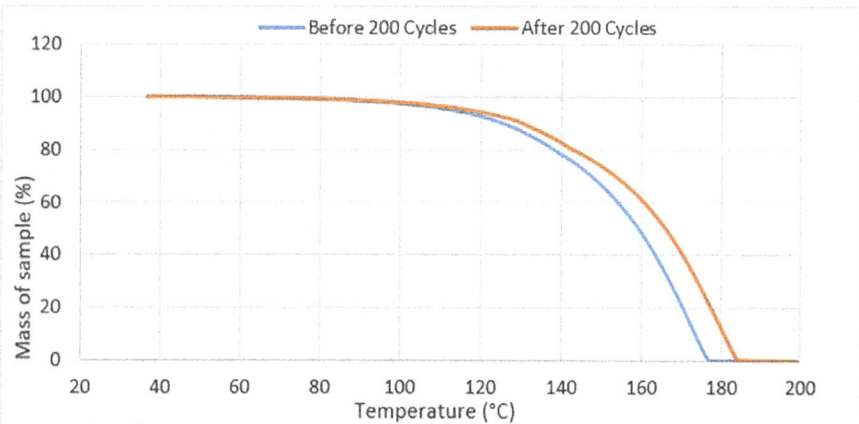

FIGURE 12.9   TGA curve for NPG + 0.5% Al_2O_3 before and after 200 thermal cycles.

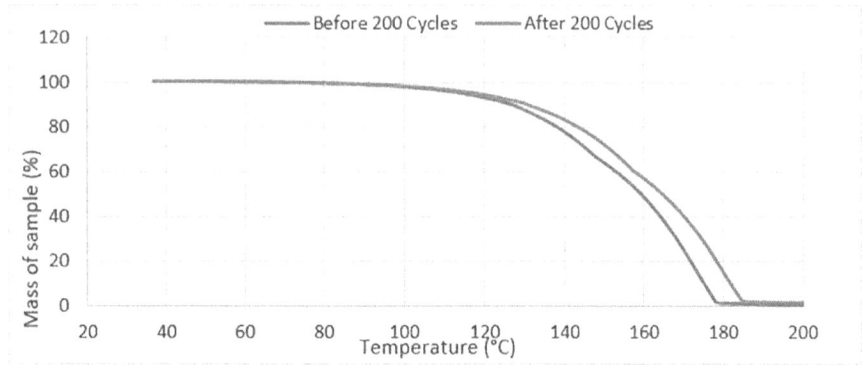

FIGURE 12.10   TGA curve for NPG + 1% Al$_2$O$_3$ before and after 200 thermal cycles.

degradation temperatures were determined to be 87.5°C, 132°C, 167°C, and 184.5°C, respectively. Following the thermal cycling process, there were slight adjustments in these degradation temperatures, with values of 88°C, 129.3°C, 169.5°C, and 187.2°C, suggesting that the incorporation of 0.1% Al$_2$O$_3$ had a minimal impact on the material's thermal stability. Ultimately, for both sets of samples, complete degradation was observed at higher temperatures. Specifically, these temperatures reached 226.8°C before thermal cycling and 191°C after thermal cycling.

In Figure 12.9, the TGA curves for NPG + 0.5% Al$_2$O$_3$ samples are presented both before and after undergoing 200 thermal cycles. The samples, with initial weights of 5.338 mg and 8.399 mg, were analyzed using TGA. It is crucial to note that the observed weight losses remained minimal, recording values below 0.6665% and 0.6353% for samples before and after the thermal cycling process, respectively. These values were well within the defined working temperature range of 75°C, ensuring the material's thermal stability during practical applications. To further elucidate the thermal degradation characteristics, specific temperature thresholds were identified for the samples. Prior to thermal cycling, the 1%, 10%, 50%, and 90% degradation temperatures were established at 82.7°C, 125.2°C, 159°C, and 173.5°C, respectively. Post-thermal cycling, slight adjustments were noted in these degradation temperatures, with values of 84.3°C, 130.3°C, 166.2°C, and 180.5°C. These findings suggest that the addition of 0.5% Al$_2$O$_3$ had a minimal impact on the material's thermal stability, with both sets of samples degrading completely at higher temperatures. Specifically, these complete degradation temperatures reached 183°C before thermal cycling and 198°C after thermal cycling.

In Figure 12.10, the TGA curves for NPG + 1% Al$_2$O$_3$ samples are displayed both before and after undergoing 200 thermal cycles. These samples, weighing 8.754 mg and 6.448 mg initially, underwent TGA analysis. Notably, the weight losses observed remained minimal, measuring below 0.4928% and 0.5224% for samples before and after the thermal cycling process, respectively. These results fall well within the defined working temperature range of 75°C, signifying the material's robust thermal stability during practical applications. Furthermore, to characterize the thermal degradation behavior, specific temperature thresholds were identified for the samples. Before thermal cycling, the 1%, 10%, 50%, and 90% degradation temperatures were observed at 86°C, 126°C, 159.1°C, and 175°C, respectively. After the thermal cycling process, slight variations were recorded in these degradation temperatures, with values reaching 86.3°C, 130°C, 164.6°C, and 181.8°C. These findings suggest that the inclusion of 1% Al_2O_3 caused only minimal effects on the material's thermal stability. Both sets of samples exhibited complete degradation at higher temperatures, reaching 247°C before thermal cycling and 297°C after thermal cycling.

In summary, the TGA results for all the samples consistently reveal that the observed degradation, which remained under 1%, occurred below the designated working temperature range of 80°C. This crucial finding strongly supports the conclusion that the examined samples, including NPG and various NPG–Al_2O_3 composites, demonstrate exceptional thermal stability within the operational boundaries of photovoltaic cells. This notable thermal resilience positions these materials as promising candidates for efficient photovoltaic cell thermal management applications, where stability under varying temperatures is of paramount importance.

12.5.1.2 Fourier-Transform Infrared (FTIR) Spectroscopy

To assess the extended chemical stability of NPG blended with Al_2O_3, FTIR spectroscopy was employed. FTIR spectra were acquired for all PCM samples within the wave number range of 400 to 4000 cm^{-1} using an infrared spectrophotometer (Frontier FT-IR/FIR, PerkinElmer). Figures 12.11–12.14 depict the FTIR spectra of NPG samples, both before and after cycling, providing insight into their chemical stability over prolonged usage.

Figures 12.11 and 12.12 show the FTIR test results for pure NPG before and after thermal cycling. In the absence of thermal cycling, FTIR analysis revealed distinct peaks in the functional group region at wavenumbers 3286, 2956, 2934, 2871, and 2715. These characteristic peaks signify the presence of two O-H and three C-H bonds. After 200 thermal cycles, the functional group region exhibited peaks at wavenumbers 3312, 2958, 2935, 2872, and 2716. Importantly, the functional

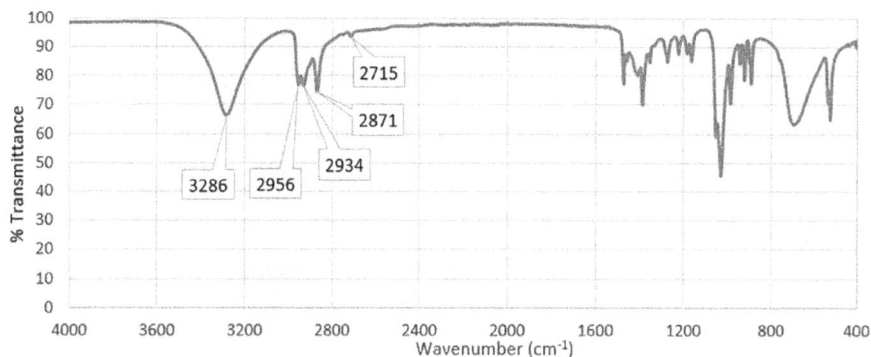

FIGURE 12.11   FTIR spectra of pure NPG without thermal cycling.

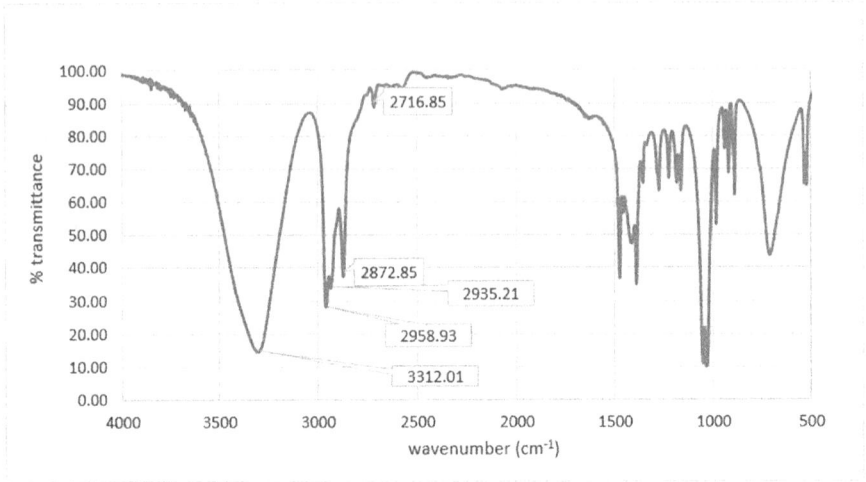

FIGURE 12.12 FTIR spectra pure NPG after 200 thermal cycles.

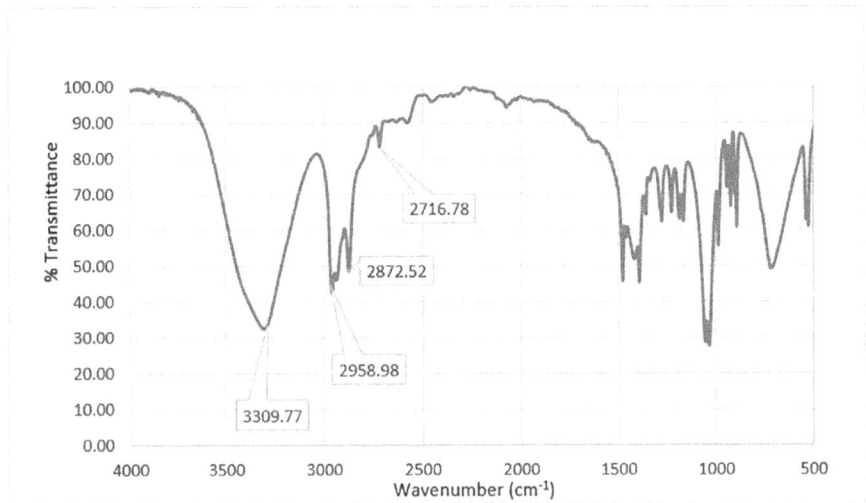

FIGURE 12.13 FTIR spectra of NPG +1% Al_2O_3 before thermal cycling.

groups responsible for these peaks, signifying the presence of two O-H and three C-H bonds, remained unaltered following the thermal cycles.

Figure 12.13 and 12.14 show the FTIR test results for NPG + 1 % Al_2O_3. In the initial state, FTIR analysis revealed distinctive peaks in the functional group region at wavenumbers 3309, 2958, 2872, and 2716. These peaks correspond to the presence of two O-H and two C-H bonds within the functional groups that absorbed the IR radiations. Following 200 thermal cycles, the functional group region maintained its integrity, exhibiting peaks at wavenumbers 3315, 2958, 2872, and 2716. These results

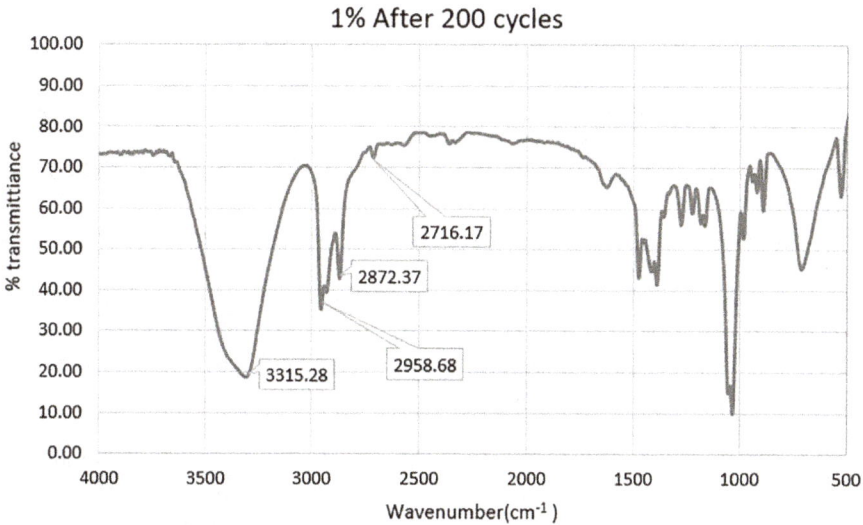

FIGURE 12.14 FTIR spectra of NPG +1% Al_2O_3 after 200 thermal cycles.

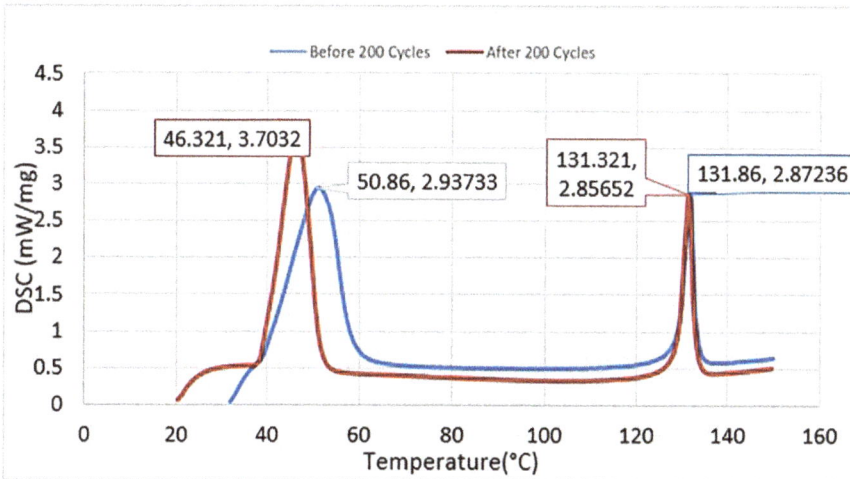

FIGURE 12.15 DSC curve of pure NPG before and after thermal cycling.

indicate that the chemical composition of the material, characterized by the presence of two O-H and two C-H bonds, remained unaltered after the thermal cycling process.

12.5.1.3 Differential Scanning Calorimetry (DSC)

In the context of DSC heating curves for pure NPG, as illustrated in Figure 12.15, the behavior of the material before and after 200 thermal cycles was evaluated. For samples prior to thermal cycling, the solid–solid phase transition commenced at

38.86°C and concluded at 65.86°C, with the peak transition occurring at 51.86°C. Simultaneously, the solid–liquid phase transition had an onset at 118.86°C, concluded at 136.86°C, and exhibited a peak transition at 131.86°C. Following 200 thermal cycles, modifications in the phase transitions were observed. The solid–solid phase transition was now initiated at 36.32°C and terminated at 55.32°C, featuring a peak transition at 46.32°C. The solid–liquid phase transition started at 121.32°C, ceased at 135.32°C, and reached its peak transition at 131.32°C.

Furthermore, it was evident that the enthalpy of phase transition, especially during the solid–solid phase transition, was higher than that of the solid–liquid phase transition for samples before thermal cycling, with values of 130.57 J/g and 42.22 J/g, respectively. However, after 200 thermal cycles, a reduction in the enthalpy values was observed for both the solid–solid and solid–liquid phase transitions. Specifically, after thermal cycling, the enthalpy of phase transition was 125.88 J/g for the solid–solid transition and 41.72 J/g for the solid–liquid transition.

The DSC heating curves of the sample containing 0.1% Al_2O_3, both before and after 200 thermal cycles, are depicted in Figure 12.16. Examination of these curves revealed variations in phase transitions. Before thermal cycling, the solid–solid phase transition was initiated at 38.68°C and concluded at 62.65°C, with the peak transition occurring at 50.65°C. Simultaneously, the solid–liquid phase transition commenced at 121.65°C, terminated at 137.65°C, and exhibited a peak transition at 131.65°C. Upon completing 200 thermal cycles, changes in phase transitions were observed. Specifically, the solid–solid phase transition began at 37.68°C, concluded at 61.68°C, and displayed a peak transition at 46.68°C. The solid–liquid phase transition now started at 121.68°C, ended at 137.68°C, and reached a peak transition at 41.27°C.

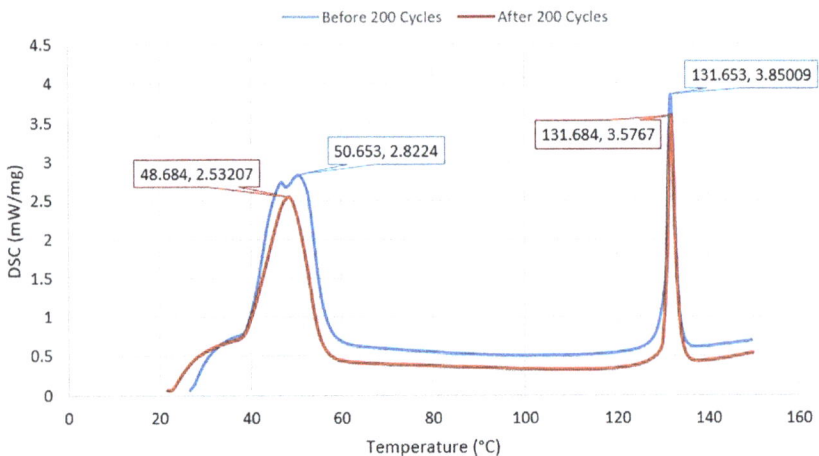

FIGURE 12.16 DSC curve of NPG + 0.1% Al_2O_3 before and after thermal cycling.

Moreover, it was noted that prior to thermal cycling, the enthalpy of phase transition, especially in the case of solid–solid phase transition, was measured at 127.40 J/g, whereas the solid–liquid phase transition exhibited an enthalpy value of 43.61 J/g. However, after 200 thermal cycles, reductions in the enthalpy values were observed for both the solid–solid and solid–liquid phase transitions. Specifically, after thermal cycling, the enthalpy of phase transition was 102.20 J/g for the solid–solid transition and 41.27 J/g for the solid–liquid transition.

The DSC heating curves for the sample containing 0.5% Al_2O_3, both prior to and after 200 thermal cycles, are visually represented in Figure 12.17. These curves revealed noteworthy changes in phase transitions. Before thermal cycling, the solid–solid phase transition was initiated at 37.35°C and terminated at 58.35°C, with its peak transition occurring at 46.35°C. Concurrently, the solid–liquid phase transition commenced at 121.35°C, ceased at 137.35°C, and exhibited a peak transition at 131.35°C. Upon the completion of 200 thermal cycles, the DSC curves demonstrated notable modifications in these phase transitions. Specifically, the solid–solid phase transition was observed to initiate at 37.82°C and cease at 60.82°C, with the peak transition occurring at 50.82°C. Similarly, the solid–liquid phase transition commenced at 121.82°C, terminated at 136.82°C, and displayed a peak transition at 131.82°C.

The enthalpy values of these phase transitions before thermal cycling were found to be 136.20 J/g for the solid–solid transition and 45.88 J/g for the solid–liquid transition. After 200 thermal cycles, there was a reduction in the enthalpy values for both the solid–solid and solid–liquid phase transitions. Specifically, after thermal cycling, the enthalpy of the solid–solid phase transition was measured at 104.40 J/g, while the solid–liquid phase transition exhibited an enthalpy value of 39.45 J/g.

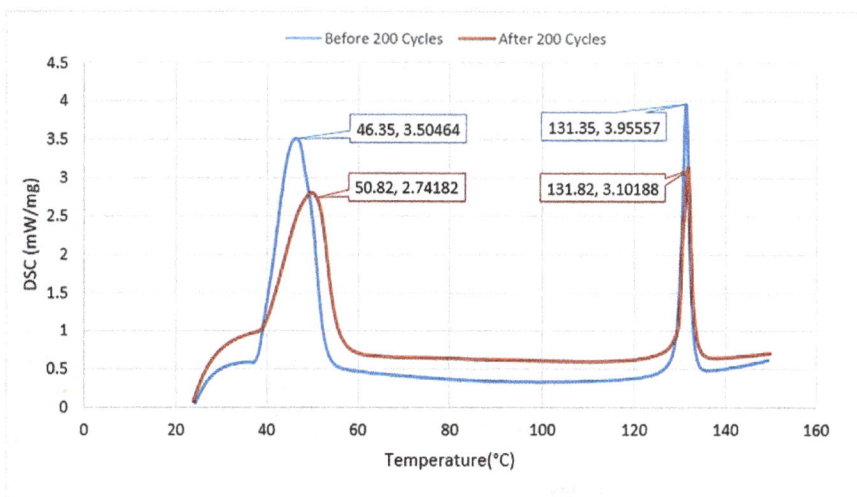

FIGURE 12.17   DSC curve of NPG + 0.5% Al_2O_3 before and after thermal cycling.

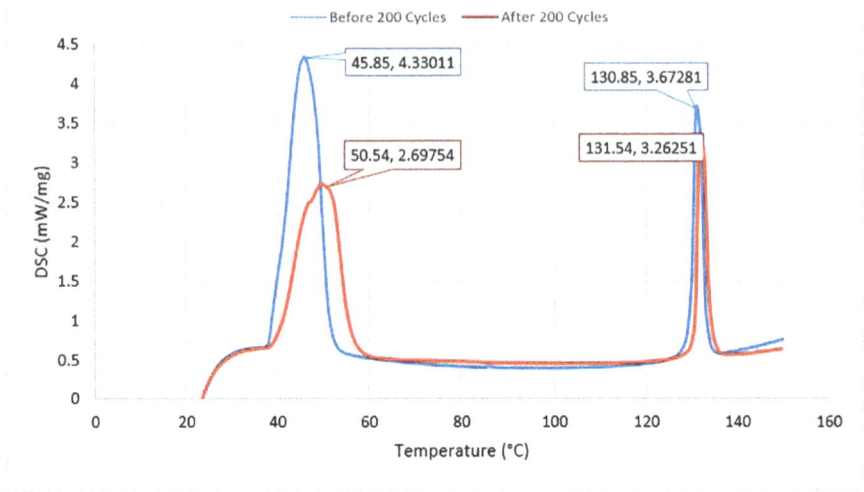

FIGURE 12.18   DSC curve of NPG + 1% Al$_2$O$_3$ before and after thermal cycling.

Figure 12.18 displays the DSC heating curves for the PCM sample containing 1% Al$_2$O$_3$ both before and after 200 thermal cycles, providing insights into the phase transition characteristics. Before thermal cycling, the solid–solid phase transition had an onset temperature of 36.85°C, concluding at 55.85°C, with a peak transition at 45.85°C. Simultaneously, the solid–liquid phase transition was initiated at 124.85°C, terminating at 135.85°C and featuring a peak transition at 130.85°C. Following 200 thermal cycles, the DSC heating curves exhibited notable changes in the phase transition behavior. The solid–solid phase transition was now initiated at 37.54°C, concluding at 60.54°C, with a peak transition at 49.54°C. The solid–liquid phase transition commenced at 124.54°C and ceased at 138.54°C, with a peak transition at 130.85°C.

Quantitatively, before thermal cycling, the enthalpy of the solid–solid phase transition was measured at 150.98 J/g, while the solid–liquid phase transition exhibited an enthalpy of 49.03 J/g. Following 200 thermal cycles, these enthalpy values decreased, with the solid–solid phase transition enthalpy measuring 120.48 J/g and the solid–liquid phase transition enthalpy recorded at 43.73 J/g. It is worth noting that the PCM with 1% alumina nanoparticles demonstrated a notably high solid–solid phase transition enthalpy of 150.95 J/g. Compared to pure NPG (130.57 J/g), this represents a substantial 15.63% enhancement in the enthalpy of the solid–solid phase transition attributed to the inclusion of 1% alumina. Furthermore, the solid–liquid phase transition enthalpy increased from 42.22 J/g to 49.03 J/g for this sample. These observations highlight the positive impact of alumina nanoparticles on the thermal properties of the PCM.

12.5.1.4   T-History Analysis

The T-history method serves as a vital tool for determining the specific heat capacity and thermal conductivity of the samples. The outcomes of the T-history experiment reveal an interesting trend: both the thermal conductivity and specific heat capacity of the samples exhibited a reduction as the number of thermal cycles increased. Moreover, the concentration of alumina played a significant role in this behavior, with thermal conductivity generally increasing while specific heat capacity decreased.

For a comprehensive understanding of the impact on thermal characteristics, Figures 12.19 and 12.20 visually depict the overall effects. Further quantification and analysis of the percentage changes in thermal properties arising from the inclusion of alumina nanoparticles and the influence of thermal cycling are presented in Tables 12.2 and 12.3, respectively. These findings shed light on the dynamic interplay between thermal conductivity, specific heat capacity, alumina concentration, and the effects of thermal cycling.

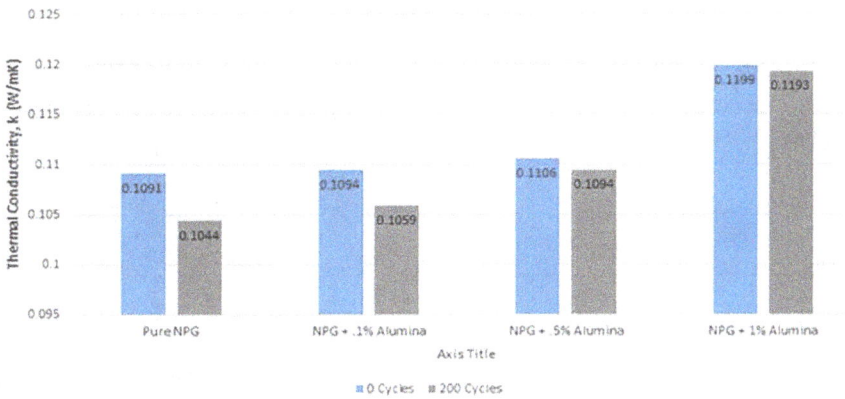

FIGURE 12.19 Thermal conductivity of NPG samples.

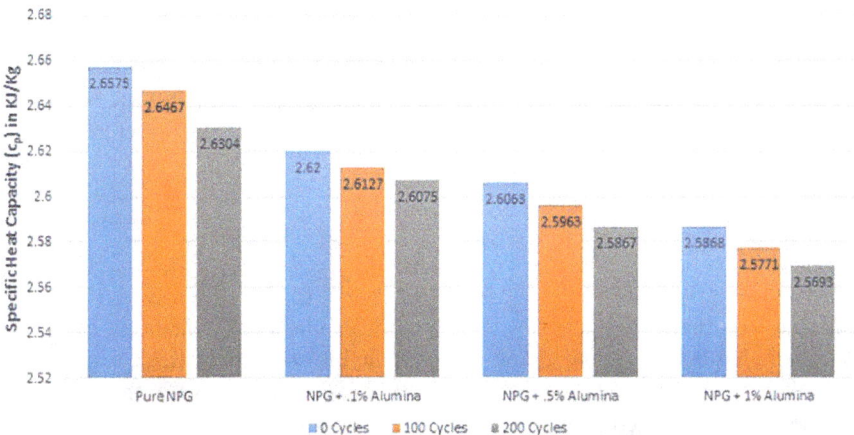

FIGURE 12.20 Specific heat capacity values of PCM samples.

TABLE 12.2
Comparison of Specific Heat and Thermal Conductivity before Thermal Cycling

Sample	% Decrease in C_p	% Increase in k
NPG + 0.1% Al$_2$O$_3$	1.411	0.275
NPG + 0.5% Al$_2$O$_3$	1.927	1.375
NPG + 1% Al$_2$O$_3$	2.660	9.899

TABLE 12.3
Comparison of Specific Heat and Thermal Conductivity after Thermal Cycling

Samples after 200 thermal cycles	% Decrease in C_p	% Decrease in k
Pure NPG	1.020	4.308
NPG + 0.1% Al$_2$O$_3$	0.477	3.199
NPG + 0.5% Al$_2$O$_3$	0.752	1.085
NPG + 1% Al$_2$O$_3$	0.677	0.500

Among the various samples, NPG with the addition of 1% alumina demonstrated the most significant variations, exhibiting a 2.66% change in specific heat capacity (C_p) and a 9.9% shift in thermal conductivity (k). Remarkably, these variations due to thermal cycling remained relatively modest, each below 0.7%. Conversely, pure NPG experienced the most substantial reduction in thermal conductivity during thermal cycling (4.3%), while NPG with 1% alumina content showed only a minimal 0.5% change in thermal conductivity.

In summary, the results underscore that NPG with 1% alumina exhibited the least thermal variation during thermal cycling compared to NPG with other alumina concentrations. Notably, the inclusion of alumina led to an enhancement in the thermal conductivity of the PCM, albeit accompanied by a reduction in specific heat. This suggests that NPG with 1% alumina holds promise for applications where consistent and reliable thermal performance is paramount.

12.5.2 PERFORMANCE TEST RESULTS

12.5.2.1 Thermal Performance of the PV Cell

The heat absorbed and released by the system with a heatsink containing NPG with 1% wt of alumina is calculated and tabulated in Tables 12.4 and 12.5, respectively. In evaluating the thermal performance of PV cells with pure NPG, the calculated heat absorbed amounted to 26.118 kJ. Conversely, for PV cells integrated with NPG containing 1% alumina, the heat absorbed totaled 30.199 kJ. Following 200 thermal

TABLE 12.4
Heat Absorbed during Charging

Sample	Heat absorbed (kJ)
Pure NPG	26.118
NPG+1% Al$_2$O$_3$	30.199
% increase in heat absorbed	15.626

TABLE 12.5
Heat Discharged during Discharging

Sample	Heat rejected (kJ)
Pure NPG	17.80
NPG+1% A$_2$O$_3$	20.31
% increase in the heat rejected	14.08

cycles, the heat absorbed by the heatsink containing pure NPG diminished to 17.8 kJ, while the heat rejected by the heatsink comprising NPG with 1% alumina decreased to 20.31 kJ. Comparing the PV-PCM setup with 1% alumina to pure NPG, it was evident that the heat absorbed increased by 15.62%, and heat release was augmented by 14.08%. This rise in both heat absorption and heat release can be attributed to the increased thermal conductivity resulting from the addition of alumina. These findings underscore the positive impact of enhanced thermal conductivity on the efficiency of heat transfer in the PV-PCM system.

12.5.2.2 Electrical Performance of the PV Cell

Table 12.6 provides an overview of various performance metrics, including voltage, current, power output, surface temperature, and the percentage change in power output concerning the reference cell. The PV-PCM module equipped with NPG and 1% alumina exhibited a voltage output of 0.53 V, a current output of 9.53 A, and a power output of 5.09 W. Notably, the percentage change in power output relative to the reference cell amounted to 3.12%. Additionally, the surface temperature of the cell, enhanced with the NPG and 1% alumina heatsink, registered at 51.31°C, marking a 7.20°C reduction compared to the reference cell devoid of a heatsink.

Conversely, the PV-PCM module employing pure NPG delivered a voltage output of 0.52 V, a current output of 9.62 A, and a power output of 4.93 W. Here, the percentage change in power output concerning the reference cell was 2.91%. Moreover, the cell's surface temperature with the pure NPG heatsink measured 51.31°C, reflecting a 6.51°C decrease in temperature relative to the reference cell without a heatsink. This temperature reduction can be attributed to the efficient heat absorption facilitated by the heatsink. Interestingly, the results showcased an average 0.44% increase in power

TABLE 12.6
Performance Matrix

Type	Voltage (V)	Current (A)	Power (W)	% Calibrated change in power from the reference	Cell temperature (°C)	Temperature difference from the reference cell (°C)
PV-PCM of 1% Al₂O₃ + NPG	0.5350	9.5223	5.0944	3.1274	51.3191	7.2064
PV-PCM of pure NPG	0.5276	9.6242	4.9334	2.9166	52.0097	6.5158
Reference cell	0.5126	9.3515	4.7936	0.0000	58.5255	0.0000

output for each degree reduction in cell temperature. These findings underscore the significant impact of heatsink integration on enhancing the electrical performance and temperature management of the PV cells.

12.6 CONCLUSION

The primary objective of the study presented in this chapter was to enhance the performance of monocrystalline silicon PV cells through the integration of PCM. Our research endeavors have successfully achieved this objective through a comprehensive experimental approach. Monocrystalline silicon PV cells were selected as our focus, given their sensitivity to surface temperature fluctuations.

Our experimental methodology involved the integration of NPG as a heat-absorbing medium on the back of the PV cells. Furthermore, we explored the augmentation of NPG with Al_2O_3 nanoparticles to enhance the cooling efficiency. We meticulously prepared three distinct combinations of NPG and Al_2O_3 nanoparticles, evaluating their suitability for experimentation. These combinations, featuring alumina nanoparticle weight proportions of 0.1%, 0.5%, and 1%, were subjected to 200 thermal cycles and rigorously tested to assess thermal and chemical stability. Notably, we also examined changes in thermal conductivity, latent heats, and transition temperatures employing DSC and T-history tests.

The key findings can be summarized as follows:

1. The results underscore the significant impact of temperature reduction on PV cell output power, indicating an average increase of 0.44% in power output for every degree reduction in cell surface temperature.
2. T-history test results demonstrated a reduction in specific heat by 2.6% and an increase in thermal conductivity by 9.89% upon the addition of Al_2O_3 nanoparticles to NPG. Importantly, these thermal properties remained relatively stable even after thermal cycling.
3. Thermogravimetric analysis revealed that pure NPG exhibited thermal instability beyond 82°C, while the introduction of Al_2O_3 nanoparticles further lowered this threshold to 65°C.

4. DSC results highlighted a 3.6% reduction in transition enthalpy for pure NPG after 200 thermal cycles. This reduction notably increased to 20.2% upon the addition of Al_2O_3 nanoparticles.
5. The FTIR spectra of both pure NPG and NPG with Al_2O_3 nanoparticles remained consistent before and after thermal cycling, signifying that thermal degradation did not induce any chemical alterations within the samples.

In conclusion, our study successfully enhances PV cell performance through PCM integration, demonstrating the considerable potential of NPG and Al_2O_3 nanoparticle blends. The findings not only underscore the pivotal role of temperature management in optimizing PV cell efficiency but also offer insights into the thermal and chemical characteristics of the materials involved. These outcomes provide valuable contributions to the field of renewable energy and the development of advanced cooling solutions for PV systems.

ACKNOWLEDGMENT

The authors gratefully acknowledge the financial support provided by the Centre for Engineering Research and Development (CERD), APJ Abdul Kalam Technological University, Thiruvananthapuram, Kerala, India under Sanction Order No: KTU/RESEARCH 2/2590/2018, dated 29.05.2018. This support was instrumental in conducting the experimental investigation detailed in this chapter.

REFERENCES

1. Al-Shahri OA, Ismail FB, Hannan MA, Lipu MH, Al-Shetwi AQ, Begum RA, Al-Muhsen NF, Soujeri E. Solar photovoltaic energy optimization methods, challenges and issues: A comprehensive review. *Journal of Cleaner Production*. 2021 Feb 15;284:125465.
2. Mohsin M, Rasheed AK, Sun H, Zhang J, Iram R, Iqbal N, Abbas Q. Developing low carbon economies: An aggregated composite index based on carbon emissions. *Sustainable Energy Technologies and Assessments*. 2019 Oct 1;35:365–374.
3. Heffron R, Halbrügge S, Körner MF, Obeng-Darko NA, Sumarno T, Wagner J, Weibelzahl M. Justice in solar energy development. *Solar Energy*. 2021 Apr 1;218:68–75.
4. Mekhilef S, Saidur R, Safari A. A review on solar energy use in industries. *Renewable and Sustainable Energy Reviews*. 2011 May 1;15(4):1777–1790.
5. Malahayati M. Achieving renewable energies utilization target in South-East Asia: Progress, challenges, and recommendations. *The Electricity Journal*. 2020 Jun 1;33(5):106736.
6. Said Z, Sohail MA, Pandey AK, Sharma P, Waqas A, Chen WH, Nguyen PQ, Pham ND, Nguyen XP. Nanotechnology-integrated phase change material and nanofluids for solar applications as a potential approach for clean energy strategies: Progress, challenges, and opportunities. *Journal of Cleaner Production*. 2023 Jun 17:137736.
7. Venkitaraj KP, Suresh S, Praveen B, Venugopal A, Nair SC. Pentaerythritol with alumina nano additives for thermal energy storage applications. *Journal of Energy Storage*. 2017 Oct 1;13:359–377.
8. Cao F, Ye J, Yang B. Synthesis and characterization of solid-state phase change material microcapsules for thermal management applications. *Journal of Nanotechnology in Engineering and Medicine*. 2013 Nov 1;4(4):040901.

9. Xing D, Chi G, Ruan D, Huo L, Li D, Zhang T, Zhang D. Solid state phase transition in binary systems of polyhydric alcohols. *Acta Energiae Solaris Sinica.* 1995;16(2):131–137.

10. Venkitaraj KP, Suresh S, Praveen B. Energy storage performance of pentaerythritol blended with indium in exhaust heat recovery application. *Thermochimica Acta.* 2019 Oct 1;680:178343.

11. Xi P, Gu X, Cheng B, Wang Y. Preparation and characterization of a novel polymeric based solid–solid phase change heat storage material. *Energy Conversion and Management.* 2009 Jun 1;50(6):1522–1528.

12. Li WD, Ding EY. Preparation and characterization of cross-linking PEG/MDI/PE copolymer as solid–solid phase change heat storage material. *Solar Energy Materials and Solar Cells.* 2007 May 23;91(9):764–768.

13. Gu X, Xi P, Cheng B, Niu S. Synthesis and characterization of a novel solid–solid phase change luminescence material. *Polymer International.* 2010 Jun;59(6):772–777.

14. Venkitaraj KP, Suresh S, Praveen B. Experimental charging and discharging performance of alumina enhanced pentaerythritol using a shell and tube TES system. *Sustainable Cities and Society.* 2019 Nov 1;51:101767.

15. Venkitaraj KP, Suresh S, Praveen B, Nair SC. Experimental heat transfer analysis of macro packed neopentylglycol with CuO nano additives for building cooling applications. *Journal of Energy Storage.* 2018 Jun 1;17:1–10.

16. Nazari MA, Maleki A, Assad ME, Rosen MA, Haghighi A, Sharabaty H, Chen L. A review of nanomaterial incorporated phase change materials for solar thermal energy storage. *Solar Energy.* 2021 Nov 1;228:725–743.

17. Ali HM. Phase change materials based thermal energy storage for solar energy systems. *Journal of Building Engineering.* 2022 Sep 15;56:104731.

18. Fleischer AS. *Thermal energy storage using phase change materials: Fundamentals and applications.* Springer; 2015 Jun 22.

19. Sharma A, Tyagi VV, Chen CR, Buddhi D. Review on thermal energy storage with phase change materials and applications. *Renewable and Sustainable Energy Reviews.* 2009 Feb 1;13(2):318–345.

20. Kumar R, Mukhtar A, Yasir AS, Eldin SM, Musa DA, Rocha CM, Le BN, Ghalandari M. Simultaneous applications of fins and nanomaterials in phase change materials: A comprehensive review. *Energy Reports.* 2023 Nov 1;10:1028–1040.

21. Arıcı M, Tütüncü E, Yıldız Ç, Li D. Enhancement of PCM melting rate via internal fin and nanoparticles. *International Journal of Heat and Mass Transfer.* 2020 Aug 1;156:119845.

22. Sharma S, Micheli L, Chang W, Tahir AA, Reddy KS, Mallick TK. Nano-enhanced phase change material for thermal management of BICPV. *Applied Energy.* 2017 Dec 15;208:719–733.

23. Liu K, Wang N, Pan Y, Alahmadi TA, Alharbi SA, Jhanani GK, Brindhadevi K. Photovoltaic thermal system with phase changing materials and MWCNT nanofluids for high thermal efficiency and hydrogen production. *Fuel.* 2024 Jan 1;355:129457.

24. Khanafer K, Al-Masri A, Marafie A, Vafai K. Thermal performance of solar photovoltaic panel in hot climatic regions: Applicability and optimization analysis of PCM materials. *Numerical Heat Transfer, Part A: Applications.* 2023 May 2:1–21.

25. Said Z, Sohail MA, Pandey AK, Sharma P, Waqas A, Chen WH, Nguyen PQ, Pham ND, Nguyen XP. Nanotechnology-integrated phase change material and nanofluids for solar applications as a potential approach for clean energy strategies: Progress, challenges, and opportunities. *Journal of Cleaner Production.* 2023 Jun 17:137736.

26. Gürbüz H, Demirtürk S, Akçay H, Topalcı Ü. Experimental investigation on electrical power and thermal energy storage performance of a solar hybrid PV/T-PCM energy conversion system. *Journal of Building Engineering.* 2023 Jun 15;69:106271.

27. Stritih U. Increasing the efficiency of PV panel with the use of PCM. *Renewable Energy.* 2016 Nov 1;97:671–679.

28. Abo-Elnour F, Zeidan EB, Sultan AA, El-Negiry E, Soliman AS. Enhancing the bifacial PV system by using dimples and multiple PCMs. *Journal of Energy Storage.* 2023 Oct 15;70:108079.

29. Yinping Z, Yi J. A simple method, the-history method, of determining the heat of fusion, specific heat and thermal conductivity of phase-change materials. *Measurement Science and Technology.* 1999 Mar 1;10(3):201.

30. Pillai PB, Suresh S, Venkitaraj KP, Ali HM. Phase change material application in heat sink: Microencapsulated PCM. In *Phase Change Materials for Heat Transfer* (pp. 155–179). Elsevier; 2023 Jan 1.

31. Midhun VC, Suresh S, Praveen B, Shiju RS. Experimental study on phase transition behaviour of shape stable phase change material for application in vacuum insulation panel. *Journal of Energy Storage.* 2020 Dec 1;32:101825.

13 Phase Change Materials for Thermal Energy Storage Systems
An Introduction

*Ankit Bisariya, Rajan Kumar, Dwesh Kumar Singh,
Shailendra Kumar Shukla, Pushpendra Kumar
Singh Rathore, and Hafiz Muhammad Ali*

HIGHLIGHTS

- Different categories of phase change materials, along with their pros and cons, are outlined.
- Various properties of phase change materials are discussed.
- Key applications of phase change materials across several fields are highlighted.

NOMENCLATURE

c_p	Specific heat
k	Thermal conductivity
L_H	Latent heat
$T_{melting}$	Melting temperature
T_{PC}	Phase change temperature

ABBREVIATIONS

LHS	Latent heat storage
PCM	Phase change material
SHS	Sensible heat storage
TE	Thermal energy
TES	Thermal energy storage

13.1 INTRODUCTION

The efficient use of energy has become a crucial issue due to the exhaustion of fossil resources and worries about greenhouse gas emissions. A useful and practical way to enhance the energy storage efficiency in several home and industrial

DOI: 10.1201/9781003331957-13

sectors is to use phase change materials (PCMs) for thermal energy storage (TES) [1–2]. The TES system is regarded as a very important technological advancement with excellent adaptability to renewable energy sources. It might be possible to minimize the difference between energy production and demand by storing extra energy that would otherwise be wasted. TES systems can be classified into three different types of heat storage systems: sensible heat storage (SHS), latent heat storage (LHS), and thermochemical heat storage. SHS is a frequently employed technique used to store energy in various fields; in the solar heating system, water and rock beds are utilized for heat storage, and in air-based heating systems, the air is sensibly heated for heat storage. Furthermore, the LHS system is thought to be a highly efficient and effective TES approach due to extensive PCM availability, higher capacity to store thermal energy, and nearly constant temperature thermal storage/retrieval [3]. Due to their high enthalpy of fusion, PCMs can quickly gain or discharge large amounts of energy in the form of latent heat (L_H) within a small space during melting and solidification [4]. The adoption of PCMs for the storage of energy shrinks the difference between the source and requirement of energy and improves the effectiveness and reliability of energy distribution networks, hence contributing substantially to overall energy conservation. [5–6].

Practical PCMs must also have a thermal conductivity (k) of higher value for effective heat transfer and congruent phase change behavior for the prevention of irreversible separation of their constituents. Also, upper and lower temperatures at which phase transition occurs in PCM should be within the operational range of temperature for any application [7]. During the course of PCM research, researchers have investigated a wide range of materials, encompassing organic and inorganic compounds, as well as polymeric materials. These materials will be further explored and analyzed in subsequent sections of this chapter.

13.2   METHODS OF TES

The process of storing extra thermal energy in materials and using it for heating and cooling when needed is known as TES. Every time there is an imbalance between the consumption and the generation of energy in terms of temperature, time, power, and location, TES improves the efficiency of thermal energy. The TES process encompasses three discrete stages, specifically charging, storage, and subsequent discharge. When an exogenous energy source is present, the material undergoes charging, whereas in the absence of an energy source, the substance undergoes discharging. The three processes of charging, storing, and discharging undergo a cyclical pattern, sequentially alternating with one another [8].

The storage medium's thermo-physical characteristics determine how well these three phases work. Since thermal energy is of low grade, it must be stored using appropriate materials, even though doing so for an extended period is either challenging or expensive. According to thermodynamics, a material's ability to store heat is closely correlated with its c_p, volume, material temperature difference, and density. The approaches generally used for TES are often categorized as SHS, LHS, and thermochemical heat storage methods. These approaches are discussed in detail in the upcoming sections.

13.2.1 SHS Method

Sensible heat is the heat that a system exchanges when it changes temperature without altering its phase. It is the simple and most straightforward type of heat storage technique, in which the material does not go through a phase change over the designated temperature limits. These materials retain heat energy for a longer period since the temperature shift happens extremely gradually. The storage medium utilizes convection, conduction, and radiation to absorb and release heat energy. It can be quantified using the relationship given in Equation 13.1:

$$Q = m \cdot c_p \cdot \left(T_f - T_i\right)$$

(13.1)

According to Equation 13.1, a material's SHS quantity is dependent on its mass (m) and c_p, as well as the difference between final temperature (T_f) and initial temperature (T_i) [9]. Heat escapes from a cool environment as the temperature gradually drops, leading to uneven heat distribution over time. The underground hot aquifers are a real-world illustration of the SHS principle. These materials' accessibility and compatibility are two highly valuable advantages. These materials, which include agricultural, industrial, and household waste, are readily accessible in our environment and can provide useful heat storage during the summer. Water, metal waste, and rock-type materials like bricks, stones, pebbles, and gravel are the most frequently used SHS materials.

13.2.2 LHS Method

The heat that a system releases or absorbs during a process without a temperature change is known as L_H. The phase transition process is a typical example of L_H. This process maintains a consistent system temperature irrespective of the absorption or release of thermal energy. The heat of fusion is the type of L_H involved in the melting or freezing of a solid or liquid, while the heat of vaporization is the L_H that is related to the vaporization of a liquid, solid, or condensing vapor. The L_H capacity of any material can be stated mathematically as:

$$Q = m \cdot c_p \cdot dT\left(s\right) + m \cdot L_f + m \cdot c_p \cdot dT$$

(13.2)

where m is the mass of PCM (in kg), L_f is the enthalpy of fusion, and dT is the temperature difference.

The sensible heat of the solid phase is the first term in the equation, followed by the L_H of fusion and the sensible heat of the liquid phase. L_H storage using phase transition materials provides a higher energy storage density (around 350 MJ/m³) than sensible heat storage, making it a more attractive thermal energy storage method [10]. Paraffin wax, salt hydrates, and fatty acids are commonly utilized PCMs in various applications.

13.2.3 Thermochemical Heat Storage Method

The thermochemical system operates based on the fundamental principle of acquiring and releasing energy through the process of breaking and reforming bonds in chemically reversible reactions. With this sort of technique for storing thermal energy, heat is supplied to reversible endothermic reactions and can be extracted when an exothermic reversible reaction takes place. A catalyst may occasionally be employed to enable reversible chemical reactions. This sort of TES system has the benefits of its high energy density, lesser loss of heat, and availability of heat for the long term [11].

13.3 CLASSIFICATION OF PCMS

The rationale behind the utilization of PCMs in TES is rooted in the thermodynamic phenomena of heat absorption and release during reversible phase transitions. These transitions encompass various states, such as solid to liquid, liquid to gas, solid to gas, and solid to solid. The possible phase transformation of materials is shown in Figure 13.1 [12]. The limited application of TES systems is attributed to the substantial volume change observed during the phase transition of solid-to-gas or liquid-to-gas PCMs, despite their high latent heat [13].

The solid-to-solid and solid-to-liquid transformations exhibit minimal changes in volume, typically amounting to approximately 10% or less. Due to this reason, although having a lower heat of phase transition, they are nevertheless desirable materials for TES systems from an economic and practical perspective [11].

The classification of PCM materials can be determined by considering their physical state prior to and subsequent to their application. This includes categories such as gas–liquid PCMs, solid–gas PCMs, solid–liquid PCMs, and solid–solid PCMs, as illustrated in Figure 13.2. In the current scenario, solid–liquid and solid–solid PCMs are commonly used due to low volume variation and high L_H capacity. In the context of solid–liquid PCMs, the temperature gradually increases until it reaches the phase change temperature (T_{PC}). At this point, a significant amount of L_H is absorbed by the material as it transforms from solid to liquid [14]. The solid–liquid PCMs can be categorized into three categories depending on their chemical composition: organic, inorganic, and eutectic. The organic PCMs can be categorized into distinct groups,

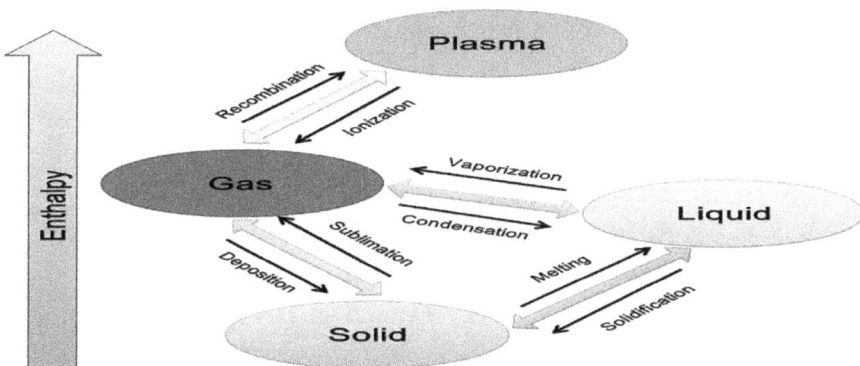

FIGURE 13.1 Possible phase transitions of materials [12].

including paraffins, alcohols, fatty acids, and esters. Conversely, inorganic PCMs are classified into two main categories: salts and salt hydrates [15]. When comparing organic PCMs with inorganic PCMs, the latter exhibit several notable advantages. These include a higher L_H per unit mass, non-flammability, and a lower cost. Eutectic PCMs comprise multiple soluble constituents that are combined and possess the characteristic of undergoing both melting and solidification concurrently, without undergoing any material separation. Solid–solid PCMs include polyols, polymeric, and organometallics, as shown in Figure 13.2.

13.3.1 Solid–Liquid PCMs

Organic, inorganic, and eutectic PCMs are the three types that are employed in solid–liquid phase changes.

13.3.1.1 Organic PCMs

Hydrocarbons make up the majority of organic compounds utilized as PCMs, which have a long chain of macromolecules that are predominantly made of carbon and hydrogen. Their optimal melting or solidification temperature, stable chemical and physical properties, high storage density, and abundance make this class of PCMs most generally used for TES [16]. Paraffins and non-paraffins are two distinct types of organic PCMs, with paraffins being further classified into two subcategories: (1) saturated hydrocarbons with several carbons ranging from 12 to 40 and having the general formula C_nH_{2n+2} of the linear alkanes, and (2) paraffin wax, which refers to a composite substance composed of alkanes and various hydrocarbons with the formula $CH_3(CH_2)$ nCH_3 [17]. Although pure alkanes are costlier than blends, the $T_{melting}$ and heat of fusion rise in both situations with the presence of more carbons. Also, the cost of paraffin increases with its purity.

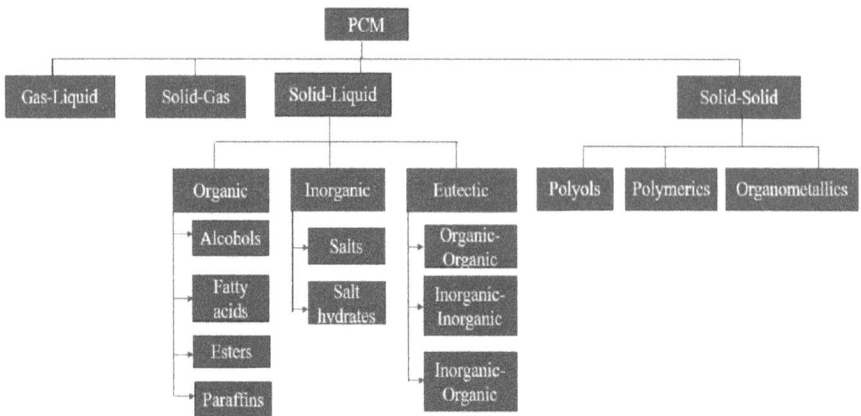

FIGURE 13.2   Types of PCMs.

In general, alcohols, esters, fatty acids, and glycols are some examples of non-paraffins. Their thermo-physical characteristics are almost identical to those of organic PCMs, which are made from paraffin. However, their higher flammability limits their suitable applications at elevated temperatures. Since the 1970s, alcohols have been studied for applications in energy storage, but more recent developments in these PCMs are linked to nanomaterials or unique composites along with distinctive features, including the addition of electrical conductivity. They can withstand some thermal shock when employed as conductive materials, which is an important characteristic. The researchers investigated a 1-tetra decanol (TD)/nano-Ag composite material's ability for energy storage and found that with an increase in the proportion of Ag nanoparticles, the k of the composite material was also enhanced [18]. TD/polyaniline (PANI) composites were also examined, which are prepared using the polymerization technique, and it was observed that these composites exhibit dual functionality, as they possess both electrical conductivity and the ability to store thermal energy [19]. When employed as conductive materials, they have characteristics that allow them to withstand high thermal stresses.

Recently, the use of organic materials for energy storage, such as fatty acids ($CH_3(CH_2)_{2n}COOH$), has increased due to their favorable thermodynamic and kinetic properties for low-temperature LHS. While saturated fatty acids have not received as much research attention as paraffin and salt hydrates, they are indeed valuable for TES. Moreover, they exhibit a broad range of $T_{melting}$, ranging from 8 °C to 64 °C, with a fusion enthalpy that varies from 149 kJ/kg to 222 kJ/kg. Also, fatty acids have a comparatively low heat conductivity, which is typically not a good thing.

The advantages of organic PCMs (including both paraffins and non-paraffins) are:

- Availability across a broad temperature range.
- High L_H of fusion.
- Long-term stability.
- Phase transition causes a slight volume shift.
- Recyclable.

Some limitations of organic solid–liquid PCMs are:

- Low k value.
- Flammable.
- Incompatible with plastic container materials.

13.3.1.2 Inorganic PCMs

A wide range of inorganic materials, eutectics, and mixtures have been extensively investigated as potential PCMs suitable for use in both low- and elevated-temperature applications. This indicates that inorganic PCMs exhibit a broad operating temperature range. Inorganic PCMs have an equivalent L_H per unit mass to organic PCMs, but because of their higher density, they have a higher L_H per unit volume.

One category of inorganic PCMs known as salt hydrates comprises inorganic salts (AB) combined with one or more water molecules (H_2O), leading to the formation of a crystalline solid structure (AB·xH$_2$O) [20]. A wide range of salt hydrates exhibiting melting points ranging from 5 to 130 °C have been identified, making them suitable for diverse applications. Rocks, concrete, stones, and other inorganic materials with a high melting point are examples of metallic materials.

The advantages of inorganic PCMs include:

- Storage density is high as compared to organic PCMs.
- Higher k value.
- Cheaper.
- Compatible with plastic containers.
- Non-flammable.

This type of PCMs has the following drawbacks:

- Corrosive in nature.
- High volume change during phase transition.

13.3.1.3 Eutectic PCMs

Eutectic PCMs comprise a mixture of two or more soluble PCMs, resulting in the acquisition of improved characteristics, including precise melting temperature and enhanced heat storage capacity [21]. The composition of two or more components should be such that these components freeze and melt simultaneously, without any segregation [11]. Inorganic and/or organic mixtures, such as organic–inorganic and inorganic–inorganic, can be found in eutectics.

13.3.2 Solid–Solid PCMs

While solid–solid PCMs often exhibit a lower L_H of phase transition, solid–liquid PCMs are now being investigated and employed in applications for energy storage. However, when compared to PCMs that transition between solid and liquid states, solid–solid PCMs offer distinct benefits such as minimal volume alteration during phase transition and the absence of leakage potential [22]. The phenomenon of reduced heat deterioration can be attributed to the phase transition exhibited by solid–solid PCMs, wherein they undergo a transformation from a crystalline structure to a semi-solid polycrystalline structure and eventually transition into an amorphous phase. As compared to form-stable PCM and solid–liquid PCMs, solid–solid PCMs have higher thermal stability. Although solid–solid PCMs have some significant advantages over solid–liquid PCMs, there are not many reviews available.

The aggressive volume growth, high vapor pressure, collection of mass, surface reactions, leakage, and slow nucleation rate are all constraints of this solid–solid phase transition mechanism. The benefit of phase transition mechanisms applies to particular applications when there is a restriction of liquid flow, which implies a faster rate of conduction heat transfer [14]. Solid–solid PCMs are further classified into organics (polyols), polymerics, and organometallics.

13.3.2.1 Organics (Polyols)

Pentaglycerine (PG), pentaerythritol (PE), 2-amino-2-methyl-1,3-propanediol (AMPL), and amino glycol are among the polyalcohols commonly employed as solid–solid PCMs. These polyalcohols are utilized either in isolation or in combination as binary, ternary, or quaternary systems. According to studies, it has been observed that polyols demonstrate a lack of symmetry and possess a layered crystal structure at lower temperatures. However, upon surpassing the T_{PC}, these polyols undergo a transformation, adopting a higher degree of symmetry and assuming a face-centered cubic structure [23].

The extent of hydrogen bonding exhibited by the hydroxyl functional groups plays a crucial role in determining the phase transition temperatures and enthalpies of polyalcohols. As PG and PE have extremely similar structural formulae, they can reliably combine to produce solutions while retaining the solid–solid phase transition properties of the original polyol [24].

13.3.2.2 Polymerics

Polymeric PCMs are made up of a solid–liquid ('soft') macromolecule that is chemically connected to a more rigid polymer. A solid–solid phase transition happens when the softer component melts and is confined by the harder component.

These solid–liquid PCMs were developed through research as form-stable PCMs that go through a solid–solid phase transition and result in a variety of polymeric solid–solid PCMs. Physically limiting the PCM in a supporting matrix of polymer to stop leakage is one such technique [25]. An alternative approach entails employing chemical processes such as grafting, blocking, and crosslinking copolymerization to affix the solid–liquid PCM onto a high-melting-point polymer substrate. Since the latter is a homogenous polymer, it is more favored. When a phase transition occurs, the PCM, such as a fatty acid or PEG, is the active component; however, the supporting polymer works for stabilizing it and keeps it away from melting [26].

13.3.2.3 Organometallics

Layered perovskites are organometallic solid–solid PCMs with the formula $(1-C_xH_{2x}+1NH_3)_2-MX_4$ where x represents the total number of carbon atoms in the hydrocarbon chain, M represents an divalent metal atom constituting the substrate, and X represents a halogen. Halide octahedra are the MX_4 layers, which have adjoining corners on a plane and are shown as the supporting structure of the layered perovskites. Covalent bonds and hydrogen bonds are used to bind NH_3 groups to the hydrocarbon chain and the halogens, respectively [27].

13.4 PROPERTIES OF PCMS

The properties necessary in PCMs for a particular application can vary from one application to another. These properties also act as a basis for selecting an appropriate PCM for the specific application. But in the broader sense PCMs should exhibit the following properties [4,11]:

13.4.1 THERMAL PROPERTIES

- Good heat transfer capability to enhance the system's effectiveness.
- High L_H capacity to accumulate large amounts of heat.
- Optimum T_{PC} to provide heat storage and extraction in a fixed temperature application.

13.4.2 PHYSICAL PROPERTIES

- Minimum volume change during phase transition to maximize the mechanical stability of the PCM-containing vessel.
- High c_p to accumulate heat within itself.
- Low vapor pressure to reduce the mechanical stability requirements of the PCM-containing vessel.

13.4.3 CHEMICAL PROPERTIES

- It should be stable for a long duration.
- It should be compatible with other materials so that it assures the longevity of the PCM-containing vessel and other surrounding materials when the PCM leakage issue arises.
- The melting and freezing cycle should be reversible so that PCM can be used several times.
- Non-toxic for safety reasons.
- Non-corrosive so that it does not decompose the PCM vessel.
- Non-explosive and non-flammable to enhance the system's safety.

13.4.4 KINETIC PROPERTIES

- There should be no subcooling to assure constant temperature that melting and solidification.
- The crystallization rate should be sufficient.

13.4.5 ECONOMIC PROPERTIES

- It should be abundantly available.
- It should be cheap so that it can be competitive with other TES methods.
- It should be easy to use so that can be easily popularized.

13.4.6 ENVIRONMENTAL PROPERTIES

- No adverse effects on the environment.
- Should have recycling potential.
- Non-polluting.

The various thermo-physical properties of different PCM materials that are used for TES in various fields are shown in Table 13.1.

TABLE 13.1
Thermo-Physical Properties of Some PCM Materials [3, 4, 28–30]

Class of PCM	PCM name	Melting point (°C)	Latent heat of fusion (kJ/kg)	Thermal conductivity (kJ/kg·K)		Density (kg/m³)	
				Solid	Liquid	Solid	Liquid
Organic	Paraffin wax	32	251	0.514	0.224	830	–
	RT-10	10	190	0.2	0.2	880	770
	Polyethylene	110–135	200			910	870
	RT-9 HC	–9	260	0.2	0.2	880	770
	Naphthalene	80	147.7	0.341	0.132	1145	976
	Erythritol	118	339.8	0.733	0.326	1480	1300
	Caprylic acid	16	149	–	0.149	981	901
	RT-28 HC	28	245	0.2	0.2	880	770
	Capric acid	32	153	–	0.153	1004	886
	n-Octadecane	27.7	243.5	0.19	0.148	865	785
	Vinyl stearate	27–29	122	–	–	–	–
	Palmitic acid	61	185	–	0.162	989	850
	Stearic acid	69	202.5	–	0.172	989	965
	Lauric acid	42–44	178		0.147	1007	862
Inorganic	Calcium chloride hexahydrate ($CaCl_2 \cdot 6H_2O$)	29.8	190.8	1.088	0.54	1802	1562
	$Ba(OH)_2 \cdot 8H_2O$	78	265–280	1.225	0.653	2070	1937
	$Mg(NO_3)_2 \cdot 6H_2O$	89	162.8	0.611	0.49	1636	1550
	Glauber's salt ($Na_2SO_4 \cdot 10H2O$)	32	254	0.554	–	1485	1458
	$Zn(NO_3)_2 \cdot 6H_2O$	36	146.9	–	0.464	1937	1828
	$Ba(OH)_2 \cdot 8H_2O$	78	265.7	1.255	0.653	2180	1937
	$Mg(NO_3)_2 \cdot 6H_2O$	89	162.5	0.611	0490	1636	1550
	$MgCl_2 \cdot 6H_2O$	117	168.6	0.694	0.570	1569	1442
	$Na_2HPO \cdot 12H_2O$	35–45	279.6	0.514	0.476	–	1520
Eutectic	61.5% $Mg(NO_3)_2 \cdot 6H_2O$ + 38.5% NH_4NO_3	52	125.5	0.552	0.494	1596	1515
	66.6% urea + 33.4% NH_4Br	76	161	0682	0.331	1548	1440
	58.7% $Mg(NO_3) \cdot 6H_2O$ + 41.3% $MgCl_2 \cdot 6H_2O$	59	132.2	0.678	0.510	1630	1550
	45.8% LiF + 54.2% MgF_2	746	–	–	–	2880	2305
	35.1% LiF + 38.4% NaF + 26.5% CaF_2	615	–	–	–	2820	2225
	48.1% LiF + 51.9% NaF	652	–	–	–	2720	2090

13.5 MAJOR APPLICATIONS OF PCMS AS PER PRESENT TRENDS

Applications of PCMs have emerged in the building industry, automobile industry, and solar energy installations. Applications in electronics and medicine are a few of them that have evolved in recent years. New, more advanced thermal energy storage materials for smart textiles and biomaterials with thermoregulation features are advancing more established industries, such as the construction sector. Some broad areas of applications of PCMs are discussed herein.

13.5.1 THERMAL STORAGE IN BUILDINGS

Energy conservation in buildings has benefited from energy storage technology using PCM. PCM's application to buildings, including both roofs and walls, boosts the building's ability to store heat, which could raise its energy effectiveness and lower the required electricity for heating and cooling of spaces. When PCMS are incorporated into the structural component, in the summer, the sun and high temperatures cause a heat wave that infiltrates the building walls during the day. The PCM delays the heat wave inside the building and even lowers its peak by absorbing extra heat through melting. The normal temperature of the room stays relaxing for most of the day, and the cooling system requires less electricity. PCMs releases the heat stored to both the internal and exterior environments throughout the night, when temperatures are lower, maintaining suitable room temperature and solidifying itself [31].

The main distinction between PCMs and more conventional thermal mass is that, as opposed to brickwork, which has a slower rate of absorption and release of heat, PCMs do it quickly. As a result, they are incredibly helpful in buildings with lightweight construction, where it is difficult to apply typical passive design principles. However, the effectiveness of a PCM is contingent upon its proper integration within a given structure.

13.5.2 HEATING OF WATER

Research has been conducted on a stratified storage tank for hot water incorporating a PCM module consisting of multiple cylinders. The study unveiled that the extended utilization of a PCM module within a water tank for residential hot water provision, even in the absence of an external energy source, can be achieved through the implementation of a granular PCM–graphite composite as the PCM [32]. The researchers also have done the charging test at the different water discharge and PCM water tanks. It was found that adding little amount of PCM in the tank resulted in a 3% increased energy capture capacity. PCM with 81 wt% of the combination of paraffin and fatty acids, etc., when used in the domestic hot water tank and experimented results showed that the temperature of 14 to 36 L of water increased by 3 to 4 °C within 10 to 15 minutes by using 3 kg of PCM, respectively [9].

13.5.3 SOLAR DESALINATION

Solar desalination is the most widely used technique to overcome water scarcity in all elements of sustainable development. Numerous prior studies have effectively illustrated the progressive augmentation of distilled water yield through the solar desalination process. However, in order to enhance their productivity and efficiency, solar desalination systems have undergone various modifications. Among these, solar heat storage methods find a solution to balance the sun's erratic activity and the demands for supply and demand.

The use of LHS and SHS materials in conjunction with solar stills has shown excellent responsiveness to ensure a steady supply of fresh water. The natural abundance of SHS materials makes them accessible and reasonably priced, whereas LHS material boosts productivity over the night and during cloudy weather [33].

13.5.4 BATTERY THERMAL MANAGEMENT SYSTEMS (BTMSs)

By maintaining the temperature within a designated safe range, a BTMS serves to maintain the functionality and safety of the battery. Due to its straightforward construction, exceptional temperature control capabilities, absence of consumption of energy, requirement of minimal components, and low cost, PCM-based BTMSs have increasingly attracted research interest in recent years [34]. However, the PCM's lower k poses a significant problem by reducing its ability to regulate the temperature variations inside the battery pack, and for enhancing the PCM's k, metals, oxides of metals, and nano-additives based on carbon are added to the PCM. This generally results in increased performance of BTMSs based on nanocomposites [35]. Figure 13.3 (a) shows the aligned layout of 25 18,650 Li-ion cells, and Figure 13.3 (b) shows this same layout surrounded by the PCM material, which efficiently keeps the maximum temperature of the cells and temperature uniformity within ideal limits by storing the surplus heat [36]. In this study, researchers used paraffin and metal foam to prepare the composite PCM because of the higher k of metal foam and because the paraffin embedded in metal foam fills the metal foam pores and forms a paraffin/metal foam composite PCM. The investigation showed that this arrangement using composite PCM lowered the battery pack's maximum temperature by 6.7 °C more than the pure paraffin. The enhancement of PCM's k helped to speed up the dissipation of heat from the battery pack to the surroundings, delayed the PCM's phase

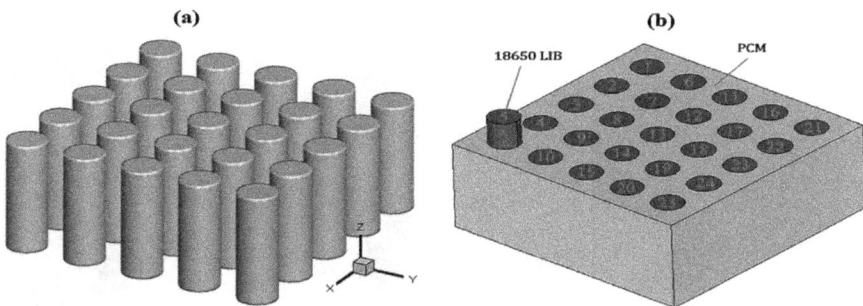

FIGURE 13.3 Li-ion battery pack with 25 18,650 cell units: (a) before packing with PCM; (b) after packing with PCM [36].

change process, and also improved the ability of PCM to regulate the battery pack's temperatures. The researchers also used fins to enhance the PCM BTMS's k, and this technique also showed positive results [37].

13.5.5 SMART TEXTILES

The utilization of PCM capsules for fabric coatings and textile fibers in commercial settings has recently become apparent. They are employed in bedding materials such as mattresses, blankets, duvets, and pillowcases, as well as in the production of outdoor apparel. Textiles made with PCMs can help to regulate body temperature in addition to providing warmth, which is why their effect is referred to as thermoregulation [38]. If a PCM is added to the textile, then it can greatly enhance its insulation ability due to its high thermal storage capacity and constant T_{PC}. The PCM having T_{PC} ranging within 18 to 35 °C is found to be most suitable for applications in clothes [14].

Microencapsulated PCMs can be added to a polymer solution before the fiber is extruded to give the fiber thermoregulating properties. Figure 13.4 shows the PCM microcapsules coated on the surface as well as embedded within the fiber now used commercially [39]. Other practical methods for including PCMs in a textile matrix include coating, laminating, melt spinning, finishing, bi-component synthetic fiber extrusion, and injection molding procedures [40].

13.5.6 BIOMATERIALS AND BIOMEDICAL APPLICATIONS

Several applications in the field of biomedicine require thermal protection, like specialized bandages or burn wound dressings, etc. For these applications in the biomedical field, PCMs are now regarded as promising materials.

Polyethylene glycol-treated textiles have demonstrated potential in biomedical contexts due to their desirable liquid absorption and antibacterial properties. These treated fabrics find applications in various medical products such as surgical gauze, diapers, and incontinence products. They can be utilized as bandages,

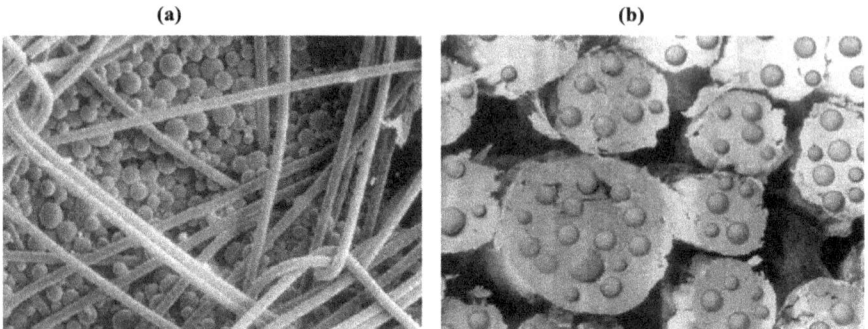

(a)   **(b)**

FIGURE 13.4 PCM microcapsules: (a) coated on the surface of the fabric; (b) embedded within fiber [39].

for burns, and in hot or cold therapies since textiles with thermoregulation can keep the temperature of skin within a desired range [41]. The researchers also found applications of material alloy PCMs in the biomedical field such as thermal protection of double-walled blood pouches packing during polymerization of poly methyl methacrylate. The PCM helped in lowering the maximum temperature of the process and could be used as bone cement composite [9]. Also, it has been noted that using microencapsulated PCM with high L_H and low k during cryosurgery maintained the healthy tissue surrounding the cancerous tumor by accumulating energy in the form of L_H [42].

13.5.7 SOLAR ENERGY STORAGE

In recent times, LHS methods have been utilized in thermo-solar systems to store energy in the daytime that can be effectively used late at night [9]. The researchers concluded that using solar collectors by incorporating PCM and water showed the enhancement of their absorption as well as insulation ability as compared to conventional hot water solar collectors [43]. The researchers also suggested a hybrid TES system that is capable of controlling heat produced by both solar and electrical sources. The simulation results of the proposed system showed 32% less energy utilization for space heating over four months in winter [44]. The researchers concluded that an efficient and economical TES system is still required for temperatures ranging from 250 to 500 °C. The researchers proposed cascaded LHS, and a concrete regenerator showed good storage capacity and needed minimal usage of the necessary material for heat storage compared to conventional LHS systems. The flux heat storage method for producing superheated steam at 350 to 400 °C by employing a zinc–tin alloy PCM produced good outcomes [45].

13.5.8 LATENT FUNCTIONAL THERMAL FLUIDS (LFTFs)

LFTFs are two-phase fluids used for heat transfer and are gaining popularity in recent times because of the inability of single-phase heat transfer fluids (HTFs) to transfer high amounts of heat at T_{PC} ranges because of their low c_p compared to LFTFs [4]. The LFTFs are made up of PCMs and HTFs in the form of a phase change emulsion. The PCM gains and removes significant heat during phase change and also exhibits high c_p. LFTFs raise the heat transfer between tube walls and fluid, which results in minimizing the pump energy requirements and mass flow. Hence, LFTFs find applications in heating, air conditioning, refrigeration, etc. [9]. The researchers examined the hydrated hydrophilic polymers by utilizing PCMs containing slurries for storage and thermal energy transfer at temperatures below the surrounding environment. The results showed that a polymeric concentration equal to or more than 10% causes substantial improvement in the heat transfer coefficient [46]. The researchers also studied the feasibility of a paraffin/water emulsion for cooling applications and found that the emulsion containing paraffin varying from 30–50 wt% is appropriate for practical usage because at a comparatively low viscosity, the proposed system has twice the energy density of pure water [47].

13.5.9 ELECTRONICS

Electronic device technical advances have enhanced functionality, decreased form factors, and crammed ever-increasing amounts of power into ever-tinier spaces. The power density of electronic components experiences a significant increase as their size decreases. The high power density intensifies the challenge of dissipating heat from components [48]. Thermal management is now more important than ever for the successful design of electronic gadgets including smartphones, laptops, tablets, and digital cameras. Cooling systems based on PCMs have a lot of potential because these devices are not often operated constantly for an extended period [49].

Heat sinks that are based on PCMs are an alternative passive cooling strategy that help to restrict the electronic equipment temperature below the critical level. The typical range of worldwide maximum permitted temperatures for various chips to prevent damage from overheating is between 85 and 120 °C.

A new PCM packaging was used in an experiment to manage the temperature of portable electronic gadgets. The results indicated that the effectiveness of a passive thermal control system relied heavily on the thermal resistance of the device as well as the power level administered to the PCM package. Another study looked into a PCM-based heat sink for plastic quad flat package (QFP) electrical device transient thermal management, as shown in Figure 13.5. The heat sink is placed on top of the QFP, and the QFP is mounted on the printed circuit board. When the level of input power was relatively high, the heat sinks with PCM in their cavities outperformed those without PCM in terms of cooling performance [50].

13.5.10 AUTOMOBILE INDUSTRY

In the automotive sector, PCMs are utilized for internal combustion engines, engine cooling, boosting passenger thermal comfort, and preheating catalytic converters [51–53]. A TES system was used in a study to lower internal combustion engine

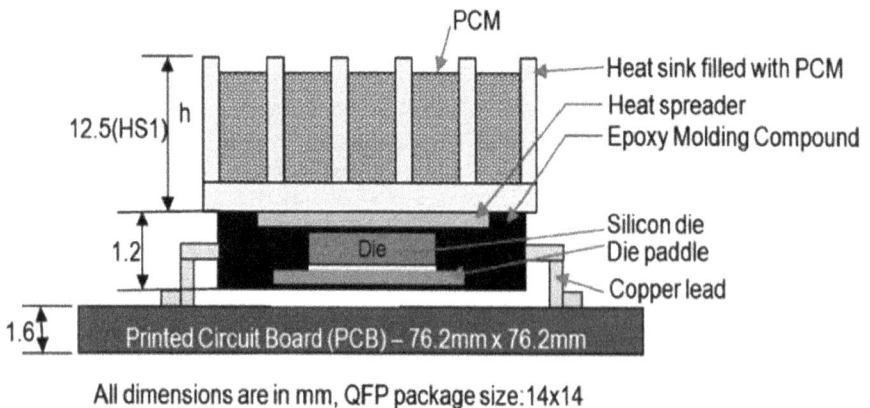

All dimensions are in mm, QFP package size:14x14

FIGURE 13.5 Schematic of a PCM-filled heat sink with a QFP [50].

cold-start emissions. With the engine preheating during cold start and warm-up time, emissions of hydrocarbons and CO dropped by almost 15% and 64%, respectively. In an alternative system, a combination of an evaporator and a pressure regulator (EPR) employed PCMs for TES in order to address the challenges associated with cold starts in vehicles powered by liquefied petroleum gas. After a period of waiting, it was discovered that the EPR with phase change material might partially resolve the issue during the cold start with LPG-fueled engines.

13.5.11 FOOD INDUSTRY

Applications of PCM for the food industry are now under investigation. Moreover, PCMs are also used to dry agricultural goods with sun dryers. With the utilization of PCM storage units in the solar dryer, the updated solar dryers can now effectively dry food, even in the evenings and nights when no sun is there, which is not possible with conventional sun dryers [54].

According to a study, a composite PCM made of nano-structured calcium silicate (NCS) and paraffin is an efficient thermal buffer for paperboard containers used to transport and temporarily store chilled perishable foods. The study determined that the NCS-PCM exhibited a satisfactory thermal buffering capability, effectively sustaining the container's temperature within 10 °C for approximately 5 hours, given an ambient temperature of 23 °C [55]. Additionally, it has been demonstrated that PCM exhibits notable efficacy in maintaining a consistent temperature for food items, while concurrently reducing the food's temperature by 1.5 °C during the defrosting procedure. Because of these findings, the PCM can also be used in cabinet shelves for temperature regulation [56]. Hence, the usage of PCMs is effective in the food industry and its usage increases with time.

13.6 CONCLUSIONS

Significant research has been carried out for utilizing PCMs as energy storage media due to their advantageous characteristics, including a high storage capacity, minimal energy losses, and reduced mass and volume of the system. Using PCM to accumulate surplus energy for future use enhances the system's performance. This chapter aims to deliver readers with a comprehensive understanding of the utilization of various PCMs as TES systems in diverse sectors, including building construction, the food industry, automotive engineering, textile manufacturing, electronics, and biomedical applications. The aforementioned observations have been noted.

- The TES in sensible and L_H form has become an effective energy management technique that emphasizes the use the waste heat and the efficient use of energy sources in various sectors.
- The utilization of PCMs for energy storage offers several advantages in comparison to sensible systems. These advantages stem from the lower mass and volume of PCM systems, the ability to store or release energy at relatively constant temperatures, and the minimal energy losses experienced when compared to conventional systems.

- The organic PCMs are inexpensive, have a moderate TES density, are non-corrosive, have a good level of chemical stability, have high capacities, etc., but because of their poor k, a lot of surface area is required to achieve the necessary TES effects. Hence methods such as incorporating carbon-based and metal-based additives to enhance the k of these PCMs are utilized.
- Inorganic PCMs, on the other hand, exhibit a higher L_H per unit volume and k as compared to organic PCMs, but these PCMs are corrosive and also show supercooling effects that may harm their phase change characteristics. Hence, techniques should be developed to minimize these adverse effects.
- TES has functioned as a thermal battery; hence, the future aspect involves suitable techniques to find out the PCM's degradation mechanisms that would enable the use of PCMs for long-term TES applications.
- To satisfy the demands of TES applications, new techniques and new materials must be developed in addition to the techniques already described for increasing the k of PCMs. The use of conductive additives enhances the k, but it also reduces the PCM capacity. Therefore, it is advisable to employ an optimization technique in order to enhance the performance of PCMs as TES systems by optimizing the quantity of additives used.

REFERENCES

[1] Kenisarin, M. and Mahkamov, K., 2007. Solar energy storage using phase change materials. *Renewable and Sustainable Energy Reviews*, *11*(9), pp. 1913–1965.

[2] Garg, H.P., Mullick, S.C. and Bhargava, A.K., 1985. *Solar thermal energy storage.* Dordrecht, Holland: Reidel Publishing Company.

[3] Khan, Z., Khan, Z. and Ghafoor, A., 2016. A review of performance enhancement of PCM based latent heat storage system within the context of materials, thermal stability and compatibility. *Energy Conversion and Management*, *115*, pp. 132–158.

[4] Pielichowska, K. and Pielichowski, K., 2014. Phase change materials for thermal energy storage. *Progress in Materials Science*, *65*, pp. 67–123.

[5] Rathod, M.K. and Banerjee, J., 2013. Thermal stability of phase change materials used in latent heat energy storage systems: A review. *Renewable and Sustainable Energy Reviews*, *18*, pp. 246–258.

[6] Reddy, M.R., Nallusamy, N., Prasad, A.B. and Reddy, H.K., 2012. Thermal energy storage system using phase change materials: constant heat source. *Thermal Science*, *16*(4), pp. 1097–1104.

[7] Lane, G.A., 1980. Low-temperature heat storage with phase change materials. *International Journal of Ambient Energy*, *1*(3), pp. 155–168.

[8] Cabeza, L.F. and Oró, E., 2016. Thermal energy storage for renewable heating and cooling systems. In *Renewable Heating and Cooling* (pp. 139–179). Cambridge: Woodhead Publishing,

[9] Kumar, N., Gupta, S.K. and Sharma, V.K., 2021. Application of phase change material for thermal energy storage: An overview of recent advances. *Materials Today: Proceedings*, *44*, pp. 368–375.

[10] Berardi, U. and Gallardo, A.A., 2019. Properties of concretes enhanced with phase change materials for building applications. *Energy and Buildings*, *199*, pp. 402–414.

[11] Sharma, A., Tyagi, V.V., Chen, C.R. and Buddhi, D., 2009. Review on thermal energy storage with phase change materials and applications. *Renewable and Sustainable Energy Reviews*, *13*(2), pp. 318–345.

[12] Kant, K., Biwole, P.H., Shamseddine, I., Tlaiji, G., Pennec, F. and Fardoun, F., 2021. Recent advances in thermophysical properties enhancement of phase change materials for thermal energy storage. *Solar Energy Materials and Solar Cells*, *231*, p. 111309.

[13] Abhat, A., 1983. Low temperature latent heat thermal energy storage: Heat storage materials. *Solar Energy*, *30*(4), pp. 313–332.

[14] Lin, Y., Jia, Y., Alva, G. and Fang, G., 2018. Review on thermal conductivity enhancement, thermal properties and applications of phase change materials in thermal energy storage. *Renewable and Sustainable Energy Reviews*, *82*, pp. 2730–2742.

[15] Milian, Y.E., Gutierrez, A., Grageda, M. and Ushak, S., 2017. A review on encapsulation techniques for inorganic phase change materials and the influence on their thermophysical properties. *Renewable and Sustainable Energy Reviews*, *73*, pp. 983–999.

[16] Lamrani, B., Kuznik, F. and Draoui, A., 2020. Thermal performance of a coupled solar parabolic trough collector latent heat storage unit for solar water heating in large buildings. *Renewable Energy*, *162*, pp. 411–426.

[17] Ukrainczyk, N., Kurajica, S. and Šipušić, J., 2010. Thermophysical comparison of five commercial paraffin waxes as latent heat storage materials. *Chemical and Biochemical Engineering Quarterly*, *24*(2), pp. 129–137.

[18] Zeng, J.L., Sun, L.X., Xu, F., Tan, Z.C., Zhang, Z.H., Zhang, J. and Zhang, T., 2007. Study of a PCM based energy storage system containing Ag nanoparticles. *Journal of Thermal Analysis and Calorimetry*, *87*, pp. 371–375.

[19] Zeng, J., Zhang, J., Liu, Y., Cao, Z., Zhang, Z., Xu, F. and Sun, L., 2008. Polyaniline/1-tetradecanol composites: Form-stable PCMS and electrical conductive materials. *Journal of Thermal Analysis and Calorimetry*, *91*(2), pp. 455–461.

[20] Luo, L. and Le Pierrès, N., 2015. Innovative systems for storage of thermal solar energy in buildings. In *Solar Energy Storage* (pp. 27–62). London: Academic Press Inc.

[21] Wang, X., Lu, E., Lin, W., Liu, T., Shi, Z., Tang, R. and Wang, C., 2000. Heat storage performance of the binary systems neopentyl glycol/pentaerythritol and neopentyl glycol/trihydroxy methyl-aminomethane as solid–solid phase change materials. *Energy Conversion and Management*, *41*(2), pp. 129–134.

[22] Jelle, B.P. and Kalnæs, S.E., 2017. Phase change materials for application in energy-efficient buildings. In *Cost-effective energy efficient building retrofitting* (pp. 57–118). Cambridge: Woodhead Publishing.

[23] Murrill, E. and Breed, L., 1970. Solid—solid phase transitions determined by differential scanning calorimetry: part I. Tetrahedral substances. *Thermochimica Acta*, *1*(3), pp. 239–246.

[24] Benson, D.K., Burrows, R.W. and Webb, J.D., 1986. Solid state phase transitions in pentaerythritol and related polyhydric alcohols. *Solar Energy Materials*, *13*(2), pp. 133–152.

[25] Benson, D.K., Webb, J.D., Burrows, R.W., McFadden, J. and Christensen, C., 1985. *Materials research for passive solar systems: solid-state phase-change materials* (No. SERI/TR-255-1828). Solar Energy Research Inst., Golden, CO (USA).

[26] Peng, S., Fuchs, A. and Wirtz, R.A., 2004. Polymeric phase change composites for thermal energy storage. *Journal of Applied Polymer Science*, *93*(3), pp. 1240–1251.

[27] Alkan, C., Ensari, Ö.F. and Kahraman, D., 2012. Poly (2-alkyloyloxyethylacrylate) and poly (2-alkyloyloxyethylacrylate-co-methylacrylate) comblike polymers as novel phase-change materials for thermal energy storage. *Journal of Applied Polymer Science*, *126*(2), pp. 631–640.

[28] Mehling, H. and Cabeza, L.F., 2007. Phase change materials and their basic properties. In *Thermal energy storage for sustainable energy consumption: fundamentals, case studies and design* (pp. 257–277). Dordrecht: Springer.

[29] Zalba, B., Marın, J.M., Cabeza, L.F. and Mehling, H., 2003. Review on thermal energy storage with phase change: Materials, heat transfer analysis and applications. *Applied Thermal Engineering*, 23(3), pp. 251–283.

[30] Wei, G., Wang, G., Xu, C., Ju, X., Xing, L., Du, X. and Yang, Y., 2018. Selection principles and thermophysical properties of high-temperature phase change materials for thermal energy storage: A review. *Renewable and Sustainable Energy Reviews*, 81, pp. 1771–1786.

[31] Arend, H., Huber, W., Mischgofsky, F.H. and Richter-Van Leeuwen, G.K., 1978. Layer perovskites of the (CnH2n+ 1NH3) 2MX4 and NH3 (CH2) mNH3MX4 families with M= Cd, Cu, Fe, Mn or Pd and X= Cl or Br: Importance, solubilities and simple growth techniques. *Journal of Crystal Growth*, 43(2), pp. 213–223.

[32] Rathore, P.K.S., Gupta, N.K., Yadav, D., Shukla, S.K. and Kaul, S., 2022. Thermal performance of the building envelope integrated with phase change material for thermal energy storage: An updated review. *Sustainable Cities and Society*, p. 103690.

[33] Cabeza, L.F., Ibanez, M., Sole, C., Roca, J. and Nogues, M., 2006. Experimentation with a water tank including a PCM module. *Solar Energy Materials and Solar Cells*, 90(9), pp. 1273–1282.

[34] Rana, S., Kumar, R. and Bharj, R.S., 2023. Lithium-ion battery thermal management techniques and their current readiness level. *Energy Technology*, 11(1), p. 2200873.

[35] Chauhan, V.K., Shukla, S.K. and Rathore, P.K.S., 2022. A systematic review for performance augmentation of solar still with heat storage materials: A state of art. *Journal of Energy Storage*, 47, p. 103578.

[36] Huang, R., Li, Z., Hong, W., Wu, Q. and Yu, X., 2020. Experimental and numerical study of PCM thermophysical parameters on lithium-ion battery thermal management. *Energy Reports*, 6, pp. 8–19.

[37] Kumar, R., Mitra, A. and Srinivas, T., 2022. Role of nano-additives in the thermal management of lithium-ion batteries: A review. *Journal of Energy Storage*, 48, p. 104059.

[38] Trinquet, F., Karim, L., Lefebvre, G. and Royon, L., 2014. Mechanical properties and melting heat transfer characteristics of shape-stabilized paraffin slurry. *Experimental Heat Transfer*, 27(1), pp. 1–13.

[39] Nelson, G., 2002. Application of microencapsulation in textiles. *International Journal of Pharmaceutics*, 242(1–2), pp. 55–62.

[40] Hansen, R.H., JP Stevens and Co Inc, 1971. *Temperature adaptable fabrics*. U.S. Patent 3,607,591.

[41] Mondieig, D., Rajabalee, F., Laprie, A., Oonk, H.A., Calvet, T. and Cuevas-Diarte, M.A., 2003. Protection of temperature sensitive biomedical products using molecular alloys as phase change material. *Transfusion and Apheresis Science*, 28(2), pp. 143–148.

[42] Lv, Y., Zou, Y. and Yang, L., 2011. Feasibility study for thermal protection by microencapsulated phase change micro/nanoparticles during cryosurgery. *Chemical Engineering Science*, 66(17), pp. 3941–3953.

[43] Kürklü, A., Özmerzi, A. and Bilgin, S., 2002. Thermal performance of a water-phase change material solar collector. *Renewable Energy*, 26(3), pp. 391–399.

[44] Hammou, Z.A. and Lacroix, M., 2006. A hybrid thermal energy storage system for managing simultaneously solar and electric energy. *Energy Conversion and Management*, 47(3), pp. 273–288.

[45] Adinberg, R., Zvegilsky, D. and Epstein, M., 2010. Heat transfer efficient thermal energy storage for steam generation. *Energy Conversion and Management*, 51(1), pp. 9–15.

[46] Augood, P.C., Newborough, M. and Highgate, D.J., 2001. Thermal behaviour of phase-change slurries incorporating hydrated hydrophilic polymeric particles. *Experimental Thermal and Fluid Science*, 25(6), pp. 457–468.

[47] Huang, L., Petermann, M. and Doetsch, C., 2009. Evaluation of paraffin/water emulsion as a phase change slurry for cooling applications. *Energy, 34*(9), pp. 1145–1155.

[48] Ali, H.M., Rehman, T.U., Arıcı, M., Said, Z., Duraković, B., Mohammed, H.I., Kumar, R., Rathod, M.K., Buyukdagli, O. and Teggar, M., 2024. Advances in thermal energy storage: Fundamentals and applications. *Progress in Energy and Combustion Science, 100*, p. 101109.

[49] Kandasamy, R., Wang, X.Q. and Mujumdar, A.S., 2007. Application of phase change materials in thermal management of electronics. *Applied Thermal Engineering, 27*(17–18), pp. 2822–2832.

[50] Kandasamy, R., Wang, X.Q. and Mujumdar, A.S., 2008. Transient cooling of electronics using phase change material (PCM)-based heat sinks. *Applied Thermal Engineering, 28*(8–9), pp. 1047–1057.

[51] Burch, S.D., Keyser, M.A., Colucci, C.P., Potter, T.F., Benson, D.K., Biel, J.P. and Beil, J.P., 1996. Applications and benefits of catalytic converter thermal management. *SAE Transactions*, pp. 839–844.

[52] Burch, S.D., Potter, T.F., Keyser, M.A., Brady, M.J. and Michaels, K.F., 1995. Reducing cold-start emissions by catalytic converter thermal management. *SAE Transactions*, pp. 348–353.

[53] Kim, K.B., Choi, K.W., Kim, Y.J., Lee, K.H. and Lee, K.S., 2010. Feasibility study on a novel cooling technique using a phase change material in an automotive engine. *Energy, 35*(1), pp. 478–484.

[54] Bal, L.M., Satya, S. and Naik, S.N., 2010. Solar dryer with thermal energy storage systems for drying agricultural food products: A review. *Renewable and Sustainable Energy Reviews, 14*(8), pp. 2298–2314.

[55] Johnston, J.H., Grindrod, J.E., Dodds, M. and Schimitschek, K., 2008. Composite nano-structured calcium silicate phase change materials for thermal buffering in food packaging. *Current Applied Physics, 8*(3–4), pp. 508–511.

[56] Lu, Y.L., Zhang, W.H., Yuan, P., Xue, M.D., Qu, Z.G. and Tao, W.Q., 2010. Experimental study of heat transfer intensification by using a novel combined shelf in food refrigerated display cabinets (Experimental study of a novel cabinets). *Applied Thermal Engineering, 30*(2–3), pp. 85–91.

.

Index

For Product Safety Concerns and Information please contact our EU
representative GPSR@taylorandfrancis.com
Taylor & Francis Verlag GmbH, Kaufingerstraße 24, 80331 München, Germany

www.ingramcontent.com/pod-product-compliance
Lightning Source LLC
Chambersburg PA
CBHW060807220326
41598CB00022B/2560

9 7 8 1 0 3 2 3 6 4 3 8 4